AF900088

Interface Engineering in Organic Field-Effect Transistors

Interface Engineering in Organic Field-Effect Transistors

Xuefeng Guo
Hongliang Chen

WILEY-VCH

Authors

Prof. Xuefeng Guo
Peking University
202 Chengfu Rd
Haidian District
100871 Beijing
China

Prof. Hongliang Chen
Zhejiang University
No. 866 Yu Hang Tang Rd
Xihu District
310027 Hangzhou
China

Cover Image: © blackdovfx/Getty Images

All books published by **WILEY-VCH** are carefully produced. Nevertheless, authors, editors, and publisher do not warrant the information contained in these books, including this book, to be free of errors. Readers are advised to keep in mind that statements, data, illustrations, procedural details or other items may inadvertently be inaccurate.

Library of Congress Card No.: applied for

British Library Cataloguing-in-Publication Data
A catalogue record for this book is available from the British Library.

Bibliographic information published by the Deutsche Nationalbibliothek
The Deutsche Nationalbibliothek lists this publication in the Deutsche Nationalbibliografie; detailed bibliographic data are available on the Internet at <http://dnb.d-nb.de>.

© 2023 WILEY-VCH GmbH, Boschstraße 12, 69469 Weinheim, Germany

All rights reserved (including those of translation into other languages). No part of this book may be reproduced in any form – by photoprinting, microfilm, or any other means – nor transmitted or translated into a machine language without written permission from the publishers. Registered names, trademarks, etc. used in this book, even when not specifically marked as such, are not to be considered unprotected by law.

Print ISBN: 978-3-527-35145-9
ePDF ISBN: 978-3-527-84046-5
ePub ISBN: 978-3-527-84047-2
oBook ISBN: 978-3-527-84048-9

Typesetting Straive, Chennai, India

Contents

Preface

The development of microelectronics into modern integrated circuits over the past 75 years was the result of continuous efforts of huge teams of scientists and engineers. The device performance of microelectronic devices is mainly determined by the properties of the materials used and their corresponding interfaces, just as Nobel laureate Herbert Kroemer coined, "the interface is the device." Understanding and governing the charge transfer phenomena occurring at the interfaces could revolutionize a number of related fields, including batteries, catalysis, data storage, energy harvesting, imaging displays, sensing devices, and even solar cells. For these reasons, interface engineering of these heterojunctions has captured the attention of a broad cross section of scientific and industrial communities.

In comparison with traditional complementary metal oxide semiconductor transistors, the advantages of organic field-effect transistors (OFETs) originate from how OFETs can be manufactured, including the considerable variety in molecular design, low cost, lightweight, mechanical flexibility, solution processability, and large-area fabrication. The key concept of this book is to use interface engineering strategies to improve the inherent properties of OFETs as well as to build novel functional devices based on OFETs. Both are of great importance for laying the foundation for the OFET research. For the first point, rapid progress has been made on each isolated component of OFETs to improve the performance, with major emphasis on organic semiconducting material design and relevant film processing methods. Although the improved electronic properties of these materials could offer device performance advantages, interfacial barriers between the semiconductors, dielectric layers, and electrodes are often treated as secondary concerns in comparison with the generation and preservation of carrier population in the semiconductor itself. In pursuit of high performance, OFETs will inevitably require an optimized carrier transport pathway that leverages all the device layers and their corresponding interfaces. This should be as important as the material design and relevant film processing methods.

The second point is that how to "use" OFETs is another important issue in OFET research. Of course, huge progress has been made to develop a variety of functional devices based on OFETs, such as chemo/biosensors, photodetectors, integrated circuits, and flexible electronics. They can be realized by using various strategies, for example, designing functional semiconducting materials, developing semiconductor processing methods, and tailoring functional interlayers. In this book, however, we will specifically survey the most recent and important works employing interface engineering strategies to develop functional OFET devices. We believe that interfacial modification is a much simpler, more efficient way to integrate functionalities.

This book will focus on interface engineering of OFETs and aim to provide fundamental understandings of the interplay between the molecular structure, assembly, and emergent functions at the molecular level and consequently offer novel insights into designing a new generation of multifunctional integrated circuits and sensors toward practical applications. Actually, the interface engineering strategy can be leveraged to improve the efficiency and stability of perovskite solar cells. We believe that this book will draw widespread interest from researchers working on organic electronics, chemo- and biosensors, solar cells, flexible electronics, and other device-related fields.

We would like to thank Wiley for offering the opportunity to prepare this book, especially Ms. Alice Qian. We would also like to express our deep gratitude to Ms. Katrina Maceda and Ms. Monica Chandra Sekar who have contributed their time and outstanding expertise to make this book possible. Also, thanks to Dr. Weining Zhang for the helpful discussions during the preparation of the book. Finally, we hope that the readers of this book will find it both useful and delightful.

Beijing, China
December 2022

Xuefeng Guo and Hongliang Chen

Author Biographies

Xuefeng Guo received his PhD in 2004 from the Institute of Chemistry, Chinese Academy of Sciences, Beijing. From 2004 to 2007, he was a postdoctoral research scientist at the Columbia University Nanocenter. He joined the faculty as a professor under the "Peking 100-Talent" Program at Peking University in 2008. In 2012, he won the National Science Funds for Distinguished Young Scholars of China. His current research is focused on functional nanometer/molecular devices. Professor Xuefeng Guo has authored over 210 scientific publications and has received numerous scientific awards, including the First prize of Ministry of Education Natural Science Award.

Hongliang Chen received his PhD in 2016 from the College of Chemistry and Molecular Engineering, Peking University, under the guidance of Professor Xuefeng Guo. From 2016 to 2018, he worked as a research scientist at the Core R&D department in Dow Chemical Company. Then he moved to Northwestern University in United States and worked as a postdoctoral research fellow in Professor Sir Fraser Stoddart's group from 2018 to 2021. He joined Zhejiang University as an assistant professor under "ZJU 100-Talent" Program in June 2021. His research interests are focused on organic functional devices and molecular electronics.

List of Acronyms and Abbreviations

1D	one-dimensional
2D	two-dimensional
3D	three-dimensional
AFM	atomic force microscopy
AZOs	azobenzenes
BCB	divinyltetramethyldisiloxane-bis(benzocyclobutene)
BTBT	[1]benzothieno[3,2-*b*][1]-benzothiophene
C_8-BTBT	2,7-dioctyl[1]-benzothieno[3,2-*b*][1]benzothiophene
CIL	charge insertion layer
CNTs	carbon nanotubes
CT	charge-transfer
D–A	donor–acceptor
DAE	diarylethene
DNTT	dinaphtho[2,3-*b*:2′,3′-*f*] thieno[3,2-*b*]thiophene
DOS	densities of state
E_A	energy gap of OSC
E_F	Fermi level of electrodes
E_{VAC}	vacuum energy level
FLG	few-layer graphene
GOs	graphene oxides
Gr	graphene
HMDS	hexamethyldisilazane
HOMO	highest occupied molecular orbital
k	relative permittivity or dielectric constant
LB	Langmuir–Blodgett
LUMO	lowest unoccupied molecular orbitals
MWCNTs	multiwalled carbon nanotubes
NDI	naphthalenediimide
NHCs	N-heterocyclic carbenes
OFET	organic field-effect transistor
OLET	organic light-emitting transistor
OSCs	organic semiconductors
P3HT	poly(3-hexylthiophene)

PA	Phosphonic acid
PCBM	[6,6]-phenyl-C_{61}-butyric acid methyl ester
PCMs	photochromic molecules
PDI	perylenediimide
PDMS	polydimethylsiloxane
PEN	polyethylene naphthalate
PHBD	photo hybrid bilayer dielectric
PLA	polylactide
PMMA	poly(methyl methacrylate)
POM	polyoxometalate
PS	polystyrene
PSA	poly(stearyl acrylate)
PSS	poly(sodium-*p*-styrenesulfonate)
PVA	poly(vinyl alcohol)
P(VDF–TrFE)	poly(vinylidene fluoride-trifluoroethylene)
PVP	poly(vinyl phenyl)
PVPyr	poly(4-vinylpyridine)
PαMS	poly(α-methylstyrene)
RGO	reduced graphene oxide
SAM	self-assembled monolayer
SAMFETs	self-assembled monolayer field-effect transistors
SEBS	poly(styrene-*b*-(ethylene-*co*-butylene)-*b*-styrene)
SEM	scanning electron microscopy
SLG	single-layer graphene
SPs	spiropyrans
SWCNTs	single-walled carbon nanotubes
TCNB	1,2,4,5-tetraacyanobenzene
TCNQ	tetracyanoquinodimethane
TEM	transmission electron microscopy
TIPS–pentacene	6,13-bis(triisopropylsilylethynyl) pentacene
TTF	tetrathiafulvalene
V_D	source–drain voltage
V_G	gate voltage
V_{th}	threshold voltage
WEG	weak epitaxy growth
W_F	work function
γ_{tot}	total surface energy
ε_0	vacuum permittivity
μ_{DT}	aligned dipole moment of the SAM
$\mu_{M\text{-}S}$	M—S bond dipole interface
φ_m	electrode work function

1

Introduction

1.1 Different Interfaces in OFETs

When Nobel laureate Herbert Kroemer coined the famous phrase that "the interface is the device" [1], he pointed out the essential importance of the interfaces between different materials in any device action, especially for semiconducting thin-film devices, which have shown astonishing successes in photonic and electronic applications. From the point of view of a typical organic field-effect transistor (OFET), the device performance is mainly determined by the properties of the materials used and their corresponding interfaces. In general, an OFET [2–5] consists of three electrodes (source, drain, and gate), a dielectric layer, and an organic semiconductor (OSC) layer (Figure 1.1). To improve the device performance and even build novel functionalities, interface engineering of these heterojunctions is currently a research focus because of its promising potential for further applications in broad areas ranging from integrated circuits and energy conversion to catalysis and chemical/biological sensors [6–14]. In fact, scientists in a variety of disciplines have been devoting great efforts to this concept, which started with simple improvement of the device performance and has branched out in different directions, indicating the interdisciplinary of these efforts [15–18]. In comparison with traditional complementary metal oxide semiconductor transistors, the advantages of OFETs originate from how they can be manufactured, including considerable variety in molecular design, low cost, light weight, mechanical flexibility, solution processability, and large-area fabrication.

There are four major contact interfaces in OFETs, namely, the interface within the semiconductor layers, semiconductor/electrode interface, semiconductor/dielectric interface, and semiconductor/environment interface (Figure 1.2). In general, the interface within the semiconductor layers [19, 20] contains traps, grain boundaries, and other defects, which provide scattering sites for carriers and are detrimental to charge transport (Figure 1.2a). Recent studies [21, 22], however, proved that semiconductor doping and blends had a positive influence on the crystallinity and morphology of the semiconductor films. Several strategies have been developed to achieve high-quality thin films, as well as to form good heterojunctions, thus affording ambipolar OFETs that are useful for building future OFET-based integrated

Interface Engineering in Organic Field-Effect Transistors, First Edition. Xuefeng Guo and Hongliang Chen.
© 2023 WILEY-VCH GmbH. Published 2023 by WILEY-VCH GmbH.

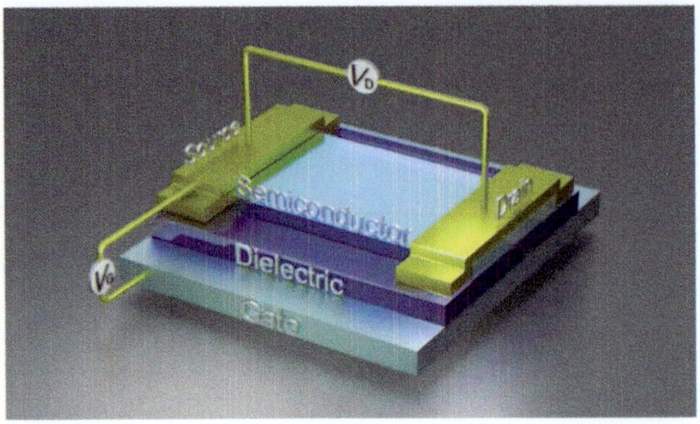

Figure 1.1 Schematic illustration of a typical OFET architecture.

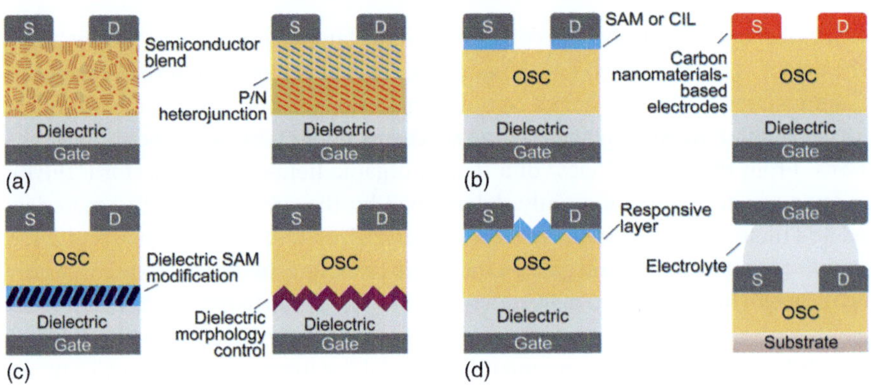

Figure 1.2 Interface engineering of interfaces in OFETs. (a) Interface within the semiconductor layers, (b) semiconductor/electrode interface, (c) semiconductor/dielectric interface, and (d) semiconductor/environment interface. SAM: self-assembled monolayer; OSC: organic semiconductor; CIL: charge insertion layer.

devices, accelerating charge separation, and improving the device performance. By designing mixed or segregated crystal heterojunctions made of OSCs, functional devices such as photodetectors, memory devices, and ambipolar transistors can be realized.

Another important interface in OFETs is the semiconductor/electrode interface (Figure 1.2b) [9, 23–25], which usually determines the efficiency of charge-carrier injection and extraction. By carefully choosing the self-assembled monolayer (SAM) or charge insertion layer (CIL) to modify metal electrodes, the charge-injection barrier can be fine-tuned at the interface, which could significantly reduce the contact resistance between semiconductors and electrodes, thus enhancing the device performance. More interestingly, by using functional layers, whose molecular conformation or electronic structure can be switched by external stimuli, the properties of OFETs can respond to external signals. This concept sets the foundation for building functional devices such as optical/electrical switches, memories, and photodetectors.

The semiconductor/dielectric interface is a vital interface [26, 27] that dominates carrier transport (Figure 1.2c) because (i) charge carriers are generated at the interface and (ii) the electrical current path is through at most the first few layers of molecules at the interface. In addition, the surface energy and microstructure of the dielectric interface have a significant impact on the local morphology of the semiconductor layer. Therefore, modification of this interface by SAMs or stimuli-responsive layers offers an important and universal methodology to improve the device performance and to even integrate new molecular functionalities such as photo-controllable memories, superconductors, and charge-trap memories into organic electrical circuits.

The semiconductor/environment interface is useful for building functional OFETs for sensing applications [12, 13, 15, 17, 28–30] (Figure 1.2d) by directly exposing the semiconductor layer to the environment being analyzed. This method offers some valuable advantages over other technologies, such as label-free capability, real-time detection, and compatibility with both the biological system and the electronics industry. Some intrinsic drawbacks, however, hamper the use of OFET-based sensors for practical sensing applications. The first is the diffusion of the analyte into grain boundaries in the semiconductors, causing either trapping or doping of charge carriers. This could result in serious problems such as device instability and irreversibility because of the physically destructive interaction between semiconductors and analytes. Second, because of the signal amplification characteristics of OFETs, any nonspecific response or noise would cause interference with the intrinsic signal of the analytes, leading to a pseudo-signal and poor selectivity. Therefore, rational control of interfacial charge transport is crucial to control the doping effects, increase the sensing selectivity/sensitivity, and even install new functionalities.

1.2 Brief Historic Overview of Interface Engineering in OFETs

We have summarized in Figure 1.3 the key developments to give a comprehensive big picture of the interface-engineered functional OFETs. Starting from the first demonstration [31, 32] of OFETs in the 1980s to their very recent applications [21, 29, 30] in skin electronics, the performance and functionality of OFETs have evolved simultaneously and promoted each other, shedding light on their potential applications in healthcare, wearable electronics, and degradable electronics.

1.3 Scope of the Book

In this comprehensive *book*, we aim to provide a systematic survey of recent significant advances in developing effective methodologies of interface engineering in OFETs, from fundamental models to experimental techniques, for creating high-performance and multifunctional optoelectronic devices. In Chapter 2, we

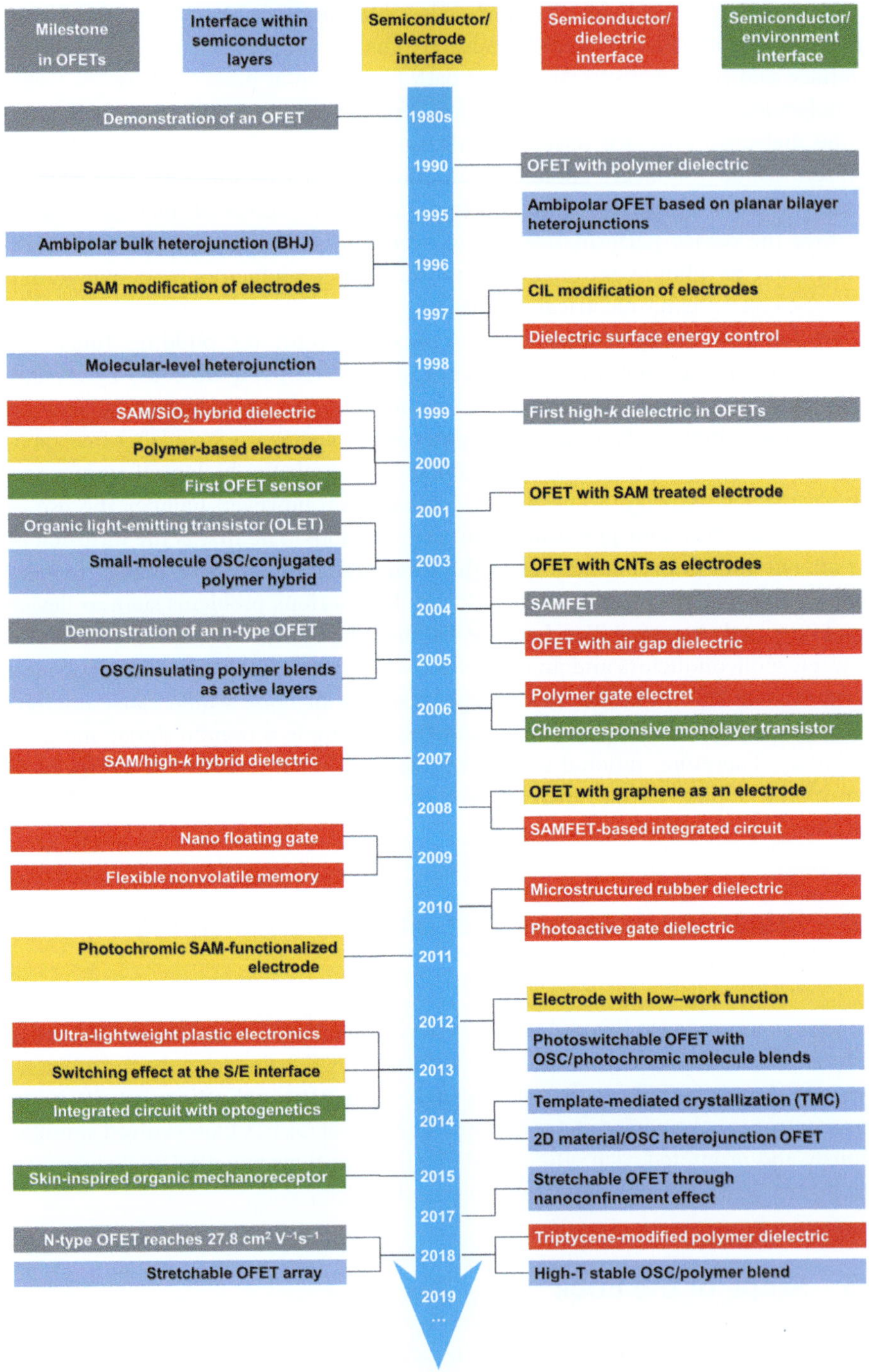

Figure 1.3 Timeline showing key developments in interface-engineered functional OFETs [10, 21, 25, 28, 29, 31, 32, 37, 57, 71, 74, 77, 81, 114, 137, 143–145, 160, 168, 172, 180, 187, 205, 237–240, 298, 312, 315, 326, 331, 333, 348, 353, 359–361, 376, 393, 397, 417, 447, 459].

will summarize interfacial modification methods in OFETs, which include physical and chemical modification methods. We will not go into details of them. Instead, summative tables and figures will be present to compare each method. Then, we focus on how to use interface engineering strategies to improve the inherent properties of OFETs at the semiconductor/semiconductor interface (Chapter 3), semiconductor/electrode interface (Chapter 4), semiconductor/dielectric interface (Chapter 5), semiconductor/environment interface (Chapter 6), as well as interfacing organic electronics with biology (Chapter 7). Both are of great importance to lie the foundation of the OFET research. We also provide a critical discussion of the limitations and main challenges that still exist for the development of practical applications in Chapter 8. Such a book will be invaluable for better understanding fundamental charge-transport mechanisms at the interfaces and will offer novel insights into the device design concept, the device fabrication process, and the development of future multifunctional practical transistors.

2

Interfacial Modification Methods

Over the past two decades, extensive efforts [6, 8, 9, 16] have been devoted to the modification of interfaces in organic field-effect transistors (OFETs), including physical modification methods, such as encapsulation or insertion on the surface, and chemical modification methods, such as covalently bonded self-assembled monolayer (SAM) modification.

2.1 Noncovalent Modification Methods

2.1.1 Charge Insertion Layer at the Electrode Surface

Because the injection barrier originates generally from the gap between the Fermi level of metal electrodes and the highest occupied molecular orbital (HOMO)/lowest unoccupied molecular orbital (LUMO) energy levels of organic semiconductors (OSCs) according to the energetic alignment, a reasonable solution is to insert a suitable additional layer to modify the electrode and function as a springboard for charge carriers, thus possibly overcoming the poor injection [70]. Initially, this method of introducing a thin buffer layer of several nanometers at metal OSC/electrode interfaces, referred to as a charge insertion layer (CIL) [71, 72], had already been applied to increase the efficiency of organic light-emitting diodes by improving the contact.

A preliminary study [73] was carried out by Hajlaoui et al., who inserted a thin film of tetracyanoquinodimethane (TCNQ) (**1**), working as molecular dopant, between Au and OSC active layers in OFETs to achieve an increased hole mobility. By using an ultrathin LiF interlayer as CIL between Al electrodes and F_{16}CuPc active layers [33], enhanced n-channel performance was achieved. Since then, CIL modification has attracted more and more attention in terms of enhancing the hole/electron injection efficiency. Various materials, such as inorganic compounds (metal oxides and inorganic salts) and organic compounds (small molecules and polymers), have been used as CILs, as shown in Figure 2.1.

Metal oxides [74] (Figure 2.1a) provide a variety of candidates for CILs because of their wide range of work functions. Transition metal oxides in particular

Interface Engineering in Organic Field-Effect Transistors, First Edition. Xuefeng Guo and Hongliang Chen.
© 2023 WILEY-VCH GmbH. Published 2023 by WILEY-VCH GmbH.

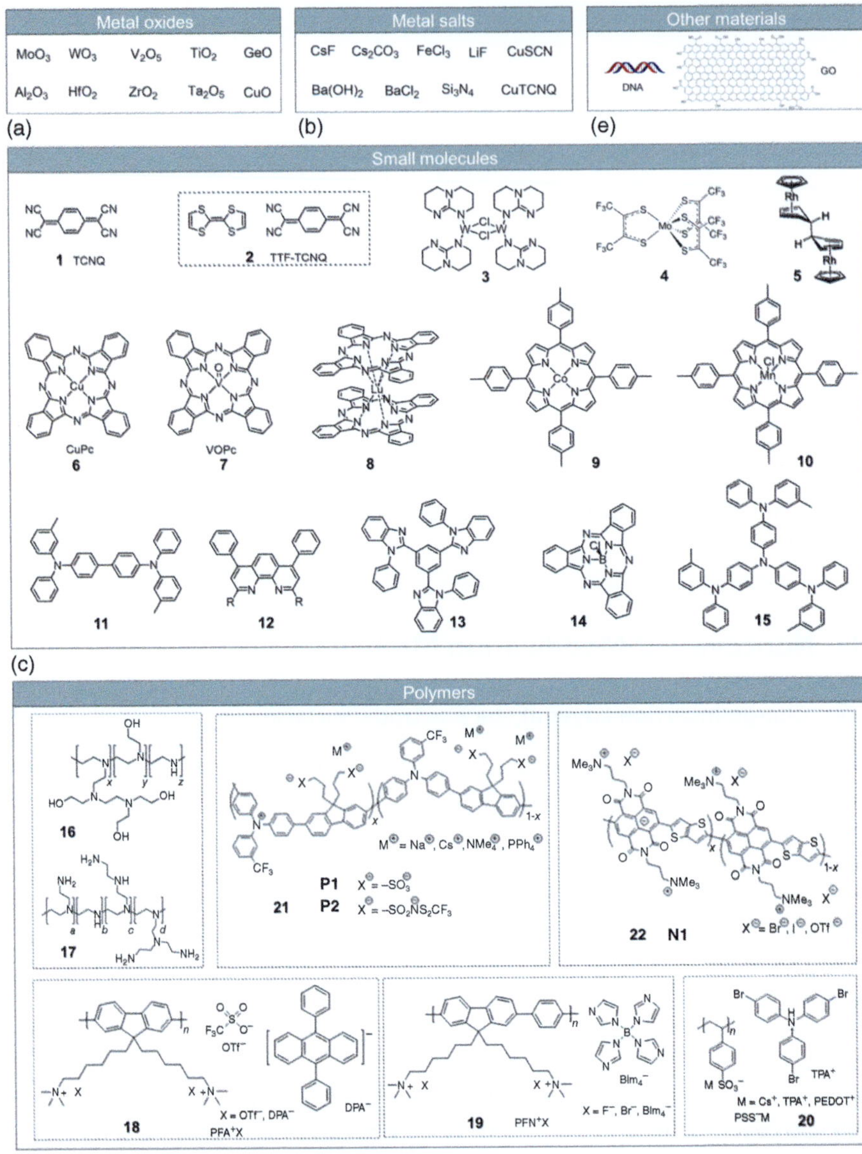

Figure 2.1 Chemical structures of the charge insertion layer (CIL). (a) Metal oxides, (b) metal salts, (c) small molecules, (d) polymers, and (e) other materials.

[75, 76], whose work functions range from very low ($\varphi \approx 3$ eV for ZrO_2) to extremely high ($\varphi \approx 7$ eV for V_2O_5), are suitable for modifying different metals with matched energy levels. After introducing the CIL, the thickness of the depletion layer between the metal and metal oxide decreases to a few nanometers, producing a reachable distance for tunneling injection, and the contact resistance could be significantly reduced once the tunneling injection is dominant. The CIL layer can

also act as a protection layer against the diffusion [77] of metals from electrodes or the permeation [78] of O_2 or H_2O into the active layer, thus improving the device stability.

Metal salts are another big family of inorganic materials working as CILs (Figure 2.1b) [79, 80]. They experience a similar mechanism as metal oxides. One of the benefits of using metal salts as CILs is their compatibility with the solution-process method. For example, CuSCN [81, 82] can be spin-casted on Au electrodes as a novel CIL with an intrinsic electron-blocking property. However, there are some drawbacks of metal salts when applying them as CILs. First, the thickness of the CIL affects electron transport. A thicker layer of CsF [83] (more than 0.6 nm) decreased the mobility owing to its insulating nature. Second, when depositing a CIL into ambipolar OFETs by using a single component, the outcome performance of p- and n-channels cannot be enhanced simultaneously. This problem could be solved by mixing two different materials to form a hybrid CIL [84]. The injection barrier for different types of carriers was tuned finely according to the molecular ratio of the two compounds.

When it comes to organic small molecules as CILs (Figure 2.1c), i.e. charge-transfer (CT) complex (**2**) [85], metal complex (**3–5**) [86, 87], metal tetraphenylporphyrin and phthalocyanine (**6–10**), [88] and N-heterocyclic compounds (**11–15**) [89], direct CT occurs at the interface between the organic CIL and the semiconductor, which helps to reduce the contact resistance by the additional interface dipole. However, a problem that still exists is the possibility of matrix–dopant hybridization [90, 91]. The organic polymers with dipolar groups [23, 92, 93] (**16** and **17**) (Figure 2.1d) can also act as electrode surface modifiers to achieve electrodes with a low work function. Polyelectrolytes (**18–22**), which comprise the groups capable of being charged with cations or anions, were proven recently to be a promising CIL to achieve ohmic contacts with a wide range of work functions (3.0–5.8 eV) [23].

Other materials (Figure 2.1e), such as DNA [94] and graphene oxides (GOs) [95–97], have also been used as CILs. The presence of phosphate groups in the DNA leads to a dipole layer, which could change the orientation in response to applied gate voltages and enhance the charge injection of both types of carriers. When GOs were applied as CILs in OFETs, the improvement in device performance could be attributed to the good contact nature of carbon-based materials with semiconductors, which will be discussed in Section 3.3.2.

2.1.2 Dielectric Surface Passivation Methods

Polymer encapsulation is a main choice for modifying the inorganic dielectric surface. Most of the polymer modifier, for example poly(imide-siloxane) [98], poly(4-vinylpyridine) (PVPyr) [99], and polystyrene (PS) [100], are used to change the surface energy of the dielectric and thus influence the grain size of OSCs growing whereon. There is still a debate on how the surface energy of polymer modifiers influence performance of OSC thin films; but as a general rule, the closer the surface energy of the dielectric layer is to that of the OSC, the higher the carrier mobility [101]. More fundamentally, it is a competition between the aggregation force within the OSC and the adsorption of OSC on the substrate.

The polymer modifiers, however, suffer from high roughness and trap density, which hamper charge transport on the surface. The same problem exists when directly using insulating polymers as gate dielectrics [102–105]. These applications call for the efficient surface modification methods of polymer films. However, the lack of functional groups makes it very hard to modify polymer dielectrics. Several strategies have been developed recently to solve this problem:

(i) By directly depositing a bilayer dielectric consisted of a bulk dielectric layer and an ultrathin interface layer (Figure 2.2a,b);
(ii) By using blends of two components to induce a phase-separated bilayer structure (Figure 2.2c,d);
(iii) By utilizing a monolayer capable of self-assembly on the polymer surface (Figure 2.2e–g).

For the first strategy, a bilayer polymer dielectric (Figure 2.2a) was employed [34], which consisted of a bulk layer of electropositive copolymer (**26**) for inducing electrostatic polarization along the channel and an ultrathin interface layer of **23** on top to form a good interface with OSCs. The polymer dielectric was prepared via a one-step, low-temperature, solvent-free chemical vapor deposition process with different ratios of two monomers **24** and **25** (Figure 2.2b) [35].

The second strategy can be realized by blending a fluorinated high-k polymer (**27**) with poly(methyl methacrylate) (PMMA) (**28**) as a dielectric [106]. Generally, the pure high-k dielectric layer of **27** can enhance the hole mobility but suppress the electron mobility because of the interfacial C–F dipole. In the blends, however, both p- and n-channel behaviors were increased simultaneously in ambipolar polymers because of the existence of vertical phase separation that induces a PMMA-rich region adjacent to the semiconductor layer (Figure 2.2c), leading to reduced interfacial dipolar disorder (Figure 2.2d). These discoveries provided new strategies for designing dielectric materials by choosing multiple components with proper chemical and morphological properties.

Last but not least, the technique of using SAMs via covalent bonding in inorganic dielectrics could not work well on polymer substrates, because of the lack of particular anchoring sites on the polymer surface whose chains are randomly oriented. To overcome this problem, Yokota et al. [108] applied a particular paraffinic triptycene (**29**) (Figure 2.2f), which can form completely oriented thin films with a high structural order featuring 2D nested hexagonal packing and 1D layer stacking for the surface functionalization of a polymer substrate. The triptycene formed noncovalently on polymer substrates, where the paraffinic side chains were arranged vertically to the substrate, thus equalizing the surface properties of these substrates (Figure 2.2e) [107]. With a few-layer triptycene film, the surface energies of different gate dielectrics were almost identical (~22.2 $mJ\,m^{-2}$), thus ensuring the high crystallinity and enhanced structural integrity of OSCs (Figure 2.2g). This is a more general strategy by depositing a few-layer film of triptycene on polymer substrates, shedding new light on interface engineering of the OSC/polymer dielectric interface.

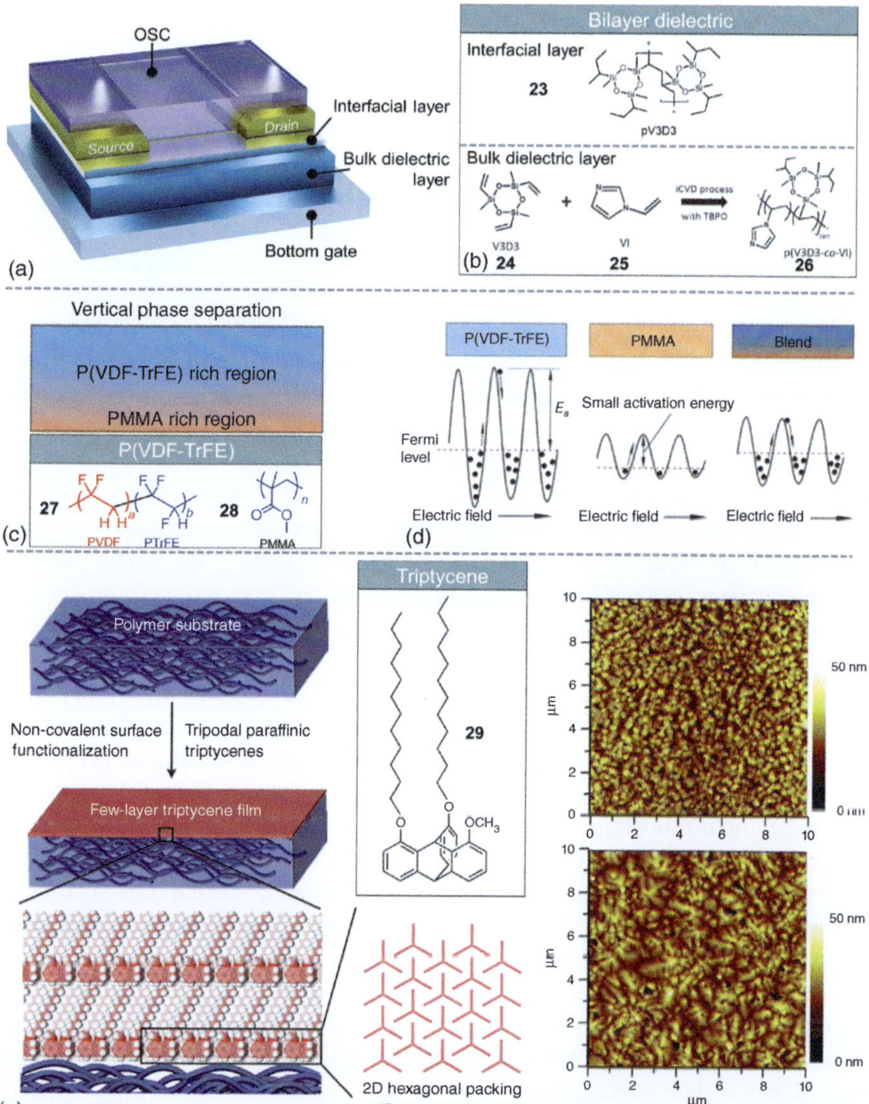

Figure 2.2 Modification of polymer dielectrics. (a) Schematic representation showing the bilayer dielectric made from an initiated chemical vapor deposition process. (b) Molecular structures of the interfacial layer and the bulk dielectric layer. (c) A schematic structure of the vertical phase separation of the blended dielectric. (d) Schematic representation showing the effect of the polaronic disorder and carrier concentration in OFETs with three dielectrics. (e) A few-layer triptycene film featuring 2D nested hexagonal packing and one-dimensional (1D) layer stacking on a parylene dielectric. (f) Molecular structure of the triptycene molecule. (g) Atomic force microscopy (AFM) height images of evaporated OSC films on parylene (top) and triptycene-coated parylene (bottom) gate dielectrics. Source: (a, b) Adapted from Pak et al. [34]. (c, d) Adapted from Khim et al. [106]. (e–g) Reproduced from Yokota et al. [107], © 2018 Springer Nature.

2.2 Covalent Modification Methods

2.2.1 SAM Modification of Electrodes

The metal electrode surface, which is capable of absorbing adventitious organic materials, offers a good platform for studying SAMs formed by chemisorption. SAM modification is not only a universal method to modify the surface properties [36, 109–112], but also a method for providing more functionalities [37, 113]. Herein, we focus on SAM modification methods applied to the metal electrodes in OFETs (Figure 2.3).

Commonly used SAM molecules are thiol derivatives (**30–53**), which have a sulfhydryl group for docking to the surfaces of metals such as Au. This forms metal–sulfur interactions through the sulfur long-pair electrons, called the thiolate—gold bond (Au—S), whose strength is close to that of the Au—Au bond [38, 114]. Organic compounds with disulfide (S–S) (**54–56**) or sulfide acetyl (-SAc) (**57–59**) bonds can also form SAMs on Au or Ag electrode surfaces [115].

Thioketone (**60**), which is structurally similar to pentacene, was used as a SAM to modify Au electrodes prior to the deposition of pentacene [116], which induced a complementary surface for the growth of pentacene and reduced the contact resistance. In addition, dithiocarbamate (DTC)-based molecules (**61–69**) have been proven to be stably adsorbed on metals under various types of environmental stresses [117], and could reduce the work function of noble metals (Au and Ag) to an ultralow value (such as ~3.1 eV for Au) [118]. Recently, *N*-heterocyclic carbenes (NHCs) (**70**) emerged as a versatile surface modification of gold surfaces [119, 120], exhibiting higher stability under harsh conditions in contrast to the most commonly used linear organosulfur ligands.

Despite the progress, great challenges remain in the SAM modification of metal electrodes. The relatively poor mechanical performance and scratch resistance hamper their practical application and commercialization. Carbon nanomaterials, i.e. carbon nanotubes (CNTs) and graphene (Gr), have emerged as potential electrodes with high conductivity, stability, and producibility. However, their chemical inertness makes them hard to functionalize without breaking the well-defined structures and scarifying their performances. Moreover, the mass production, purification, and integration of carbon nanomaterials as electrodes is a huge systematic project. There is still a long way to go.

2.2.2 SAM Modification of Dielectrics

The SAM modification strategy is worth being highlighted at the OSC/dielectric interface. Based on different dielectric materials, we summarize three aspects to discuss the topic in detail: (i) SAM/SiO_2 dielectrics, (ii) SAM/high-*k* dielectrics, and (iii) self-assembled monolayer field-effect transistors (SAMFETs). In terms of different types of molecules that have been used [121], two categories of the most-commonly-used SAM molecules are discussed:

(i) Silane-based SAM molecules [122], which prefer to react with the hydroxyl group of -SiOH on the silicon-based dielectrics (SiO_2 and Si_3N_4) to form SAMs via covalent linkage of Si—O—Si bonds.

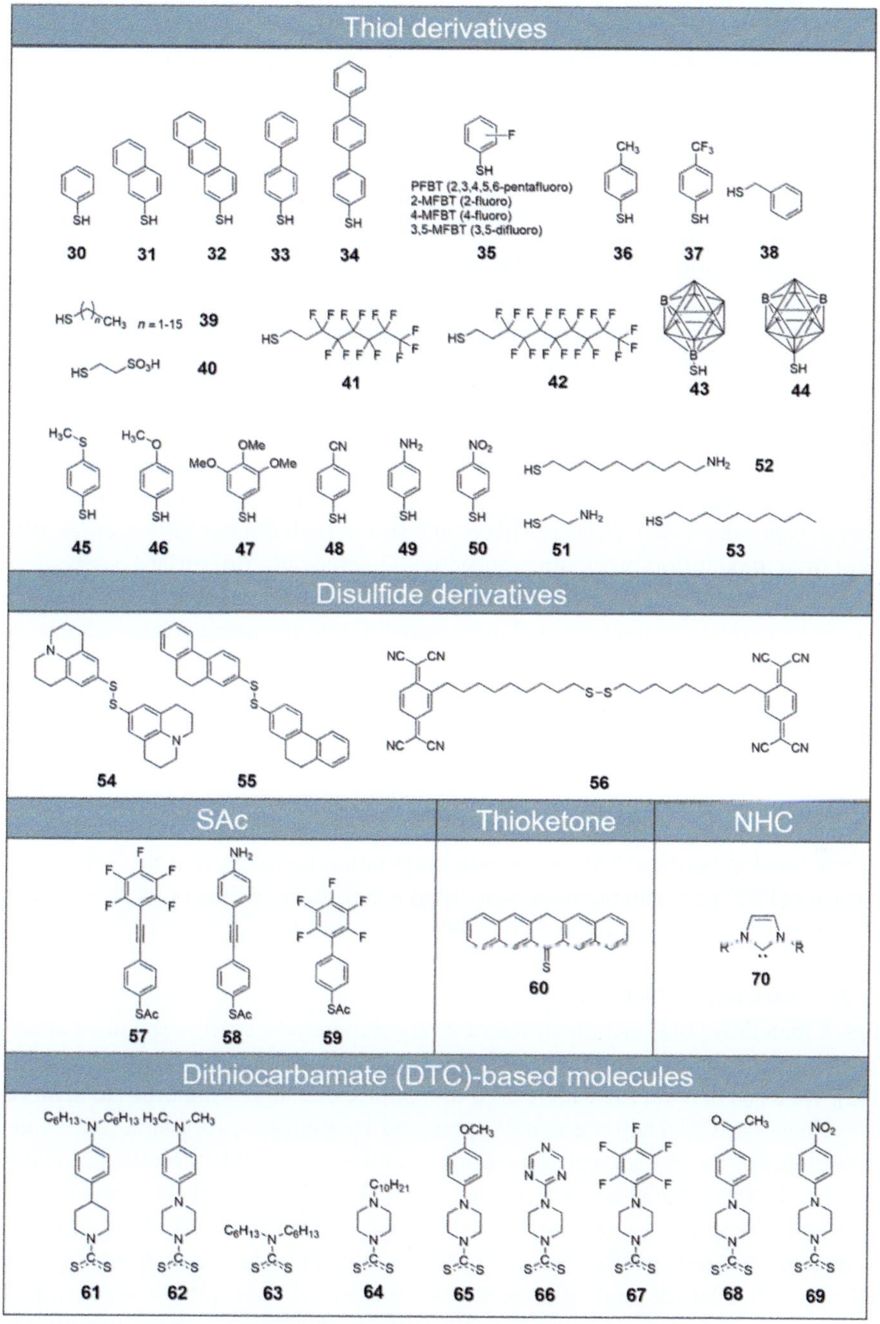

Figure 2.3 Chemical structures of SAMs used to modify metal electrodes.

(ii) The phosphonic acid (PA) and its phosphonate ester derivative-based SAM molecules[111], which can form PA–SAMs on a large range of inorganic oxide dielectrics, including high-*k* materials, by multidentate binding of strong covalent P–O–M anchoring (M stands for metal atom). Under normal ambient conditions, phosphonic acid molecules do not react chemically on the surface of SiO_2 [123].

2.2.2.1 SAM/SiO_2 Dielectrics

Surface modification with organosilanes (**71–102**) (Figure 2.4) has been a common method for a long time to prepare SAMs on a wide range of surfaces bearing –OH groups [124], such as silica, aluminum oxide, and zinc oxides. The modification can be done by reactions in solution or in the vapor phase. The challenge is that uniform silane monolayers are still difficult to obtain [121], thus limiting the device performance to some extent.

The processing conditions have a great impact on the quality of the resulting SAM. The SAM coverage was found to be very sensitive to the pretreatment conditions. In general, vapor-deposited SAMs are likely to result in high but not full coverage, different from the solution-processing procedure. Theoretically, full surface passivation and zero-onset voltage are only possible at near 100% SAM coverage, and if the coverage is incomplete or uncontrolled, high variability in the onset voltage occurs. A partially covered gate-dielectric SAM would cause two impacts on the transport of an OSC [125]. The first impact is the surface-induced electronic disorder in the adjacent semiconductor, which broadened the distribution of charge-transport states. The second impact is the rigid shift of the transport levels of the semiconductor owing to electrostatic interactions between the SAM and the semiconductor, inducing either hole or electron accumulation accordingly. To get a comprehensive understanding of the relationship between SAM materials, preparation methods, and the field-effect transistor (FET) performances, we have listed below recent works of SAM-modified dielectric surfaces, which are shown in Table 2.1.

2.2.2.2 SAM/High-*k* Dielectrics

High-*k* materials [144] include different kinds of materials, such as the most abundant inorganic oxides [145] and polymers [146, 147], as well as hybrid dielectrics [148]. We focus here on the widely used inorganic oxide high-*k* materials to achieve low-voltage OFETs. They are usually prepared by solution processing and vapor deposition, such as the sol–gel and atomic layer deposition (ALD) methods. Owing to some distinct characteristics of high-*k* metal oxides, i.e. (i) the unavoidable charge-carrier trap sites on the surface (usually associated with –OH groups), (ii) the induced ionic polarization owing to the interaction between charge carriers and the high-*k* lattice, and (iii) the surface roughness, they can adversely affect the device performance. In order to mitigate the problems, surface modification is needed for controlling the surface properties of high-*k* dielectrics.

PA–SAMs (**107–134**) (Figure 2.5) are prepared easily in air and are able to form well-ordered, strongly covalently bonded films on high-*k* oxide surfaces [149]. PA–SAMs show better stability to moisture and less tendency to homocondensation

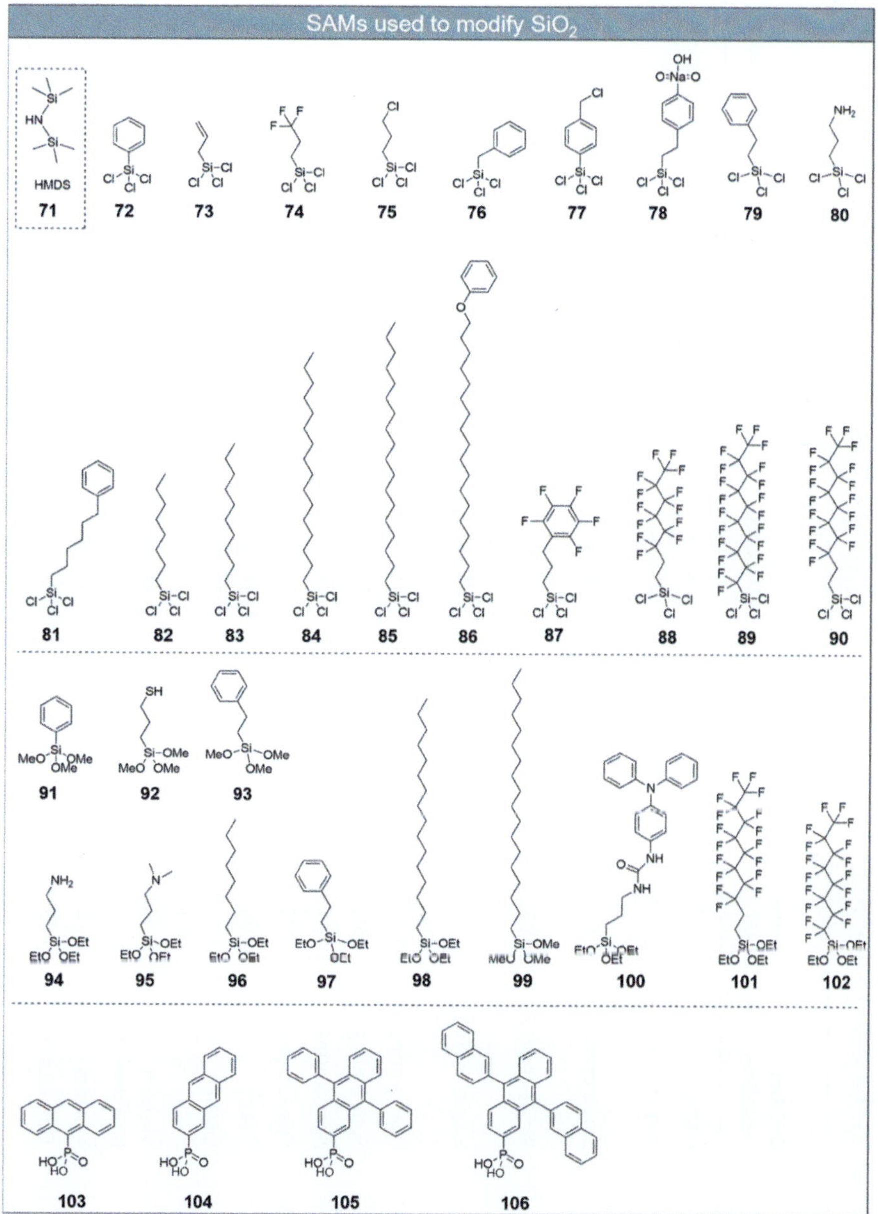

Figure 2.4 Molecular structures of SAMs used to modify the SiO_2 surface. Surface modification with organosilanes (**71–102**) is the most common method. Phosphonic acid (**103–106**) can also be used in some cases.

Table 2.1 The device performance of the devices with SAM-modified SiO_2 dielectrics.

Type	OSC material	OSC film deposition	Silane SAM	SAM preparation	Electrode	Average mobility (highest)	On/off ratio	References
p	Pentacene	OMBD[a)]	ODTS	Dipping	TC Au	1.2	10^6	[126]
p	Pentacene	Single crystals	ODTES	PDMS stamp printing	BC Au	~0.3 (0.36)	10^5	[127]
p	Rubrene	Single crystals	ODTES	PDMS stamp printing	BC Au	0.6 ± 0.5 (2.4)	10^7	[127]
n	C60	Single crystals	ODTES	PDMS stamp printing	BC Au	0.03	10^2	[127]
n	TCNQ	Single crystals	ODTES	PDMS stamp printing	BC Au	10^{-4}	10^1	[127]
p	MEH-PPV	Spin-coating	ODTS	Vapor	TC Au	1.2×10^{-4}	10^2	[128]
p	MEH-PPV,	Spin-coating	HMDS	Vapor	TC Au	1.3×10^{-3}	10^2	[128]
p	OC1C10-PPV	Spin-coating	ODTS	Vapor	TC Au	1.6×10^{-4}	10^2	[128]
p	OC1C10-PPV	Spin-coating	HMDS	Vapor	TC Au	1.3×10^{-3}	10^2	[128]
p	Pentacene	OMBD	ordered ODTS	Dipping	TC Au	~0.6	10^6	[129]
p	Pentacene	OMBD	disordered ODTS	Dipping	TC Au	~0.3	10^6	[129]
p	Pentacene	Vacuum sublimation	HDMS	Vapor	BC Au	0.1–0.6	10^5	[130]
p	Pentacene	Vacuum	β-PhTS	Immersion	TC Cu	0.4	—	[131]
p	8QT8	Drop-casting	β-PhTS	Coating	TC Au	0.014	10^6	[132]

p	8QT8	Drop-casting	β-PhTS	Coating	TC Au	0.17	—	[132]
p	pBTCT	Spin-coating	ODTS	Immersion	Au	0.021	10^7	[133]
p	pBTCT	Spin-coating	OTS	Immersion	Au	0.018	10^6	[133]
p	pBTCT	Spin-coating	HMDS	Refluxing	Au	0.005	10^5	[133]
p	pBTCT	Spin-coating	PTS	Immersion	Au	1.6×10^{-3}	10^5	[133]
p	pBTCT	Spin-coating	APTES	Immersion	Au	5×10^{-5}	10^6	[133]
p	PB16TTT	Spin-coating	β-PhTS	Immersion	BC Au	0.24 (0.25)	10^7	[134]
p	PB16TTT	Spin-coating	HMDS	Immersion	BC Au	0.28 (0.30)	10^7	[134]
p	PB16TTT	Spin-coating	OTS	Immersion	BC Au	0.48 (0.51)	10^7	[134]
p	PB16TTT	Spin-coating	FOTS	Immersion	BC Au	0.56 (0.74)	10^7	[134]
p	PB16TTT	Spin-coating	β-PhTS	Immersion	TC Au	0.35 (0.38)	10^6	[134]
p	PB16TTT	Spin-coating	HMDS	Immersion	TC Au	0.41 (0.46)	10^7	[134]
p	PB16TTT	Spin-coating	FOTS	Immersion	TC Au	0.59 (0.65)	10^7	[134]
p	PB16TTT	Spin-coating	β-PhTS	Immersion	TC Au	0.84 (1.0)	10^6	[134]
p	Pentacene	Thermal evaporation	ODTS	Immersion	TC Au	0.2	10^5	[135]
p	Pentacene	Thermal evaporation	ODTMS	Vapor	TC Au	0.6 (0.9)	10^6	[136]
p	Pentacene	Thermal evaporation	ODTMS	LB technique	TC Au	2.1 (2.3)	10^6	[136]
n	C60	Thermal evaporation	ODTMS	Vapor	TC Au	0.2 (0.27)	10^6	[136]
n	C60	Thermal evaporation	ODTMS	LB technique	TC Au	4.1 (5.3)	10^7	[136]

(Continued)

Table 2.1 (Continued)

Type	OSC material	OSC film deposition	Silane SAM	SAM preparation	Electrode	Average mobility (highest)	On/off ratio	References
p	PTAA	Spin-coating	FPPTS	Immersion	BC Au	$5.2\ (\pm 1.3) \times 10^{-4}$	10^4	[137]
p	PTAA	Spin-coating	MPTMS	Vacuum	BC Au	$5.2\ (\pm 1.5) \times 10^{-4}$	10^5	[137]
p	PTAA	Spin-coating	APTES	Vacuum	BC Au	$1.4\ (\pm 0.7) \times 10^{-4}$	10^4	[137]
p	PTAA	Spin-coating	TAATS	Immersion	BC Au	$2.0\ (\pm 0.1) \times 10^{-4}$	10^5	[137]
p	PTAA	Spin-coating	ODTS	Immersion	BC Au	—	—	[137]
p	Pentacene	Thermal evaporation	ODTMS	Spin-casting	TC Au	2.8 ± 0.2	10^6	[138]
p	Pentacene	Thermal evaporation	ODTS	Vapor	TC Au	0.52 ± 0.04	10^5	[138]
n	C60	Thermal evaporation	ODTMS	Spin-casting	TC Au	4.7 ± 0.41	10^7	[138]
n	C60	Thermal evaporation	ODTS	Vapor	TC Au	0.27 ± 0.15	10^5	[138]
n	PTCDI-C4F7	Thermal evaporation	ODTMS	Spin-casting	TC Au	1.4	10^6	[138]
n	PTCDI-C4F7	Thermal evaporation	ODTS	Vapor	TC Au	0.72	—	[138]
p	Pentacene	Thermal evaporation	9-Ant-PA	T-BAG treatment	TC Au	0.8 (0.9)	10^7	[139]
p	Pentacene	Thermal evaporation	2-Ant-PA	T-BAG treatment	TC Au	1.6 (2.4)	10^7	[139]

p	Pentacene	Thermal evaporation	DiPh-2-Ant-PA	T-BAG treatment	TC Au	1.8 (3.6)	10^7	[139]
p	Pentacene	Thermal evaporation	DiNaph-2-Ant-PA	T-BAG treatment	TC Au	2.5 (4.7)	10^7	[139]
p	P-BTDT	Thermal evaporation	High humidity (HH)-ODTS	Immersion	TC Au	0.22 (0.250)	10^8	[125]
p	P-BTDT	Thermal evaporation	anhydrous (AA)-ODTS	Immersion	TC Au	0.085 (0.090)	10^8	[125]
p	Pentacene	Thermal evaporation	ODTS	Vapor	TC Au	0.517 ± 0.015	—	[140]
p	Pentacene	Thermal evaporation	FDTS	Vapor	TC Au	0.449 ± 0.002	—	[140]
p	C10–DNF–VV	Edge-cast	DTES	Vapor	TC Au	1.3	—	[141]
p	C10–DNF–VW	Edge-cast	DTES	Vapor	TC Au	1	—	[141]
p	C10–DNF–VW	Edge-cast	β-PhTS	Vapor	TC Au	1.1	—	[141]
p	Pentacene	OMBD	ODTS	Dipping at −30 °C	TC Au	0.46 (0.61)	—	[142]
p	Pentacene	OMBD	ODTS	Dipping at −5 °C	TC Au	0.31	—	[142]
p	Pentacene	OMBD	ODTS	Dipping at 20 °C	TC Au	0.19	—	[142]
p	P3HT	Spin-coating[b)]	OTS	Casting	BC Au	0.156 ± 0.010	10^3	[39]
Ambipolar	PNDTI-BT-DT	Spin-coated	ODTS	Immersion	TC Au	0.2 (p)/0.28 (n)	10^3(p)/10^3(n)	[143]
Ambipolar	PNDTI-BT-DT	Spin-coated	FDTES	Vapor	TC Au	0.43 (p)	10^6	[143]
Ambipolar	PNDTI-BT-DT	Spin-coated	MAPTES	Immersion	TC Au	0.24 (n)	10^6	[143]

a) OMBD: organic molecular-beam deposition.
b) Spin-coating from ultrasonicated solution for four minutes.

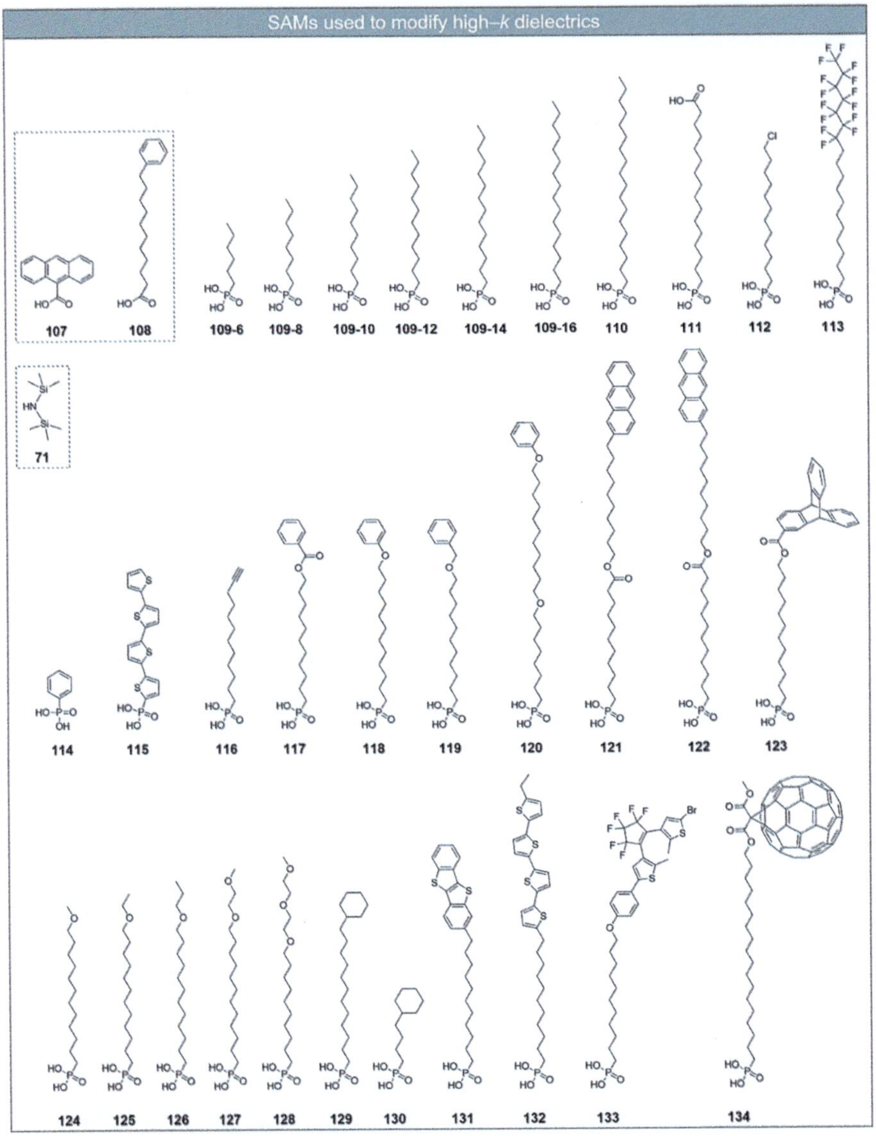

Figure 2.5 Chemical structures of SAMs used to modify high-*k* inorganic dielectrics.

between phosphonic acids. Once the monolayer is formed, the SAM [150, 151] shows more resistance of phosphonate to hydrolysis than siloxane. PA-SAMs [152, 153] can also be used to modify transparent electrodes based on metal oxides, such as indium tin oxide (ITO), with almost no impact on the transparency. To get a comprehensive understanding of the relationship between SAM materials, preparation methods, and FET performances, we have listed below recent works of SAM-modified high-*k* dielectrics, which are shown in Table 2.2.

Table 2.2 The device performance of the devices with SAM-modified high-*k* dielectrics.

Material	High-*k* preparation	PA-SAM	OSC material	OSC type	OSC deposition	V_{th}	Average mobility (highest mobility)	On/off ratio	References
CeO_2–SiO_2	Electron-beam deposition	HMDS	Pentacene	p	Vacuum	—	0.32	10^3	[154]
Al_2O_3–SiO_2	Magnetron sputtering	COOH-anthracene	Pentacene	p	Thermal evaporation	−18	0.17	—	[155]
Al_2O_3–SiO_2	Magnetron sputtering	Ba-C10	Pentacene	p	Thermal evaporation	−9	0.35	—	[155]
HfO_2–SiO_2	Sol–gel	ODPA	Pentacene	p	Thermal evaporation	−0.53	0.15	10^5	[156]
HfO_2–SiO_2	Sol–gel	π-σ-PA1	Pentacene	p	Thermal evaporation	−0.41	0.22	10^5	[156]
HfO_2–SiO_2	Sol–gel	π-σ-PA2	Pentacene	p	Thermal evaporation	−0.41	0.15	10^5	[156]
Al_2O_3	O_2 plasma treated Al	ODPA	Pentacene	p	Thermal evaporation	—	0.4	—	[157]
Al_2O_3	O_2 plasma treated Al	ODPA	F16CuPc	n	Thermal evaporation	—	0.01	—	[157]
HfO_2	Sol–gel	ODPA	PTzQT-14	p	Spin-coating	−0.32	0.10 ± 0.01 (0.11)	10^5	[158]
HfO_2 flexible	Sol–gel	ODPA	PTzQT-14	p	Spin-coating	−0.85	0.061 ± 0.0045 (0.068)	10^5	[158]
HfO_2	Sol–gel	π-σ-PA	Pentacene	p	Thermal evaporation	−0.5	0.32	10^5	[159]

(Continued)

Table 2.2 (Continued)

Material	High-*k* preparation	PA-SAM	OSC material	OSC type	OSC deposition	V_{th}	Average mobility (highest mobility)	On/off ratio	References
HfO_2	Sol–gel	ODPA	Pentacene	p	Thermal evaporation	−0.6	0.3	10^5	[159]
HfO_2	Sol–gel	π-σ-PA	TIPS-PEN	p	Drop cast	−0.5	0.38	10^5	[159]
HfO_2	Sol–gel	ODPA	TIPS-PEN	p	Drop cast	—	—	—	[159]
HfO_2–SiO_2	Sol–gel	C18-PA	Pentacene	p	Thermal evaporation	−1.06	0.25 ± 0.08	10^6	[160]
HfO_2–SiO_2	Sol–gel	C16-PA	Pentacene	p	Thermal evaporation	−0.88	0.30 ± 0.03	10^6	[160]
HfO_2–SiO_2	Sol–gel	C14-PA	Pentacene	p	Thermal evaporation	−0.86	0.82 ± 0.03	10^6	[160]
HfO_2–SiO_2	Sol–gel	C12-PA	Pentacene	p	Thermal evaporation	−0.69	1.09 ± 0.02	10^6	[160]
HfO_2–SiO_2	Sol–gel	C10-PA	Pentacene	p	Thermal evaporation	−0.59	0.84 ± 0.13	10^6	[160]
HfO_2–SiO_2	Sol–gel	C8-PA	Pentacene	p	Thermal evaporation	−0.58	0.74 ± 0.03	10^6	[160]
HfO_2–SiO_2	Sol–gel	C6-PA	Pentacene	p	Thermal evaporation	−0.56	0.56 ± 0.04	10^6	[160]
Al_2O_3–SiO_2	O_2 plasma treated Al on Si/SiO_2	PhO-19-PA	TIPS-PEN	p	Spin-cast	−0.7 to −0.9	0.02–0.11	10^6	[161]
Al_2O_3–SiO_2	O_2 plasma treated Al on Si/SiO_2	PhO-19-PA	PC61BM	n	Spin-cast	0.4–0.6	0.02–0.06	10^6	[161]

Al_2O_3–SiO_2	O_2 plasma treated Al on Si/SiO_2	PhO-19-PA	Pentacene	p	Thermal evaporation	−0.8	0.9–0.11	10^6	[161]
Al_2O_3–SiO_2	O_2 plasma treated Al on Si/SiO_2	C10-PA	DH4T	p	Thermal evaporation	−1.5	0.024	10^6	[162]
Al_2O_3–SiO_2	O_2 plasma treated Al on Si/SiO_2	PHDA	DH4T	p	Thermal evaporation	−1.3	0.018	10^6	[162]
Al_2O_3–SiO_2	O_2 plasma treated Al on Si/SiO_2	F15C18-PA	DH4T	p	Thermal evaporation	−0.4	0.024	10^6	[162]
AlO_x/banknote	O_2 plasma treated Al	ODPA	DNTT	p	Thermal evaporation	−1.0 to −1.4	—	10^4 (10^5)	[163]
AlO_x/banknote	O_2 plasma treated Al	ODPA	F16CuPc	n	Thermal evaporation	—	0.005	—	[163]
AlO_x	O_2 plasma treated Al	Phenyl PA	DH6T	p	Thermal evaporation	−1.100 ± 0.24	0.011 ± 0.003	—	[164]
AlO_x	O_2 plasma treated Al	Phenyl PA	Pentacene	p	Thermal evaporation	−1.780 ± 0.07	0.046 ± 0.03	—	[164]
AlO_x	O_2 plasma treated Al	Phenyl PA	F16CuPc	n	Thermal evaporation	0.31 ± 0.01	0.061 ± 0.005	—	[164]
AlO_x	O_2 plasma treated Al	C14-PA	DH6T	p	Thermal evaporation	−0.860 ± 0.08	0.034 ± 0.007	—	[164]
AlO_x	O_2 plasma treated Al	C14-PA	Pentacene	p	Thermal evaporation	−1.610 ± 0.05	0.833 ± 0.106	—	[164]

(Continued)

Table 2.2 (Continued)

Material	High-*k* preparation	PA-SAM	OSC material	OSC type	OSC deposition	V_{th}	Average mobility (highest mobility)	On/off ratio	References
AlO_x	O_2 plasma treated Al	C14-PA	F16CuPc	n	Thermal evaporation	0.330 ± 0.180	0.164 ± 0.056	—	[164]
AlO_x	O_2 plasma treated Al	4T-C12-PA	DH6T	p	Thermal evaporation	-0.80 ± 0.14	0.014 ± 0.005	—	[164]
AlO_x	O_2 plasma treated Al	4T-C12-PA	Pentacene	p	Thermal evaporation	-1.37 ± 0.07	0.001 ± 0.0001	—	[164]
AlO_x	O_2 plasma treated Al	4T-C12-PA	F16CuPc	n	Thermal evaporation	0.81 ± 0.070	0.014 ± 0.008	—	[164]
AlO_x	O_2 plasma treated Al	BTBT-C12-PA	DH6T	p	Thermal evaporation	-0.75 ± 0.02	0.015 ± 0.002	—	[164]
AlO_x	O_2 plasma treated Al	BTBT-C12-PA	Pentacene	p	Thermal evaporation	-1.23 ± 0.240	0.029 ± 0.022	—	[164]
AlO_x	O_2 plasma treated Al	BTBT-C12-PA	F16CuPc	n	Thermal evaporation	0.66 ± 0.04	0.031 ± 0.025	—	[164]
AlO_x	O_2 plasma treated Al	C60C18-PA	DH6T	p	Thermal evaporation	0.450 ± 0.04	0.016 ± 0.001	—	[164]
AlO_x	O_2 plasma treated Al	C60C18-PA	Pentacene	p	Thermal evaporation	-0.51 ± 0.07	0.155 ± 0.076	—	[164]
AlO_x	O_2 plasma treated Al	C60C18-PA	F16CuPc	n	Thermal evaporation	0.83 ± 0.09	0.004 ± 0.002	—	[164]
AlO_x	O_2 plasma treated Al	F15C18-PA	DH6T	p	Thermal evaporation	0.640 ± 0.01	0.022 ± 0.006	—	[164]

AlO_x	O_2 plasma treated Al	F15C18-PA	Pentacene	p	Thermal evaporation	-0.60 ± 0.04	0.097 ± 0.067	—	[164]
AlO_x	O_2 plasma treated Al	F15C18-PA	F16CuPc	n	Thermal evaporation	1.890 ± 0.120	0.043 ± 0.014	—	[164]
AlO_y/TiO_x	Spin-cast	ODPA	TIPS-TAP	n	Thermal evaporation	0.83 ± 0.08	1.3–1.8	—	[40]
AlO_y/TiO_x	Spin-cast	PA1	TIPS-TAP	n	Drop cast	1.1 ± 0.22	1.2 (2.5)	10^5	[40]
AlO_y/TiO_x	Spin-cast	PA2	TIPS-TAP	n	Drop cast	0.9 ± 0.25	0.88	—	[40]
AlO_y/TiO_x	Spin-cast	PA3	TIPS-TAP	n	Drop cast	—	—	—	[40]
AlO_y/TiO_x	Spin-cast	PA4	TIPS-TAP	n	Drop cast	1.2 ± 0.25	0.94	—	[40]
AlO_y/TiO_x	Spin-cast	PA5	TIPS-TAP	n	Drop cast	1.2 ± 0.24	0.97	—	[40]
AlO_y/TiO_x	Spin-cast	PA1	TIPS-TAP	p	Drop-cast	—	1.13	—	[40]
AlO_y/TiOx	Spin-cast	PA4	TIPS-TAP	p	Drop-cast	—	0.95	—	[40]
Al_2O_3–SiO_2	O_2 plasma treated Al on Si/SiO_2	BA-11-PA	Pentacene	p	Thermal evaporation	26.0 ± 12.3	3.5 ± 0.28	10^6	[165]
Al_2O_3–SiO_2	O_2 plasma treated Al on Si/SiO_2	12-PD-PA	Pentacene	p	Thermal evaporation	14.0 ± 7.3	3.15 ± 0.31	10^6	[165]
Al_2O_3–SiO_2	O_2 plasma treated Al on Si/SiO_2	DDPA	Pentacene	p	Thermal evaporation	18.6 ± 3.8	3.51 ± 0.39	10^6	[165]
Al_2O_3–SiO_2	O_2 plasma treated Al on Si/SiO_2	PhO-19-PA	Pentacene	p	Thermal evaporation	22.5 ± 7.5	0.91 ± 0.04	10^6	[165]

(Continued)

Table 2.2 (Continued)

Material	High-*k* preparation	PA-SAM	OSC material	OSC type	OSC deposition	V_{th}	Average mobility (highest mobility)	On/off ratio	References
Al_2O_3–SiO_2	O_2 plasma treated Al on Si/SiO_2	ODPA	Pentacene	p	Thermal evaporation	27.4 ± 6.8	0.98 ± 0.21	10^6	[165]
Al_2O_3–SiO_2	O_2 plasma treated Al on Si/SiO_2	Trip-12-PA	Pentacene	p	Thermal evaporation	42.1 ± 6.7	0.51 ± 0.15	10^5	[165]
AlO_x/PEN foil	Potentiostatic anodization of Al	C14-PA	DNTT	p	Thermal evaporation	−0.61 ± 0.13	0.88 ± 0.13 (1.6)	10^7	[166]
AlO_x/SiO_2	Potentiostatic anodization of Al	C14- PA	DNTT	p	Thermal evaporation	2	−0.76	10^7	[166]
AlO_y/TiO_x	Spin-coating	CDPA	Pentacene	p	Thermal evaporation	—	3.86 ± 0.47 (5.7)	—	[167]
AlO_y/TiO_x	Spin-coating	CDPA	Pentacene	p	Thermal evaporation	—	2.21 ± 0.46 (2.9)	—	[167]
AlO_y/TiO_x	Spin-coating	CDPA	C60	n	Thermal evaporation	—	2.98 ± 0.83 (5.1)	—	[167]
AlO_y/TiO_x	Spin-coating	CDPA	C60	n	Thermal evaporation	—	0.66 ± 0.28 (1.1)	—	[167]
AlO_y/TiO_x	Spin-coating	CDPA	TIPS-PEN	p	Drop-cast	—	1.64 ± 0.55 (2.7)	—	[167]
AlO_y/TiO_x	Spin-coating	CDPA	TIPS-TAP	n	Drop-cast	—	0.78 ± 0.32 (1.44)	—	[167]
Al_2O_3	Atomic layer deposition (ALD)	HC18–PA	DNTT	p	Thermal evaporation	−0.6	2	—	[168]
Al_2O_3	Atomic layer deposition (ALD)	HC18–PA	DNTT	p	Thermal evaporation	−0.4	2.3	—	[168]

Al_2O_3	Atomic layer deposition (ALD)	FC18–PA	DNTT	p	Thermal evaporation	1.2	1.8	—	[168]
Al_2O_3	Atomic layer deposition (ALD)	FC18–PA	DNTT	p	Thermal evaporation	1.3	2.1	—	[168]
AlO_x/PEN foil	O_2 plasma treated Al	HC14-PA	DNTT	p	Thermal evaporation	—	2.9 ± 0.3	10^6	[169]
AlO_x/PEN foil	O_2 plasma treated Al	FC18-PA	DNTT	p	Thermal evaporation	—	2.2 ± 0.2	10^6	[169]
AlO_x/PEN foil	O_2 plasma treated Al	HC14-PA	DPh-DNTT	p	Thermal evaporation	—	4.1 ± 0.1	10^6	[169]
AlO_x/PEN foil	O_2 plasma treated Al	FC18-PA	DPh-DNTT	p	Thermal evaporation	—	1.9 ± 0.1	10^6	[169]
AlO_x/PEN foil	O_2 plasma treated Al	HC14-PA	C10-DNTT	p	Thermal evaporation	—	3.6 ± 0.1	10^6	[169]
AlO_x/PEN foil	O_2 plasma treated Al	FC18-PA	C11-DNTT	p	Thermal evaporation	—	1.1 ± 0.04	10^6	[169]
AlO_x/PEN foil	O_2 plasma treated Al	HC14-PA	N1100	n	Thermal evaporation	—	0.5 ± 0.04	10^6	[169]
AlO_x/PEN foil	O_2 plasma treated Al	FC18-PA	N1100	n	Thermal evaporation	—	0.8 ± 0.1	10^6	[169]
AlO_y/TiO_x	Spin-coating	CDPA	TIPS-TAP	n	Drop-cast	11–15	7.6 ± 1.6	10^6	[41]
AlO_y/TiO_x	Spin-coating	CDPA	TIPS-TAP	n	Dip-coated	11–15	7.6 ± 1.9	10^6	[41]
AlO_y/TiO_x	Spin-coating	CDPA	TIPS-TAP	n	Thermal evaporation	18–22	5.1 ± 1.2	10^6	[41]

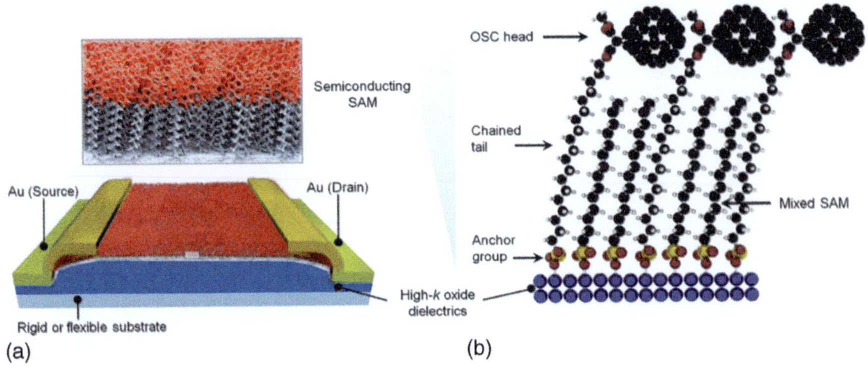

Figure 2.6 Organic semiconducting molecular design for SAMFEFs. (a) Schematic illustration showing a SAMFET device configuration. (b) Molecular design strategy of a typical SAMFET semiconductor. Source: (a, b) Adapted from Schmaltz et al. [42], © 2017 American Chemical Society.

2.2.2.3 Self-Assembled Monolayer Field-Effect Transistors (SAMFETs)

The covalent attachment of SAMs containing semiconducting cores onto an inorganic high-*k* dielectric surface provides a direct method for the development of miniaturized devices of SAMFETs, with low operation voltages and high electrical performance by reducing the dielectric thickness and improving the dielectric quality.

For OSC monolayer growing on rigid or flexible substrates to construct a SAMFET (Figure 2.6a), three golden rules (Figure 2.6b) should be taken into consideration:

(i) Anchor group was chosen based on the dielectric used;
(ii) Chained tail was chosen by considering interactions;
(iii) Head group was chosen to integrate the functionality.

We have summarized in Figure 2.7 and Table 2.3 some widely used OSCs and their performances in SAMFETs.

Early attempts to fabricate SAMFETs resulted in relatively poor performance at channel lengths of more than 1 μm because of the lack of long-range order of the OSC films. These molecules were designed by attaching the anchoring group directly to the rigid semiconductor core with no flexible chained tails [43, 170] which misses the second golden rule. For example, the Nuckolls' group [43] used a tetracene derivative (**135**) that had one of its terminal rings functionalized with *o*-hydroxyl as a catechol to form upright monolayers on the AlO_x surface. The SAM of semiconductor showed p-type operation. The current–voltage curves, however, did not allow quantitative extraction of the transistor parameters.

The flexile chained tails are necessary to tune the self-assembly process of monolayers. As described by Mottaghi et al. [44], who used a molecule (**137**) comprising a short alkyl chain (C6) linked to an oligothiophene moiety and a –COOH terminal group as an anchor on AlO_x, an estimated mobility of ~0.0035 $cm^2\ V^{-1}\ s^{-1}$ was

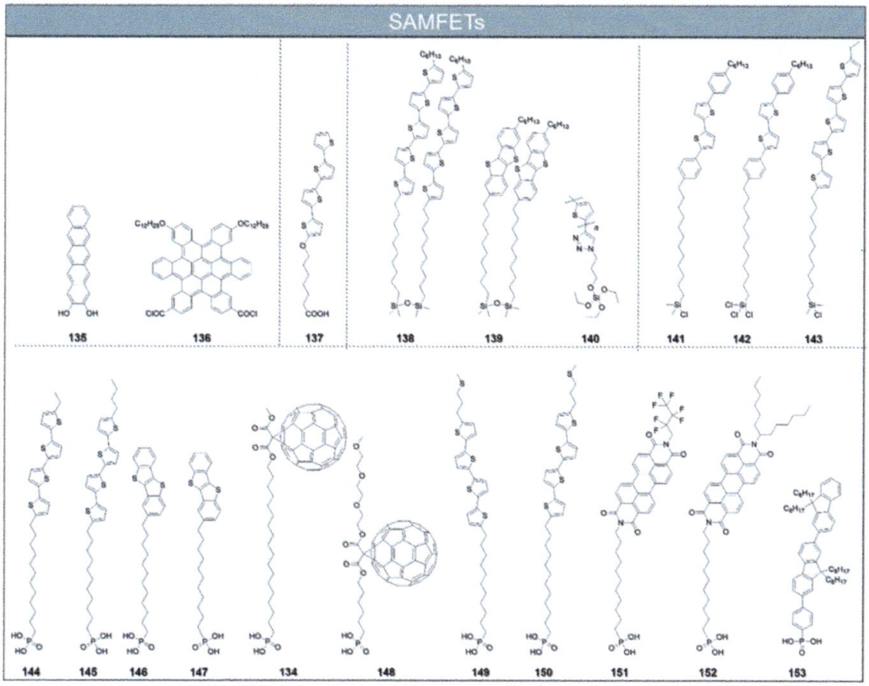

Figure 2.7 Molecular structures of OSCs for SAMFETs.

achieved in a submicrometer channel. A remarkable breakthrough was achieved by Smits et al. [171] who used a liquid-crystalline molecule (**143**) consisting of an oligothiophene core separated by a long aliphatic chain (C11) from a monofunctionalized anchor group. The SAMFET exhibited bulk-like carrier mobility (~0.04 $cm^2\ V^{-1}\ s^{-1}$) for a 40-μm channel with high reproducibility. Another good example of the use of long chained tail was demonstrated by Parry et al. [179] using monochlorosilane (**141**) or trichlorosilane (**142**) anchors.

N-type SAMFETs were demonstrated by employing the perylene diimide (PDI) derivates (**151** and **152**) [176, 177]. Because of the formation of a nanocrystalline monolayer, highly reproducible transistors were demonstrated with channel lengths up to 100 μm, showing electron mobilities of ~$10^{-3}\ cm^2\ V^{-1}\ s^{-1}$ and on/off current ratios up to 10^5.

In summary, by holistic molecular design, SAMFETs can achieve the FET property as good as bulk OFET devices. Considering their monolayer thickness, low-voltage operation, and high-sensitive OSC/environment, SAMFETs may have promising applications in chemo-/biosensors and flexible electronics. The operation stability and reproducibility are main concerns hindering their development. It calls for rational design of the molecules as well as the engineering of device architectures. A deep understanding of surface self-assembly mechanism [184], including the complicated inter-/intramolecular interactions as well as SAM–dielectric interactions, should be taken in the preparation of high-quality semiconducting SAMs.

Table 2.3 Summary of the device performance of SAMFETs.

SAM OSC	Anchor	Spacer	Dielectric	Mobility ($cm^2\ V^{-1}\ s^{-1}$)	On/off ratio	V_{th} (V)	Work voltage (V)	References
135	Ortho-hydroxyl	—	Al_2O_3	—	—	—	−2.5	[43]
136	Ortho-hydroxyl	—	SiO_2	1	10^4	—	−2	[170]
143	Monochlorosilane	C11	SiO_2	0.04	10^8	—	−2	[171]
143	Monochlorosilane	C11	SU8	0.02	10^6	—	−2	[172]
144	Phosphonic acid	C12	SiO_2	10^{-5}	10^3	−2.15	−4	[173]
148	Phosphonic acid	C6	SiO_2	10^{-4}	10^4	3	4	[173]
145	Phosphonic acid	C11	Al_2O_3	8.0×10^{-6}	10^2	−32 to −53	—	[174]
146	Phosphonic acid	C12	SiO_2	1.2×10^{-2}	10^2	—	−3	[175]
146	Phosphonic acid	C12	PEN	1.7×10^{-3}	10^3	—	−3	[175]
152	Phosphonic acid	C11	Al_2O_3	1.5×10^{-3}	10^5	0	20	[176]
151	Phosphonic acid	C11	Al_2O_3	10^{-3}	10^6	10	2	[177]
138	Tetramethyldisiloxane	C10	SiO_2	3×10^{-3}	10^5	15	−20	[178]
142	Trichlorosilane	C12	SiO_2	1.7×10^{-2}	10^6	−20	−1	[179]
149	Phosphonic acid	C11	HfO_2/SiO_2	3.56×10^{-2}	10^3	−65	−3	[180]
150	Phosphonic acid	C12	HfO_2/SiO_2	7.37×10^{-2}	10^3	−65	−3	[180]
153	Phosphonic acid	—	Al_2O_3	$(3.6 \pm 0.3) \times 10^{-5}$	10^4	—	−5	[181]
139	Monochlorosilane	C10	SiO_2	1.4×10^{-2}	10^4	−30	—	[182]
147	Phosphonic acid	C11	Al_2O_3	3×10^{-2}	Variable	—	−5	[42]
140	Triethoxysilane	—	SiO_2	1.8×10^{-3}	10^4	−20	—	[183]

2.3 Efforts in Developing New Methods

In organic electronics, the functionalization of dielectric substrates with SAMs or polymers is regarded as an effective surface modification strategy that may significantly improve the resulting device performance. This method, however, is not suitable for some latest substrates typically used in flexible electronics, e.g. stretchable substrates, paper substrates, and degradable substrates. The pioneer work has been demonstrated by Yokota et al. [107] by using a paraffinic tripodal triptycene, which self-assembles into a completely oriented 2D hexagonal triptycene array on polymer surfaces. Such few-layer films are analogous to conventional SAMs on inorganic substrates in that they neutralize the polymer surface to provide better electrical performance.

The result, however, is still far from perfect. The development of interfacial modification methods lags far behind the progress of active materials, namely, OSCs, electrodes, and dielectrics. To name but a few of the drawbacks:

(i) The ultrahigh stretchability of OSCs, electrodes, and substrates calls for ultrahigh density of SAMs, which is impossible to achieve under thermodynamic self-assembly process;
(ii) For portable use, the surface conditions of substrates are more serve than regular flat substrates, how can we expect a perfect interfacial property based on general methodologies?

There is still plenty of room to develop efficacious surface modification methodologies to optimize the surfaces on flexible substrates and thus improve the device performance.

3

Semiconductor/Semiconductor Interface

Organic semiconductors (OSCs), consisting of molecular subunits with a conjugated backbone, serve as the active layer of organic field-effect transistors (OFETs) and govern charge transport from the source electrode to the drain electrode. A variety of OSCs, from small molecules to polymers, have been synthesized by chemists with p- or n-type carriers over the past 30 years to obtain the desired high electrical performance [185, 186]. Among them, some semiconductor molecules have been found to show field-effect mobilities that are comparable to or even higher than those of amorphous silicon (>1 $cm^2 V^{-1} s^{-1}$) and polycrystalline silicon (>10 $cm^2 V^{-1} s^{-1}$) [187–189], for example, [1]benzothieno[3,2-*b*][1]-benzothiophene (BTBT)-based small molecules [190] and diketopyrrolopyrrole (DPP)-based conjugated polymers [45, 191]. The abundant diversity of the molecular structures of organic molecules, which can be changed via flexible chemical designs, provides us with plenty of choices in a broad chemical development space. In addition, it was found that not only are the distinct characteristics of OSCs determined by their diverse molecular structures, which affect the charge-transport properties (at the molecular scale), but also the packing microstructure and morphology of the channel films (at the micro-/nano-scale) influence the device performance. This can be attributed to the existing "interface within the semiconductor layers," an interface formed through the π–π stacking behaviors of adjacent molecules or the grain boundaries between crystal domains.

Various mechanisms explaining the origin of charge carriers and the charge transport/transfer between molecules and related physical processes in organic solids have been put forward, but there is no uniform model capable of comprehensively describing charge transport in all OSCs [46, 47, 192–201]. Among different charge-transport mechanisms, hopping and band transport are regarded as two extreme cases, depending on the disorder degree of OSCs [48, 202–207]. In brief, thermally activated hopping transport of charges between localized states is the most common phenomenon in the less-ordered, even amorphous, OSCs at near room temperature, in comparison to charge transport in their highly ordered inorganic counterparts via band-like conduction. Nevertheless, a few examples of band-like transport in OSCs under certain conditions in their highly ordered crystalline structures have been reported [208, 209]. This calls for overall control of

Interface Engineering in Organic Field-Effect Transistors, First Edition. Xuefeng Guo and Hongliang Chen.
© 2023 WILEY-VCH GmbH. Published 2023 by WILEY-VCH GmbH.

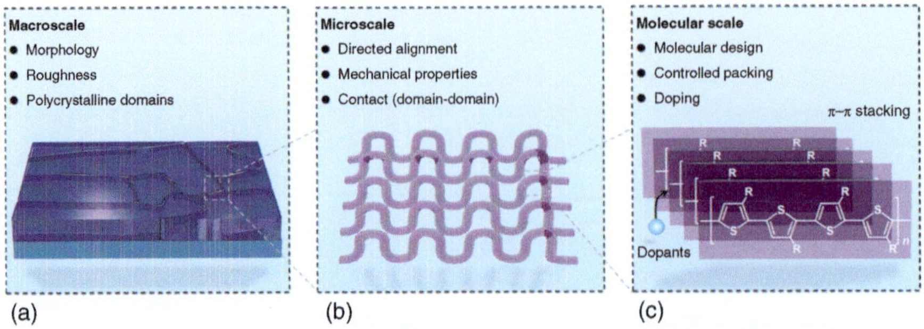

Figure 3.1 Interfaces within the semiconductor layers. (a) On the macroscale level, polycrystalline domains in OSC films can be from tens of nanometers to micrometers in size. The orientation of these ordered domains and the morphology and roughness of the films can strongly affect effective macroscale electrical transport. (b) On the microscale level, the strategies to direct chain alignment, including mechanical stretching and the use of co-solvents, can enhance chain alignment, crystallinity phase, and electronic coupling within domains. (c) On the molecular level, the molecular conjugation tends to form π–π stacking, for the strongest coupling. The molecular design of the planar π-conjugated backbone, side-chain engineering, core motif, and the dopant molecule that introduces charges into the assembly provide useful strategies for efficient charge transfer between molecules.

interfaces within semiconductors from macroscale to microscale to molecular scale (Figure 3.1).

First, the bulk transport behavior along the channel in an OFET is the overall result of combining the charge transport within a single-crystal-like domain with that between adjacent domains [200, 210–214]. Therefore, a rather general requirement for an OSC layer is thought to yield large-size grains with few domain boundaries (Figure 3.1a). Second, disorder can strongly hinder charge transport for almost all semiconductors because high molecular ordering and desirable molecular orientation occur to facilitate high mobility (Figure 3.1b). Last but not least, the packing behaviors, especially the π–π packing distance and crystal axis orientation, which determine the transfer integral and reorganization energy, have a significant influence on the carrier-transport behavior in organic semiconducting solids, no matter whether the organic crystals stack into a highly ordered structure or not (Figure 3.1c). In this section, we discuss interfacial engineering strategies focused on the interfaces within the semiconductor layers, by which considerable progress has been made thanks to an ever-deepening understanding of the charge-transport mechanism between semiconductor organic molecules.

Although the molecular structure has the most impact on determining the morphology, the variable packing structures of certain organic solids are significantly influenced by external conditions, such as solvent, temperature, processing and post-processing, and substrate. This is because external conditions can affect the solidification process. In fact, polymorphism has become an emerging design strategy for high-performance organic electronics [215]. Based on a solution-sheared deposition method [216], Giri et al. [217] found that with increasing the shearing speed of a 6,13-bis(triisopropylsilylethynyl) pentacene (TIPS–pentacene)

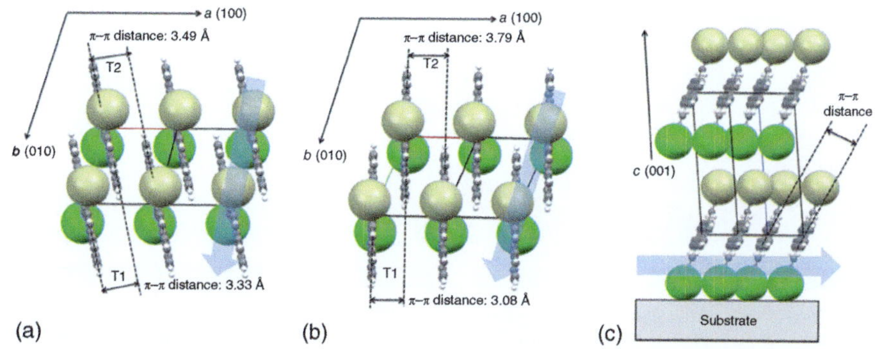

Figure 3.2 Molecular packing arrangements of TIPS–pentacene prepared under different conditions. (a) Evaporated thin films, (b) solution-sheared thin films, and (c) solution-sheared thin films viewed along the cofacial axis. The blue arrows represent the high charge-transport direction as well as the direction of shearing. Source: Reproduced with permission from Giri et al. [217]. © 2011 Springer Nature.

solution on the substrate when fabricating thin films, the obtained OSC films presented different molecular packings, with a decrease in the cofacially arranged π–π stacking distance from ~3.33 to ~3.08 Å (Figure 3.2). This was induced by strained lattices, resulting in a mobility as high as ~4.6 $cm^2\ V^{-1}\ s^{-1}$, two times higher than the highest value previously reported [218]. On the contrary, with a larger shearing speed, the mobility decreased because of the formation of boundaries, and it was unnecessary to increase the overall charge transfer integral by decreasing the distance of π–π stacking along one direction. To control the morphology of solution-shearing films with fewer defects, Bao and coworkers demonstrated an approach termed fluid-enhanced crystal engineering (FLUENCE) by using a micropillar-patterned printing blade [219]. The yielded millimeter-wide, centimeter-long, single-crystalline TIPS–pentacene thin films, which were highly ordered and had non-equilibrium domains with excellent molecule alignment and fewer grain boundaries, showed a mobility of $8.1 \pm 1.2\ cm^2\ V^{-1}\ s^{-1}$ along the coating direction. For small-molecule semiconductors, it is unquestionable that the ability to form an aligned and ordered structure guarantees high electric performance.

Things begin to get complicated when talking about charge transport within less-ordered polymer films, as the conjugated polymer semiconductors with long chains are connected by chemical bonds between small molecular units. Routinely, it is believed that two-dimensional (2D) charge transport occurs in self-organized conjugated polymers [220]. The Salleo's group reported a general relationship between aggregation, disorder, and charge transport in conjugated polymers (Figure 3.3a–c) [221]. The molecular weight has an obvious influence on the crystallinity and packing behavior of a given polymer. The presence of long conjugated chains is crucial for high carrier mobility with a poor structural order in high-molecular-weight polymers because a chain that is long enough can result in improved interconnectivity that links ordered crystalline domains together like bridges to ensure a fast charge-transport path along the chain backbones of different aggregates (here, charge transport along π–π stacking and lamellar-stacking are

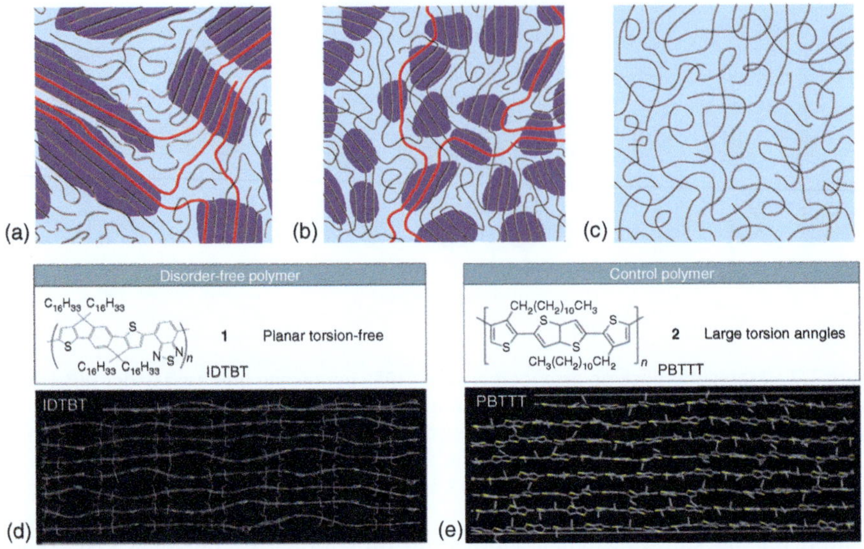

Figure 3.3 Microstructures of conjugated polymer films. (a) A semicrystalline polymer film, (b) disordered aggregates, and (c) a completely amorphous film. The figures show the coexistence of both ordered (dark shadowed areas) and spaghetti-like amorphous regions. If the molecular weight is high enough and the density of the ordered materials is large enough, long polymer chains (highlighted in red) can connect ordered regions, significantly improving charge transport. (d) The simulation shows that IDTBT adopts a wavy, yet planar, torsion-free backbone conformation, while (e) PBTTT chains maintain a linear conformation with large torsion angles between thiophene and thieothiophene. Source: (a–c) Adapted from Noriega et al. [221]. (d, e) Adapted from Venkateshvaran et al. [222].

"fast" and "slow," respectively). This well explains the high mobility extracted from amorphous conjugated polymer films. In other words, paracrystallinity governs charge transport, and thus the interconnection of different aggregates is a prerequisite. Therefore, a critical strategy for designing high-mobility polymers is to improve their ability to tolerate the inevitable disorder within the aggregates by allowing more efficient intra- and intermolecular charge transport. The Sirringhaus's group reported that "disorder-free" transport could take place in less crystalline, even amorphous donor–acceptor copolymers (Figure 3.3d,e) [222]. Transport in such polymers approached intrinsic limits, where all molecular sites along the backbones were thermally accessible. Owing to the formation of a planar, torsion-free backbone conformation (Figure 3.3d), the "disorder-free" conjugated polymer IDTBT (**1**) showed remarkable resilience to side-chain disorder with a low degree of energetic disorder and without broadening the densities of state (DOS) in comparison with the polymer PBTTT (**2**) with large torsion angles. This theory opens new insights into the development of "disorder-free" polymer semiconductors with high carrier mobilities.

Because of the diverse kinds of OSC molecules, composites, and even mixtures employed in organic electronics as active and functional components, the study of the interface within the semiconductor layers is attracting more and more interest.

As discussed above, there are various significant aspects that determine the carrier mobility of OSCs, ranging from nanoscale to mesoscale and from the impact of energy to morphology. Tremendous efforts have been made toward optimizing the packing motifs and the processing techniques [223]. Moreover, small-molecule and polymer-based OFETs have achieved field-effect mobilities of more than $10\,cm^2\,V^{-1}\,s^{-1}$ by intelligently engineering the molecular structure of soluble OSCs and controlling their microstructures and morphologies.

3.1 Influence of Additives on a Material's Nucleation and Morphology

The solution processability of OSCs during deposition strongly influences the microstructure and morphology of the resulting films. Effective control of the nucleation process offers a convenient method to optimize the performance of OFETs by making full use of crystallization kinetics to tune the microstructure and morphology [49, 224, 225]. For example, the process history of P3HT, including nucleation, growth, and alignment, has a significant effect on the electrical properties of the corresponding OFETs [226]. In addition, the proper post-processing of OSC films, such as thermal annealing and solvent annealing, may also result in a more ordered structure and thus an improvement in device performance [227]. Current strategies, including using solvent additives, adding small molecular nucleating agents, or blending with polymers, have proven highly useful to optimize the crystallinity of the matrix in the active layer of OFETs, thus improving the performance.

3.1.1 Solvent Additives

Besides deposition methods, the solvent that is utilized to dissolve OSC is another key element to tune the crystallinity, change the nano/microstructure, and enhance molecular ordering in organic electronic devices [228–230]. High boiling point solvent additives and nonsolvent additives are two of the most important solvent additives used in OFET with remarkable results (Figure 3.4).

When introducing appropriate solvent additives (such as 1-chloronaphthalene (CN, **3**), diphenyl ether (DPE, **4**), and 1,8-diiodooctane (DIO, **5**)) with the high boiling points during device fabrication, the drying time of OSCs increases, which can help in the formation of a uniform film [231, 232]. In this aspect, An et al. [233] presented a strategy by introducing CN (b.p. 259 °C) as a solvent additive to the chloroform solution of a DPP-based polymer to enhance the molecular packing order. The high boiling point of CN could significantly reduce the evaporation speed of the mixed solvent (95 wt% chloroform/5 wt% CN). This resulted in more time for the polymer chains to align, facilitating the formation of extended π–π interchain stacking and thereby significantly increasing the mobilities. But CN additives only functioned well for quasi-crystalline polymer films; no drastic effect was observed on the film structure of the highly crystalline or highly amorphous polymer.

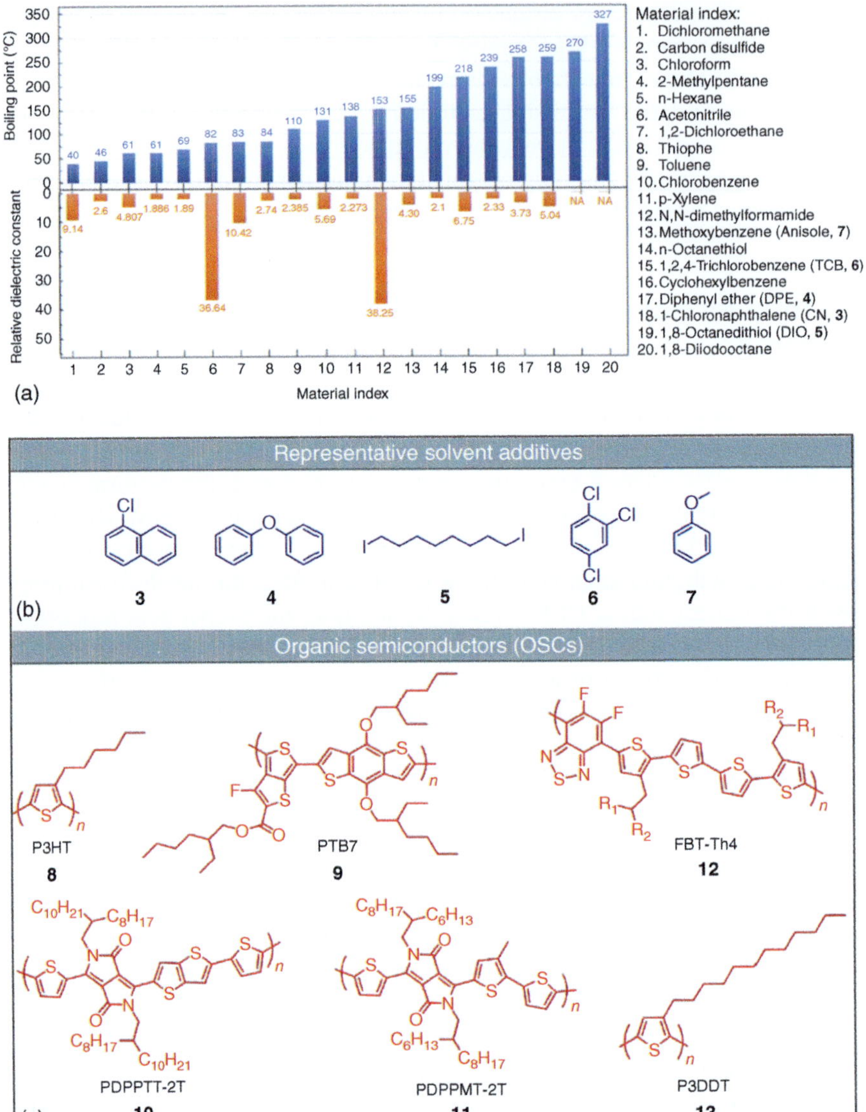

Figure 3.4 High-boiling-point solvent additives and nonsolvent additives. (a) Bar charts showing their boiling points and relative dielectric constants. (b) Chemical structures of representative solvent additives used in (c) organic semiconductors.

To understand the aggregate formation kinetics, Reichenberger et al. [234] carried out *in situ* experiments to investigate the impact of the additive DIO on the morphology of P3HT (**8**) films. Although the number of aggregates from P3HT films was only slightly higher for the samples with DIO than that without, the phase separation between aggregated and nonaggregated chains was significantly different, that is, much longer conjugation length in the aggregates was obtained with DIO. The *in situ* grazing-incidence wide-angle X-ray (GIWAXS) study of the solvent additive effect on PTB7 (**9**) films showed that not only the solvent boiling point influences the crystallite formation times but also the specific additive–polymer interactions, determined by molecular properties of additives and preferential interactions between them, played important roles, as reported by Manley et al. [235]. The significant DIO–PTB7 interactions led to a much longer crystalline formation time of 1.5 hours than that adding CN or DPE for a few minutes in the "solvent + additive" systems. The authors fabricated OFETs with these films and found that their mobilities were within a close range, due to the similar film morphologies. These studies pointed out the importance of rational control of fabrication conditions.

However, in some cases, things are much more complicated. Recently, Ji et al. [236] synthesized asymmetric DPP-based polymers (PDPPTT-2T (**10**) and PDPPMT-2T (**11**)), which showed good solubility in most common solvents. The DPP-based polymers were less soluble in toluene than in chloroform; spin-coating in toluene resulted lower mobilities (i.e. 4.68 $cm^2\ V^{-1}\ s^{-1}$ from toluene and 5.22 $cm^2\ V^{-1}\ s^{-1}$ from chloroform for PDPPMT-2T). However, when processed from their toluene with the addition of 3% DPE, OFETs showed a dramatic increase in hole mobility compared with that from pure toluene (from 1.96 to 5.24 and from 4.68 to 10.3 $cm^2\ V^{-1}\ s^{-1}$ for PDPPTT-2T and PDPPMT-2T, respectively). These were attributed to the highly ordered crystalline films with a preferentially edge-on molecular packing orientation from toluene/DPE solutions, which is usually desired for high-performance OFETs. Li et al. [237] reported that the organization of conjugated D–A polymers could be controlled from edge-on to face-on by solvent additives. The driving force was attributed to the pre-aggregation of solutions when depositing films. The D–A polymer FBT-Th_4 (**12**) in TCB (**6**) solution formed a lower aggregation and led to a face-on arrangement due to the trend of maximizing interaction between polymer and substrate by lying flat (Figure 3.5a,b) [238]. Whereas the intermolecular π-stacking interaction of polymer in chloroform was strong enough to overcome the barrier that reorients face-on into an edge-on organization (Figure 3.5c,d). But a small amount of TCB (1/240) in the chloroform solution of the co-polymer could result in face-on organization because TCB gradually became the final solvent and de-aggregation occurred. Especially, when using pure TCB as the solvent, the deposited films only exhibited mobilities of $10^{-6}\ cm^2\ V^{-1}\ s^{-1}$, far more less than that of ~2 $cm^2\ V^{-1}\ s^{-1}$ from pure chloroform. In conclusion, the high-boiling point solvent could extend the time for chain alignment, facilitating the high ordered packing, but a precondition is always needed to satisfy that the surface organization ought to be edge-on, which demands a full consideration of both the polymer solubility and the substrate-polymer/solvent-polymer interaction.

Moreover, DPE and DIO have also been used to enhance the performance of small molecule OFETs, such as TIPS–pentacene device [240] and isoindigo-derivative device [241], with very low dosage (0.2–0.5%). AFM study showed that they changed the morphologies of OSC films to either more crystalline or smaller sized grains, which correlated well with dramatic changes in device performance.

In addition to the use of high-boiling-point solvent additives, nonsolvent additives are also capable of tuning the morphology of OSC films. The choice of nonsolvent

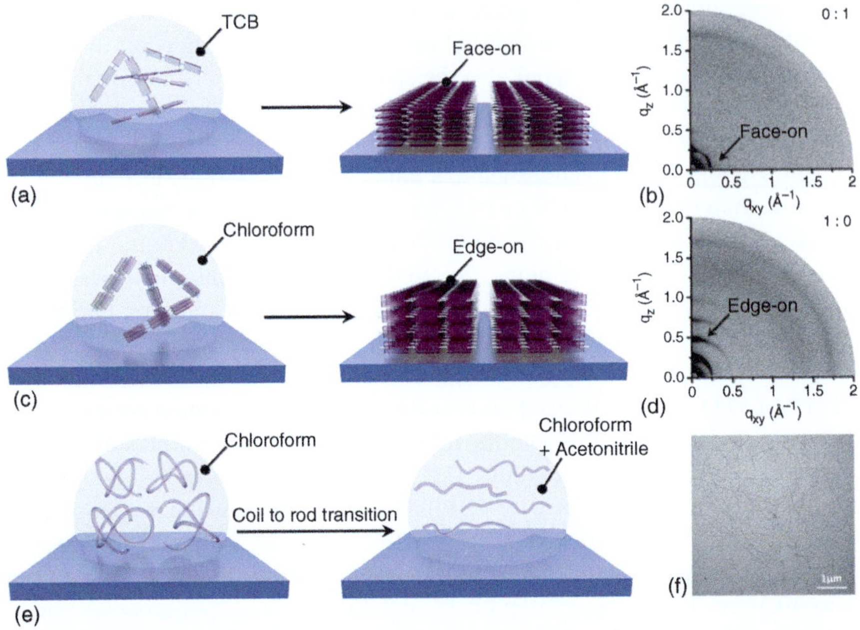

Figure 3.5 Additive-assisted crystallization. (a–d) Schematic illustrations and GIWAXS patterns of thin films showing that the pre-aggregation of OSC polymers in solution influences the surface organization in deposited thin films. (a, b) TCB solution leads to a face-on arrangement. (c, d) Chloroform solution results in an edge-on organization. (e) Self-assembly of P3DDT into nanowhiskers via addition of the marginal solvent anisole to its well-dissolved solution. (f) TEM images showing the round-like P3DDT nanowhiskers. Source: (a–d) Adapted from Li et al. [237], © 2015 John Wiley & Sons. (e, f) Adapted from Xu et al. [239], © 2011 American Chemical Society.

additives can refer to the polarity of the solvent (dielectric constant). For example, upon the addition of the marginal solvent anisole (**7**) into well-dissolved P3DDT (**13**) solution, highly ordered one-dimensional (1D) or 2D aggregation of P3DDT occurred (Figure 3.5e,f), because anisole was a nonsolvent for P3DDT [239, 242]. By using small-angle neutron scattering, Keum et al. [243] studied the effect of the solvent combination conditions (chloroform/hexane) on the nucleation and growth processes of P3HTs. As the fraction of the poor solvent hexane increased over a specific value, the aggregation of P3HT chains underwent a transition from coiled to high crystallinity rectangular parallelepipeds (RPs) phase. The Reichmanis' group demonstrated a controlled, solution-based self-assembly of P3HTs from disordered to ordered on the nano- through meso-scales, forming a rod-like P3HT nanostructure in solution under ultrasonication [244, 245]. P3HT films composed of long nanorod structures and large crystallite sizes formed after spin coating. The authors hypothesized a two-step crystallization mechanism to describe the P3HT self-assembly process. Ultrasonication of the chloroform solution of P3HTs promoted a degree of nucleation that was well balanced for subsequent growth, and the degree of supersaturation was adjusted through the addition of the nonsolvent

2-methylpentane. Therefore, the two-step crystallization may have enhanced the nanostructure growth and formed significantly longer assemblies.

3.1.2 Nucleating Agents

Nucleating agents (Table 3.1) were first used to regulate the solidification process of (semi-)crystalline polymers and to improve the mechanical properties of common plastic materials. They were applied by Treat et al. [126] to accelerate the crystallization process of OSCs and to assist in the formation of uniform films without influencing the crystallinity degree (Figure 2.2b). The authors showed that two commercially available nucleating agents, 1,3:2,4-bis(3,4-dimethylbenzylidene) sorbitol (DMDBS) and tris-*tert*-butyl-1,3,5-benzenetrisamide (BTA), could induce the nucleation of a wide range of OSC materials. Both DMDBS and BTA are inert molecules and were initially dissolved at the molecular level with the host semiconductor in the molten state or in solution. Then, fast heterogeneous nucleation occurred after cooling and solvent evaporation. The addition of only 1 wt% DMDBS to molten poly(3-dodecylthiophene) (P3DDT) resulted in an increase in the melting temperature, indicating a dramatic increase in the crystallite nucleation rate. An increase of up to 50% in carrier yield was observed in the P3DDT mixture film versus the neat polymer owing to the creation of larger interfacial areas between crystalline and amorphous domains. The introduction of a nucleating agent into a polymer solution also worked well, with no change in the crystallinity degree. Time-of-flight (TOF) photoconductivity measurements showed that the hole mobility of P3HT films obtained from *p*-xylene with BTA was not affected. These nucleating agents were also used in solutions of small organic conjugated molecules such as phenyl-C_{61}-butyric-acid-methyl-ester (PCBM) and TIPS–pentacene. The optical micrographs and X-ray diffraction (XRD) data showed that the domain size of PCBM films composed of DMDBS after annealing was greatly reduced to less than 1 μm, with no influence on the molecular ordering and crystallinity, in comparison with annealed neat samples. Again, this change in crystallite dimensions had no negative effect on the electron mobility of PCBM. The beneficial feature of using nucleation agents is the shorter annealing time because a much faster crystallization process can be achieved, minimizing exposure of the films to high temperature. Also, the addition of nucleating agents strongly helps form uniform films when inkjet printing, resulting in large, continuous coverage over the entire channel of the OFETs.

Benefiting from its wide applicability, DMDBS can also be used to effectively control the morphological development of organic photovoltaics (OPVs) [129, 130]. 1,2,3,4-bis(*p*-methylbenzylidene) (MDBS) can also accelerate the crystallization of P3HTs. Yuan et al. [131] presented the preparation of spin-coating P3HT/MDBS hybrid films with a shish-kebab nanostructure from a hot *o*-dichlorobenzene solution of the blend. The central "shish" is the self-assembled MDBS nanowire, promoted by intermolecular hydrogen bonding, supplying a heterogeneous surface for P3HTs to grow on. Because of π–π intermolecular interactions between P3HT and MDBS, P3HT crystallized into nanofibrils perpendicular to the long axis

Table 3.1 Common nucleating agents.

Nucleating agent	**Chemical structure**	ΔT_c^{nuc} **(°C)**	**NE (%)**
C_{60}		9	26
C_{70}		0, 6	0, 18
SWNT		0, 13, 24	0, 38, 71
NaBz	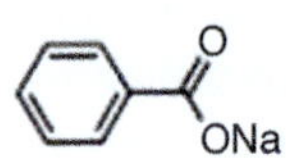	0, 14, 33	0, 41, 97
NaBz*	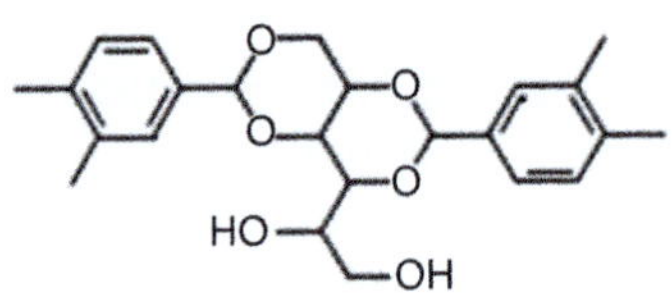	8, 17	24, 50
DMDBS		7	21
TBPMN	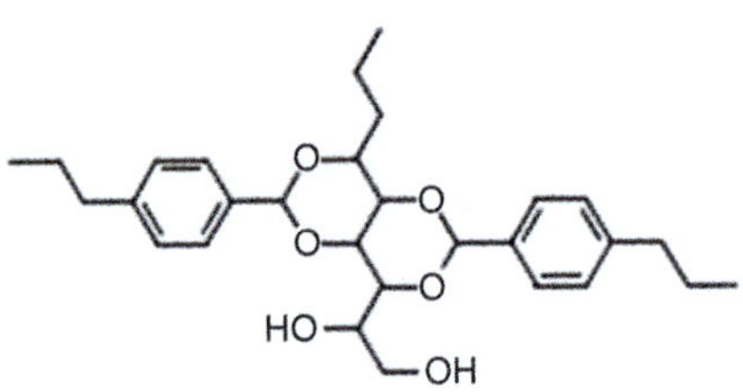	4	12
BTA	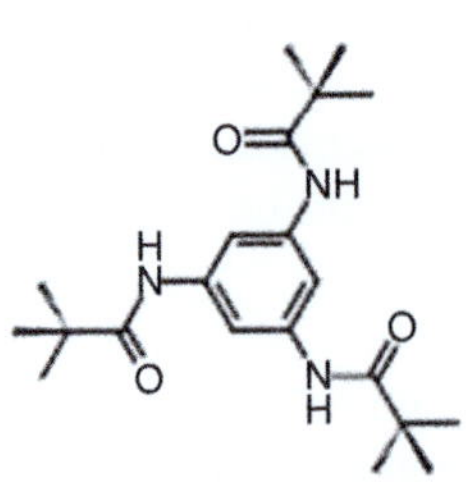	1	3

of nanowire to be "kebab". By using the hybrid P3HT/MDBS with shish-kebab morphology as the channel layer, the mobility of the corresponding OFETs is $\sim 3.80 \times 10^{-2}\ cm^2\ V^{-1}\ s^{-1}$, an order of magnitude higher than the device with neat P3HTs [132].

3.1.3 Template-Mediated Crystallization

In general, molecular ordering of OSCs is controlled by non-covalent bond interactions, mainly via the relative weak van der Waals (vdW) force (~10 kcal mol^{-1}) [246], thus resulting in a limitation for OSC to form large single crystalline domains in films and a dependence on external conditions. Usually short-range-ordered polycrystalline films with a large amount of grain boundaries are formed, which is unfavorable for charge transport. For π-conjugated polymers, it is believed that the charge dynamics is dominated by the structure disorder and/or the amorphous portions at the mesoscopic length scale [221, 247–249]. In this regard, it is crucial to achieve long-range ordered crystalline domains.

In order to achieve highly efficient charge transport over a long range, the formation of interconnected pathways along the different crystallite domains of semiconductor polymers is desired. Kwon et al. [250] reported a template-mediated crystallization (TMC) method, which resulted in the formation of nano-crystalline networks in a semiconducting polymer, P3HT (**8**), by employing a polymer template via polaron formation and electrostatic interaction (>20 kcal mol^{-1}) (Figure 3.6a). Polystyrene sulfonate (PSS) (**14**) can serve as a bifunctional agent with the sulfonic acidic side-chain group acting as the p-dopant and the polystyrene backbone acting as a template. The authors believed that the mechanism of light doping and TMC formation in the P3HT:PSS film was related to the deprotonation of the acidic side group of PSS, which happened after the initial formation of an S···H hydrogen bond between the two different molecules (the sulfur atom from thiophene and the H atom from the acidic group). The protons transferred from PSS to the sulfur atoms in P3HT, which was vital because the negative charge (ion) carried along the PSS polymer induced an electrostatic interaction with pronated P3HTs. With repetitive units of sulfonic acid side-chains attached to PSS, many electrostatic interactions can readily occur with electropositive P3HTs, manipulating the orientation of P3HTs and further promoting nucleation because of the presence of more energetically favorable crystal planes for neighboring P3HT to assemble on. By employing this TMC method, P3HT:PSS (1 : 1 w/w) films showed an evolution of favorable edge-on orientation with enhanced in-plane π–π stacking, whereas the pristine P3HT preferred a weak face-on orientation (Figure 3.6b). Physically interconnected fibril nanostructures with long-range order were formed, leading to a three-dimensional (3D) network in the TMC-treated films (Figure 3.6c). With interconnected pathways beneficial for charge transport on both lateral and vertical directions, the improved structural orientation and interconnection of the P3HT:PSS films led to a dramatical increase in the device performance with an average mobility value and on/off ratio of the films improved to ~0.1 $cm^2\ V^{-1}\ s^{-1}$ and ~10^6, respectively, values that were much greater than those of the pristine P3HT (~8.8×10^{-4} $cm^2\ V^{-1}\ s^{-1}$ and ~10^4, respectively).

In a further study [251], the Lee's group reported that the PSS template could serve as a robust organic crystal seed on a nanometer scale (~70 nm) for efficient nucleation and growth of the semiconductor molecule 2,7-dioctyl[1]-benzothieno[3,2-*b*][1]benzothiophene (C_8–BTBT) from a supersaturated solution via electrostatic

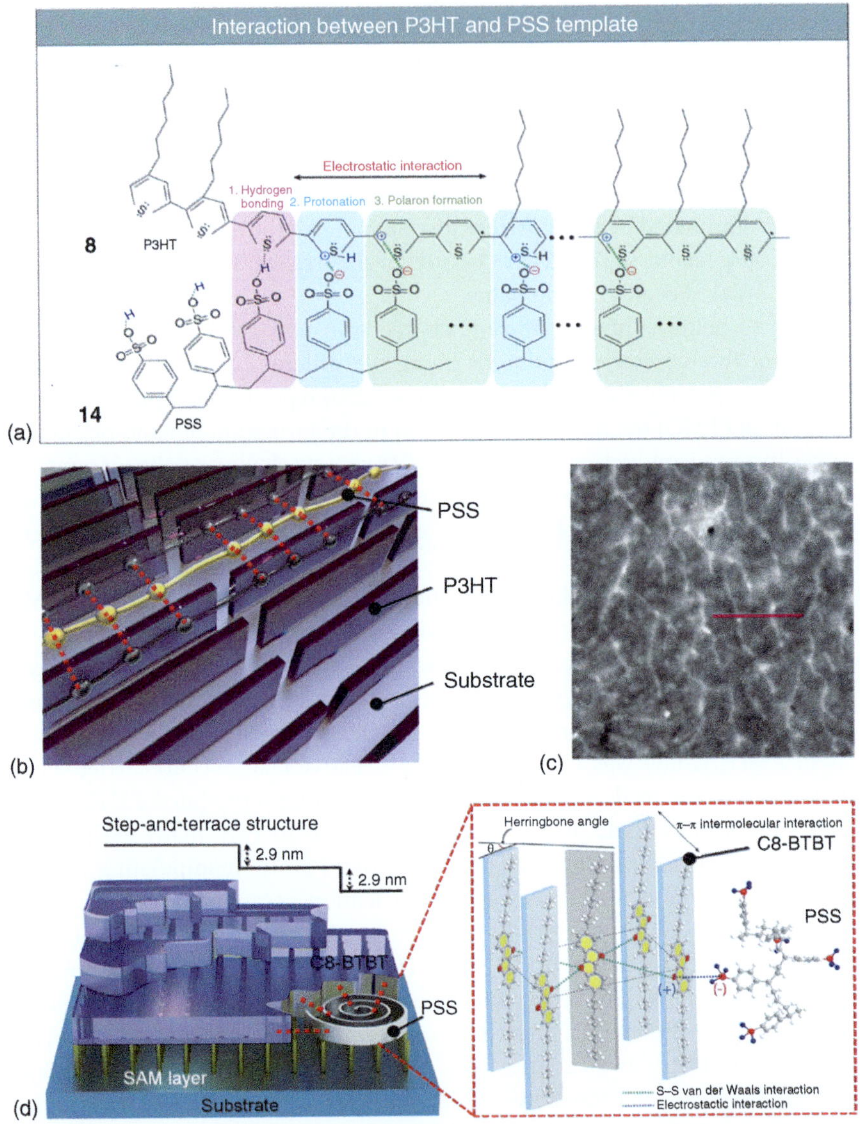

Figure 3.6 Template-mediated crystallization. (a) Schematic representation showing that P3HT crystallizes along the PSS template via electrostatic interaction between P3HT and the template. (b) Detailed mechanism of the electrostatic interaction between P3HT and PSS template followed by hydrogen bonding, protonation, and polaron formation in the solution state. (c) Electron dispersive spectroscopy showing the TMC-induced fibril structure of P3HT:PSS film. (d) Schematic representation showing a C_8–BTBT film and a PSS template on a substrate where intermolecular electrostatic interactions are induced between them. Inset shows the arrangement of C_8–BTBT and PSS groups. Source: (a–c) Reproduced with permission from Kwon et al. [250], © 2014 Springer Nature. (d) Adapted from Kwon et al. [251].

interactions, referred to as the template-mediated molecular crystal seed (TMCS) method (Figure 3.6d). In fact, the strong electrostatic interactions existing in the C_8–BTBT and PSS composites offered favorable nucleation dynamics at a lower undercooling for the crystal growth of C_8–BTBT molecules by reducing the nucleation energy barrier. Long-range-ordered large single-crystal films on the millimeter scale (2–3 mm) were obtained. The resulting C_8–BTBT:PSS crystal films showed a more ordered and compressed packing structure with a decreased π–π stacking distance than the neat C_8–BTBT film, ensuring the co-facial charge-transport path (Figure 3.6d, inset). OFET devices were fabricated within the millimeter-scale organic single-crystal domain, from which an excellent charge-carrier mobility ($\sim$2.30 $cm^2\ V^{-1}\ s^{-1}$) that was nearly 30-fold higher than that of the pristine C_8–BTBT device was obtained.

3.1.4 Blending with Insulating Polymers

Blending semiconducting small molecules or polymers with insulating polymers has recently been demonstrated to be a successful route for manufacturing high-performance OFETs capable of combining the favorable properties of each material or even exceeding the performance of individual components [252–254]. The blending of organics can be achieved by simply dissolving each material in a common solvent and then coating them onto the substrates (Figure 3.7a). The rapid

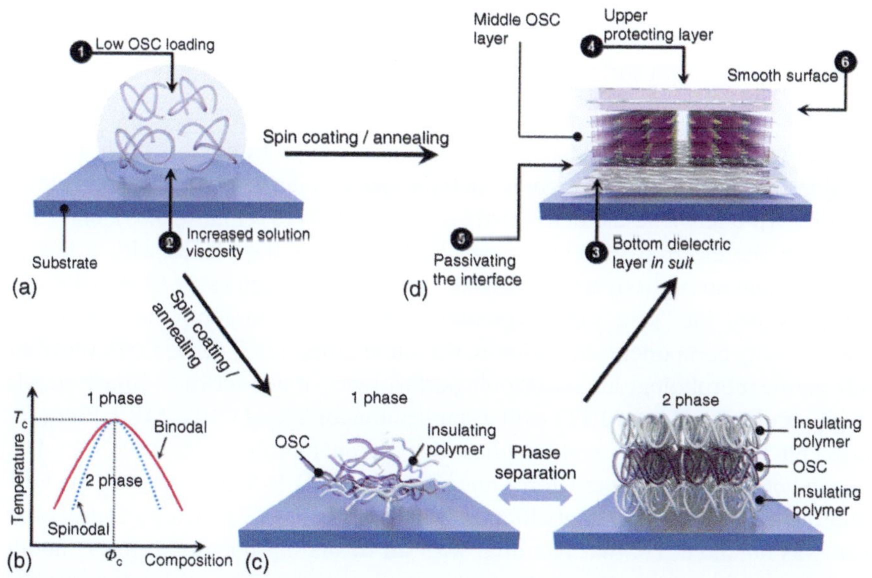

Figure 3.7 Phase separation of multicomponent systems. (a) Benefits of phase separation. (b) The thermodynamic phase diagrams describing a simple two-component polymer system. (c) Schematic representation showing vertical phase separation between OSC and insulating polymers. (d) Schematic representation showing the layer structure after phase separation with OSC molecules as channel layer, insulating polymers as dielectric layer or protecting layer.

solvent evaporation during this process did not leave enough time for equilibrium structures to develop. Immiscible blends of OSCs and insulating polymers with phase separation formed due to the negligible entropy of mixing between the two components (Figure 3.7b), resulting in an OSC-rich phase containing a fair quantity of insulating polymers and vice versa (Figure 3.7c) [50, 51]. In fact, phase separation is a complex phenomenon, and many factors have an impact on the resulting degree of disorder and morphology, such as the molecular characteristics of the two materials, solution conditions, thermodynamic parameters, and processing methods. For more detailed information about the nucleation and evolution process, please refer to reviews by Prud'homme et al. [52, 53, 255].

Phase separation of OSC/insulator polymer blends can occur in both the lateral and vertical directions. Vertically phase-separated structure is preferred to fabricate OFETs, which can be realized by choosing an appropriate insulating polymer, solvent, substrate, and processing condition. We listed below several benefits of this method in comparison with the traditional fabrication process (Figure 3.7d):

(i) Decreases the loading of semiconducting materials;
(ii) Increases solution viscosity to help form larger grain-size domains and continuous films;
(iii) Helps overcome the dewetting challenges associated with OSC material processing;
(iv) Provides semiconductor films with a passivation layer and/or a dielectric layer *in suit*;
(v) Passivates the interface between OSCs and substrates;
(vi) Provides a smooth surface to contact with electrodes.

The application of blends of conjugated polymers in OFETs was first reported by the Jenekhe's group in 2002 [256]. They built n-type transistors with a series of binary semiconductor blends poly(benzobisimidazobenzophenanthroline) (BBL)/poly(p-phenylene-2,6-benzobisthiazole) (PBZT) (**15**) (Figure 3.8a), and they found that the electron mobilities were fairly high for the blends with a highly disordered and amorphous structure over a wide composition range (5–60% PBZT). This was called the "stair step" dependence of mobility, in comparison with the lower-mobility component. In addition, the same group [257] carried out a further study on the morphology and electronic performance of two series of binary blends of regioregular P3HT (rr-P3HT) with a semiconductor, blend P3HT/MEH-PPV (**16**) and blend P3HT/PS (**17**). Similar phase-separation morphologies were found to occur in both blends by atomic force microscopy (AFM). When comparing the hole mobilities of blends **16** and **17** at different weight fractions of P3HT, a very surprising result was obtained, i.e. that blending with an insulating polymer brought about a higher mobility over almost the entire composition range, in contrast to the reduced performance of blends with semiconductor, which was believed to be the consequence of dipolar effects [258]. These findings shed light on the use of organic blends as active layers for high-performance OFETs.

A group led by Sirringhaus reported that by applying bicomponent blends rr-P3HT/PS (**17**) as channels, the BGBC OFETs showed good performance,

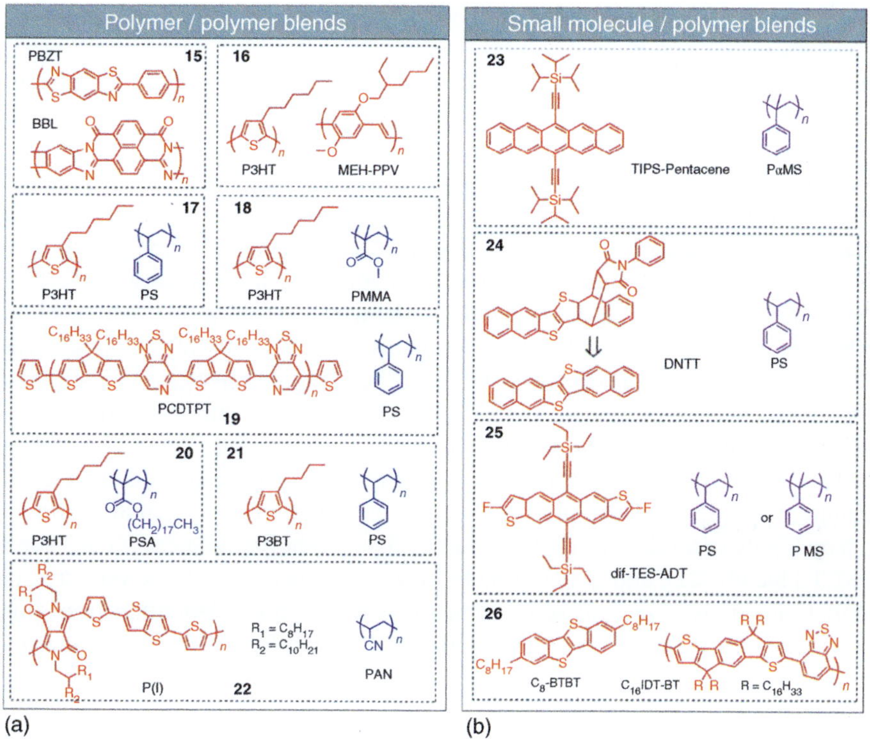

Figure 3.8 Molecular structures of (a) polymer/polymer blends and (b) small molecule/polymer blends. Semiconductors are highlighted in red, and insulators are highlighted in blue.

even using as little as 3 wt% of semiconducting materials [259]. Owing to the crystallization-induced phase segregation of the two materials, a vertically stratified structure was obtained in a one-step process, which predominantly expelled P3HT to the surface of the substrate in the resulting films. The environment stability was also improved owing to the encapsulation of the insulating polymer on the top. Different from this insulator-top/semiconductor-bottom structure, Cho et al. [260] reported a bilayer with an inverse structure formed via vertical phase separation. They made use of films formed on hydrophilic SiO_2 substrates (even directly on silicon without other dielectrics) by spin casting a chlorobenzene solution of blend P3HT/PMMA (**18**) to fabricate improved OFETs. Only 5 wt% P3HT was enough to show better hole mobility than the neat P3HT film under the same conditions. It is worth mentioning that the phase-separation-induced thin and defect-free PMMA bottom layer can act as a modifier material at the semiconductor/dielectric interface. The formation of an ultrathin layer of insulating polymers at the bottom could eliminate the scattering/trapping sites from the dielectric surface. In this regard, Ford et al. [261] deposited a blend poly[4-(4,4-dihexadecyl-4H-cyclopenta[1,2-b:5,4-b′]-dithiophen-2-yl)-alt-[1,2,5]thiadiazolo-[3,4- c]pyridine] (PCDTPT)/PS (**19**) on a nanogroove substrate to achieve a mobility as high as $2.7\,cm^2\,V^{-1}\,s^{-1}$ for a blend containing 90 wt% PS.

By changing the preparation methods of the blends, semiconducting polymer nanofibers would form embedded in the insulating polymer matrix, quite different from vertical phase separation. Qiu et al. [262] have obtained this structure by depositing P3HT/PS blends from a marginal solvent with temperature-dependent solubility. This method offered an efficient charge-transport path and allowed good encapsulation of active materials. Lin et al. [263] used semicrystalline poly(stearyl acrylate) (PSA) or amorphous PMMA as a polymer matrix to prepare P3HT/PSA (**20**) films containing well-defined P3HT nanowires, resulting in high mobility with good environmental stability in comparison to P3HT–PMMA films or neat P3HT films from the corresponding OFETs. They pointed out that the comb-like semicrystalline PSA facilitated the crystallization of P3HT into larger nanowires and the crystalline alkyl chains of PSA acted as an efficient kinetic barrier in ambient atmosphere. Highly ordered structures are preferred to exhibit high mobility. Lu et al. [264] reported a direct-writing method by a capillary array to prepare highly ordered striped films of aligned poly(3-butylthiophene) (P3BT) and P3BT/PS blends (**21**), and a higher mobility was obtained in both the vertical and parallel directions than neat semiconductor polymer films. Benefiting from this alignment method to tune the FET mobility and threshold voltage, they realized a high gain inverter up to 80. Lei et al. [141] reported that the crystalline structural order of polymer semiconductors can be enhanced via polymer matrix-mediated molecular self-assembly by using polyacrylonitrile (PAN) as a matrix. They first studied blend P(I)/PAN (**22**) with various weight ratios on an OTS-treated substrate. XRD measurements showed that the (100) diffraction peak gradually intensified, along with shifting to a higher angle, with PAN loading up to 60 wt%, indicating an enhancement of molecular self-organization to a highly ordered crystalline structure of P(I). Cross-sectional TEM images and element mapping of carbon and sulfur showed a large, even distribution of both P(I) and PAN in the layer, and careful examination confirmed that the OTS-modified SiO_2 dielectric surface had a higher concentration of P(I) semiconductor.

An in-depth study on the high performance of semiconductor/insulator polymer blend transistors was carried out by Lu et al. [265], who found that the high FET performance of P3HT/PS blend-based transistors only took place after exposure to ambient air, whereas only very poor performance was obtained when undoped. They attributed this dramatic improvement to the moderate doping in polymer blends as a general mechanism, quite different from the superior stability caused by the encapsulation. In addition, the doping effect of OSC channels is another important factor that determines the performance of the corresponding optoelectronic devices as it is known that conductive polyacetylene was initially achieved by chemical doping. The effect of doping on organic transistors has recently been well reviewed by Lüssem et al. [22] and Jacobs et al. [266].

Blending small-molecule OSC materials with insulating polymers has also been demonstrated to build high-performance OFETs with high reproducibility and low performance spread (Figure 3.8b) [252, 259, 267–272]. Kang et al. [273] fabricated

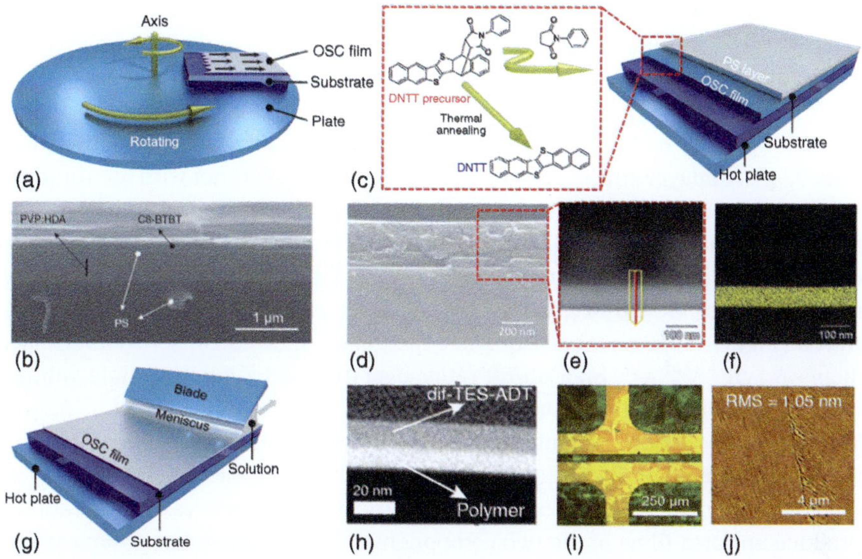

Figure 3.9 Three representative methods to blend small-molecule OSC materials with insulating polymers. (a) Schematic illustration of the off-center spin-coating process, in which the substrates are located away from the axis of the spin-coater. (b) Cross-sectional scanning electron microscopy (SEM) image of thick C_8-BTB/PS film drop coated on the PVP surface, demonstrating the PS segregation is energetically more favorable than the C_8-BTBT segregation to the PVP surface. (c) Chemical structures of the soluble DNTT precursor and its thermal conversion process. (d–f) Cross-sectional images obtained from SEM (d), STEM (3), and element mapping of S (f) in the blend film after thermal annealing. (g) Blade coating set-up. (h) Cross-sectional TEM image shows the bilayer structure of the blend with PS at the bottom and diF-TES-ADT on top. (i) Polarized optical micrographs (POM) and (j) atomic force micrographs (AFM) show the smooth morphology of the film. Source: (a, b) Adapted from Yuan et al. [274], © 2014 Springer Nature. (c, d) Adapted from Kimura et al. [275], © 2015 John Wiley & Sons. (g–j) Adapted from Niazi et al. [276], © 2015 Springer Nature/CC BY 4.0.

OFETs by using small molecule–polymer blend semiconductors, TIPS–pentacene, and poly(α-methylstyrene) (PαMS) as the channel layer (**23**). They found that highly crystalline TIPS–pentacene was formed with the π–π stacked direction parallel to the film surface, which ensured a higher mobility. Yuan et al. [144] described the use of an off-center spin-coating method to grow highly aligned meta-stable films of a blend of C_8–BTBT and PS (Figure 3.9a), reaching hole mobilities up to ~43 $cm^2\ V^{-1}\ s^{-1}$, the highest value at the time of the report. Furthermore, the authors used cross-sectional transmission electron microscopy (TEM) to study the vertical phase separation between C_8–BTBT and PS. It was clearly shown that an additional thinner layer (≤5 nm) was inserted between the C_8–BTBT layer and the substrate, which could be attributed to the PS layer (Figure 3.9b).

Kimura et al. [275] reported the fabrication of a polycrystalline dinaphtho[2,3-*b*:2′,3′-*f*] thieno[3,2-*b*]thiophene (DNTT) structure by thermally annealing films of a DNTT precursor/PS blend (**24**), which were obtained by simply coating the blends mixed in chloroform onto OTS-treated substrates in ambient air (Figure 3.9c).

With a distinct well phase-separated DNTT/PS structure (Figure 3.9d–f), the solution-processed OFETs showed hole mobilities up to ~1.1 $cm^2 V^{-1} s^{-1}$ with high environmental stability.

Recently, Niazi et al. [276] presented a method based on blade coating of a blend of diF-TES-ADT/PS or PαMS (**25**) to induce large-scale directional crystallization (Figure 3.9g). A bilayer stratification was obtained in the channel with an ultrathin (~10 nm thick) single crystal-like OSC layer on top and the insulating polymer at the bottom (Figure 3.9h). In addition, the sizes of the domains were on the order of millimeters with few boundaries (Figure 3.9i). A smooth, pinhole- and crack-free topography of the OSC surface both within the domains and at their boundaries was achieved (Figure 3.9j). As a consequence, the device showed carrier mobilities as high as ~6.7 $cm^2 V^{-1} s^{-1}$, high on/off ratios over 10^6, and a low threshold voltage (<1 V) in comparison with its single-crystal counterparts.

In addition to blending small-molecule OSC with insulating polymer matrix, blending with semiconducting polymers is also capable of forming improved films, combining the advantages of high electrical characteristics and the ability to produce uniform films of the two components. The Anthopoulos' group carried out such studies [268, 277], and recently they reported their third generation of small molecule/polymer blend-based high-mobility organic transistors with active channels consisting of a blend of C_8–BTBT/C_{16}IDT-BT) (**26**) [278]. In this work, they chose C_{16}IDT-BT as the conjugated polymer binder because of its high hole mobility and comparable highest occupied molecular orbital (HOMO) energy level with C_8–BTBT. Prepared from a suitable solution (C_8–BTBT:C_{16}IDT-BT=1 : 4 wt ratio) with moderate molecular doping (1% $C_{60}F_{60}$), the blended films showed improved layer microstructures with a strong vertical-separation phase and C_8–BTBT-rich regions on the surface, ideal for top-gate OFETs, exhibiting mobilities exceeding 13 $cm^2 V^{-1} s^{-1}$ after annealing.

In summary, besides those mentioned earlier, several other insulating polymers have been utilized to blend with numerous organic semiconducting molecules to pursue better performance. We have listed some of them in Table 3.2. The general requirements for the insulating polymers are that they be highly soluble in solvents but rather immiscible with semiconductors, thus capable of forming good phase separation within the blended films. Because the major insulating polymers in the matrix can maintain their mechanical properties, the flexibility and/or stretchability of the semiconductor/insulating polymer blends will provide more possibilities in terms of versatile integrated platforms for high-performance wearable devices with improved environmental and electrical stability. Together with proper material choice and processing technology for electrodes and dielectric materials, the polymer blend approaches have already proven to be one of the most ideal technologies for the development of stretchable electronics [212, 279–282].

3.1.5 Blending with Polymer Elastomer: Nanoconfinement Effect

Nowadays, flexible and deformable electronics have drawn a lot of attention owing to their wide potential applications, including artificial skin, wearable/implantable

Table 3.2 Summaries of the performance by using OSC/Insulating polymer blends.

OSC	Insulating polymer	OSC ratio (wt%)	Average (highest) mobility ($cm^2 V^{-1} s^{-1}$)	On/off ratio	References
169 TIBS–pentacene	**171** APC	20	$0.53 \pm 0.07(0.68)$	10^6	[272]
169 TIBS–pentacene	**172** DVS-BCB	60	5×10^{-4}	10^4	[283]
rubrene	**173** P4VP	80	0.09	10^4	[51]
161 P(I)	**174** PAN	40	11.43(15.93)	10^7	[284]
162 P(II)	**174** PAN	40	2.21(2.58)	10^7	[284]
154 P3HT	**174** PAN	50	0.068(0.076)	10^5	[284]
164 PBTTT	**174** PAN	50	0.061(0.066)	10^5	[284]
154 P3HT	**175** PE	10	6×10^{-2}	10^5	[259]
159 diF-TES-ADT	**176** PMMA	50	$(1.21 \pm 0.53) \times 10^{-3}$	10^5	[271]
154 P3HT	**176** PMMA	5	2×10^{-3}	10^5	[260]
154 P3HT	**176** PMMA	20	$(2.97 \pm 0.59) \times 10^{-3}$	10^3	[263]
154 P3HT	**176** PMMA	5	0.1–0.2	10^6	[285]
165 PQT-12	**176** PMMA	40	0.04	10^8	[286]
167 TES-ADT	**176** PMMA	20	0.47 ± 0.09	10^6	[50]
167 TES-ADT	**176** PMMA	20	0.47 ± 0.09	10^6	[50]
158 C_8–BTBT	**177** PS	95	25(43)	10^3	[274]
159 diF-TES-ADT	**177** PS	50	$(1.75 \pm 0.88) \times 10^{-4}$	10^3	[271]
159 diF-TES-ADT	**177** PS	50	3.8	10^6	[276]
159 diF-TES-ADT	**177** PS	50	6.7	10^6	[276]
160 DNTT	**177** PS	66.6	0.76–1.1	10^6	[275]
163 P3BT	**177** PS	10	0.01	10^6	[264]
154 P3HT	**177** PS	10	1×10^{-2}	10^4	[259]
154 P3HT	**177** PS	3	4.0×10^{-3}	10^5	[262]
154 P3HT	**177** PS	5	0.1–0.5	10^3	[285]
164 PBTTT	**177** PS	5	0.1–0.4	10^3	[285]
166 PCDTPT	**177** PS	75	5.5 ± 0.3	10^4	[261]
165 PQT-12	**177** PS	5	0.1–0.3	10^3	[285]
rubrene	**177** PS	80	0.4(0.52)	10^5	[51]
154 P3HT	**178** PSA	10	$(6.82 \pm 0.74) \times 10^{-3}$	10^5	[263]
154 P3HT	**179** PVCn	10	0.015	10^5	[287]
170 P1	**180** PVK	55-65	2.5	10^6	[288]
159 diF-TES-ADT	**181** PαMS	50	0.094 ± 0.042	10^5	[271]
159 diF-TES-ADT	**181** PαMS	50	1.2	10^6	[276]
167 TES-ADT	**181** PαMS	70	0.040 ± 0.011	10^6	[270]
169 TIBS–pentacene	**181** PαMS	50	0.44	10^4	[268]
169 TIBS–pentacene	**181** PαMS	50	0.5(0.54)	10^5	[273]

health monitors, and flexible/stretchable displays [281, 289, 290]. Conjugated polymer semiconductors, with the advantage of some level of softness, are regarded as ideal candidates for active channel layers in stretchable transistors. However, high-mobility semiconducting polymers are often fabricated by incorporating a large amount of planar fused aromatic rings as repeating units, together with strong intermolecular interactions, thus limiting the stretchability and mechanical plasticity. To realize a highly flexible and/or stretchable semiconductor polymer, there are two approaches in principle. The first strategy is to synthesize a brand-new, distinct stretchable polymer with a carefully and reasonably designed molecular structure. The second practical approach is to make good use of the abovementioned strategy by blending a semiconductor with a polymer elastomer to enable good stretchability with maintaining high electrical characteristics (Figure 3.10a) [291–293]. Moreover, solution-processing techniques have great effects on the stretchability of the resulting composites [165].

Recently, Gumyusenge et al. [288] employed the strategy of making semiconducting polymer blends, which were composed of interpenetrating conjugated polymers and a high glass-transition temperature insulating matrix, to realize stable charge transport at high temperatures. The hole mobility of thin-film transistors exceeded ~2.0 $cm^2 V^{-1} s^{-1}$ across a wide temperature range ranging from room temperature up to 220 °C, thus broadening the applications of OFETs in harsh thermal conditions.

Shin et al. [295] reported the use of a stretchable P3HT rubber composite blended with an elastomer styrene-ethylene-butylene-styrene (SEBS) (**27**) as the semiconductor channel layer in OFETs (Figure 3.10b). The stretchable films were made by spin coating a P3HT/SEBS blend solution onto a substrate, where high-quality P3HT nanofibers induced by a cooling-and-heating process aggregated into bundles and formed networks by side-to-side self-assembly on the top surface of the SEBS because of the *in-situ* phase separation during the solvent evaporation process (Figure 3.10c). The resulting films showed hole mobilities of ~10^{-2} $cm^2 V^{-1} s^{-1}$ and reliable stretchability at high strain up to $\varepsilon = 0.3$–0.5. In another work, Song et al. [294] described an intrinsically stretchable organic semiconducting material based on a P3HT/PDMS blend with P3HT nanowire networks embedded in the PDMS (**28**) matrix (Figure 3.10d). The blended films showed comparable electrical properties with the homo P3HT over a large range of semiconductor ratios, even with only 1 wt% P3HT. The mobilities of the 10% P3HT blended films at 100% strain under stretching degraded very little after 5000 stretching cycles.

Yu et al. [296] introduced a polymer blend system showing a high FET mobility and perfect transparency approaching 100% by blending a semiconducting polymer DPP2T (**29**) into an inert PS (**30**) matrix (Figure 3.10e). It is interesting that transistors based on the DPP2T/PS (15/85 wt% ratio) composites showed a higher mobility than that of neat DPP2T, with an average linear-regime mobility of ~1.0 $cm^2 V^{-1} s^{-1}$ (Figure 3.10f). This improvement was attributed to the morphological changes. In the neat DPP2T films, although the rigid π-conjugated backbone of DPP2T led to high intermolecular stacking and provided strong interchain charge transfer, its crystallization was unavoidably disturbed during the solution process owing

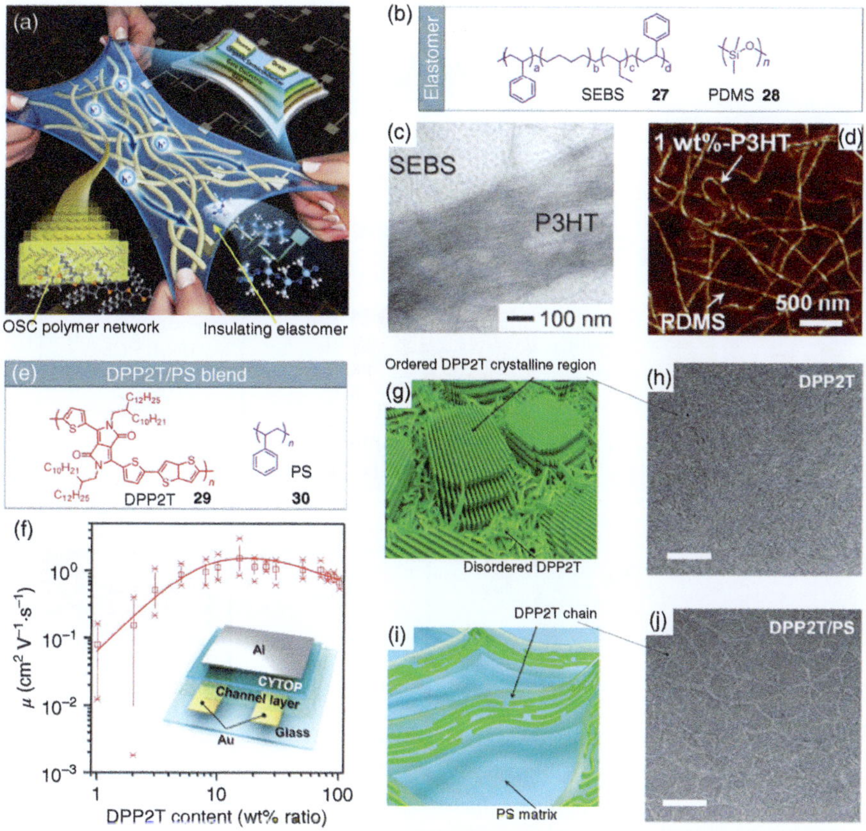

Figure 3.10 Blending with insulating polymer elastomer. (a) Schematic illustration showing the strategy by blending a semiconductor with polymer elastomer to make flexible stretch electronics. (b) Chemical structure of insulating elastomer. (c) TEM image showing a P3HT nanofibril bundle indented in the SEBS surface. (d) AFM phase image showing 1D nanowire network of P3HT embedded in transparent PDMS elastomeric matrix. (e) Chemical structure of DPP2T/PS blend. (f) Hole mobility of OFETs fabricated with DPP2T/PS blend films of various concentration ratios. (g–j) Schematic illustrations (g and i) and TEM images (h and j) showing the structural morphologies of pure DPP2T (g and h) and DPP2T/PS blend films (i and j). Source: (a) Reproduced with permission from Song et al. [294], © 2016 John Wiley & Sons. (c, d) Reproduced with permission from Shin et al. [295], © 2015 John Wiley & Sons. (f–j) Reproduced with permission from Yu et al. [296], © 2016 Proceedings of the National Academy of Sciences PNAS.

to the limited time and space for polymer chains to align, and thus the complex chain entanglements resulted in disordered regions (Figure 3.10g,h). However, for a DPP2T/PS blend, the PS matrix diluted the polymer chains of the small number of semiconductors, confined them within narrow, connected web-like structures, and provided a more flexible surrounding environment to prevent entanglement of the rigid DPP2T chains, thus producing continuously connected pathways with reduced structural/energetic disorder to guarantee efficient charge transport (Figure 3.10i,j). The authors further built a prototype of an FT-FET-integrated

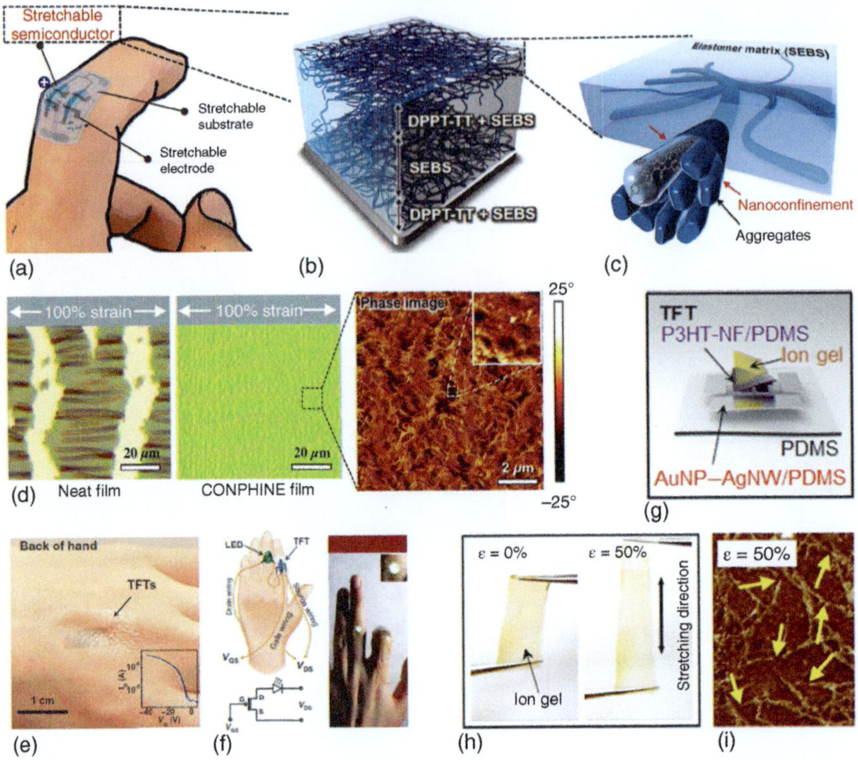

Figure 3.11 Nanoconfinement effect enhances the stretchability of a polymer semiconducting films through the CONPHINE method. (a) Schematic illustration showing a highly stretchable and wearable TFT. (b) Schematic illustration showing the morphology of the CONPHINE film. (c) Schematic illustration showing desired morphology composed of OSC aggregates embedded in SEBS nanoscale networks to achieve high stretchability. (d) Optical microscope images of a neat OSC film (left) and a CONPHINE film (middle) under 100% strain; the AFM phase image (right) showing that the color variation in a CONPHINE film is not due to cracks. (e) Optical image showing the transistor's transparency and flexibility when attached to skin. (f) Demonstration of the fully stretchable OFET to drive an LED. (g) Schematic illustrations showing an OFET consisting of P3HT-NF/PDMS semiconductor blend on a flexible substrate. (h) Image showing the stretchability of free-standing ion gel dielectric. (i) AFM phase image showing P3HT-NF/PDMS blend coated on a PDMS substrate upon uniaxial stretching. Source: (a–f) Reproduced from Xu et al. [21], © 2017 American Association for the Advancement of Science – AAAS. (g–i) Reproduced with permission from Kim et al. [297], © 2017 American Association for the Advancement of Science – AAAS.

polymer light-emitting diode (PLED) by directly fabricating transparent DPP2T/PS FETs atop PLEDs to drive them. The top FET was able to modulate the emitted light even during bending.

A remarkable work was reported recently by a group led by Bao [21] based on the nanoconfinement effect to achieve a highly stretchable polymer semiconductor with high mobility (Figure 3.11a–c). They demonstrated a polymer film with highly improved stretchability without affecting its charge-transport mobility through the conjugated polymer/elastomer phase separation-induced elasticity (CONPHINE) method. They first examined the microstructure of the films formed by the blend of a high-performance DPP-based conjugated polymer, DPPT-TT (**29**), with a soft elastomer, SEBS (**27**). The obtained CONPHINE films exhibited vertical phase separation, with semiconductors appearing on the top and bottom of the films (Figure 3.11b). Under optimum conditions with 70 wt% SEBS in the CONPHINE film, significantly improved softness and ductility were realized with an unaffected high mobility (>1 $cm^2 V^{-1} s^{-1}$ in a hard substrate device) stretched up to 100% strain. Because of the comparable surface energies of the two components, the nanoscale mixed morphology appeared to induce nano-confinement of the polymer semiconductors, forming nanofibers inside the deformable elastomer matrix (Figure 3.11c). This effect altered the physical properties, such as reducing the mechanical modulus and glass transition temperature, as well as increasing the mechanical ductility, which were attributed to the increase of polymer chain dynamics and suppressed crystallization. The nanofiber networks of conjugated polymers formed from phase separation reduced the crystallinity but still ensured good charge-transport properties by connecting the aggregations across the films; in the meantime, the elastomer matrix prevented crack propagation and thus ensured enhanced stretchability (Figure 3.11d). By using the CONPHINE films, the authors fabricated fully stretchable OFETs with BGTC configuration by using CNT networks as the electrodes and SEBS as the dielectric layer, substrate, and encapsulation layer. The OFETs exhibited an average mobility of ~0.59 $cm^2 V^{-1} s^{-1}$ (lower because of higher contact resistance) and good transparency and conformability, suitable for constructing e-skins on human epidermis (Figure 3.11e). The highly robust stretchable transistor was further demonstrated to drive a finger-wearable LED (Figure 3.11f).

In another example of stretchable devices, Kim et al. [297] fabricated rubbery electronics and sensors with intrinsically stretchable elastomeric composites composed of semiconductors and conductors, that is, P3HT nanofiber (P3HT-NF)/PDMS composites and Au nanoparticles with conformally coated silver nanowires (AuNP–AgNW)/PDMS acting as the channel layer and source/drain electrodes, respectively. The corresponding ion gel-gated transistors (Figure 3.11g) constructed on PDMS substrates retained more than 55% of their electrical performance upon 50% stretching (Figure 3.11h), with one of the highest P3HT-based FET mobilities of ~1.4 $cm^2 V^{-1} s^{-1}$ owing to the improved crystallinity of the P3HT-NFs (Figure 3.11i). Moreover, by incorporating the elastomeric semiconductors and conductors to build rubbery sensors with two-terminal configurations as strain/pressure and temperature sensors, all exhibited good sensing capabilities.

3.2 Enhancing the Performance Through Semiconductor Heterojunctions

A heterojunction (p–n junction) is the interface between two kinds of semiconductors (p- and n-type) with an energy gap, leading to a space-charge region with various

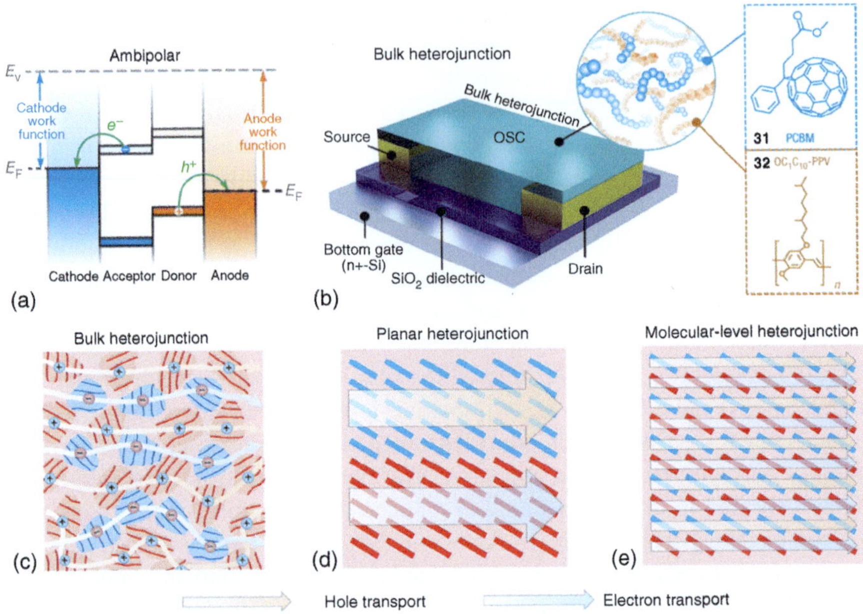

Figure 3.12 Semiconductor heterojunctions. (a) Schematic energy-level diagram showing the hole/electron transport at heterojunction. (b) Schematic illustration showing the geometry of a bulk heterojunction OFET device. Inset shows the interpenetrating networks of the two semiconductors: PCBM (blue) and OC_1C_{10}−PPV (orange). (c–e) Schematic illustrations showing ambipolar charge transport in (c) bulk heterojunction, (d) planar heterojunction, and (e) molecular-level heterojunction. Source: (b) Adapted from Meijer et al. [298].

electron states determined by the work function of each conductivity type. When two layers or regions of OSCs with proper energy alignment, that is, the work function of the p-type semiconductor is smaller than that of the n-type semiconductor, come into contact with each other, an accumulation heterojunction is obtained with excess free carriers and band bending, resulting in a built-in electric field that guarantees effective carrier separation (Figure 3.12a).

OSC heterojunctions are widely applied in fabricating high-performance optoelectronic devices, especially OPVs and organic light-emitting diodes (OLEDs), which can provide an efficient separation of photoexcited excitons into free electron and hole carriers or recombination of excitons to emit light. In the case of OFET, one of the most important applications of heterojunction is to build balanced ambipolar transport. The ambipolar transport behaviors of OSCs have shown obvious advantages in terms of building OFET-based memories [299], as well as integrated circuits (like inverters), with good electronic properties and simplified structures [300–303]. However, while the development of high-performance single-component ambipolar OSCs seems more challenging, the fabrication of organic heterojunctions comprising both hole transport and electron transport materials has been demonstrated to be a useful approach to achieving this goal [4, 54, 304–306].

Initially, ambipolar transport has been achieved in one active layer by blending p- and n-type OSCs together to form a bulk heterojunction (BHJ), first presented by Meijer et al. [298] by using a blend of PCBM (**31**) and OC_1C_{10}-PPV (**32**) in 2003 (Figure 3.12b). Other pairs of BHJs, such as the blend of P3HT and PCBM [307–309] have also been applied in the fabrication of ambipolar OFETs [305]. From then on, a variety of methods have been developed to fabricate high-quality heterojunctions between OSCs, showing satisfying electrical performance enhancement, which includes BHJ (Figure 3.12c), planar heterojunction (Figure 3.12d), and molecular-level heterojunction (Figure 3.12e).

3.2.1 Planar Bilayer Heterostructures

By applying the idea of an organic p–n bilayer junction to OFETs, many works have been presented that demonstrate OFETs with ambipolar transport behaviors (Figure 3.13a, **23–29**) [306, 310–313]. Planar bilayer heterojunctions were first demonstrated by Dodabalapur et al. [310] to act as active materials of OFETs in 1995, with electron and hole carriers accumulated in thermally evaporated C_{60}/α-6T (**33**) layers, respectively. These heterojunctions exhibited ambipolar transport with a hole mobility of ~0.004 $cm^2\ V^{-1}\ s^{-1}$ and an electron mobility of ~0.005 $cm^2\ V^{-1}\ s^{-1}$ when the α-6T layer was in direct contact with Au electrodes. In order to further enhance the charge injection process from the electrode to the OSC layer, asymmetrical electrodes were used by Rost et al. [312] (Figure 3.13b), who chose to use the low-work-function metal Mg as the contact electrode with n-type OSC and an Au electrode in contact with p-type OSC in heterojunction pentacene/PTCDI-$C_{13}H_{27}$ (**34**) to fabricate OFETs showing ambipolar properties over a wide range of biases. Herein, we mainly focus on the approaches to building organic planar heterostructures, starting with the gas deposition method.

Among the early studies, small OSC molecule pairs were chosen to form heterojunctions by sequential vapor deposition of the two components under high vacuum (owing to their poor solubility in common solvents), such as CuPc/F_{16}CuPc (**35**) [55, 314–316], pentacene/C_{60} (**36**) [56, 317, 318] and pentacene/perfluoropentacene (**37**) [319]. Besides that, Wang et al. [320] demonstrated a heterojunction with two classical p-type OSCs (terphenanthrenyl and vanadyl-phthalocyanine) and observed the surprising ambipolar transport behavior, which was attributed to the ambipolar behavior of the heterojunction caused by the interfacial electronic structures.

Furthermore, in order to achieve higher performance, molecular template growth (MTG) [321] such as the weak epitaxy growth (WEG) method (Figure 3.13c) [322], is widely used for fabricating high-quality thin-layered OSC films with controllable molecular orientations and morphologies that modify the interface properties and electronic structures. The WEG method, developed by first introducing an ultrathin ordered template (**47–50**) onto the substrate to induce/guide the ordered growth of a variety of OSCs (**43–46**), worked well for both p- and n-type materials to present mobilities approaching those of single-crystal devices [57, 323–327], as well as forming the complex layer-by-layer structure, even organic superlattices [328, 329]. The Yan's group further used this WEG method to build organic bilayer [314, 327, 329],

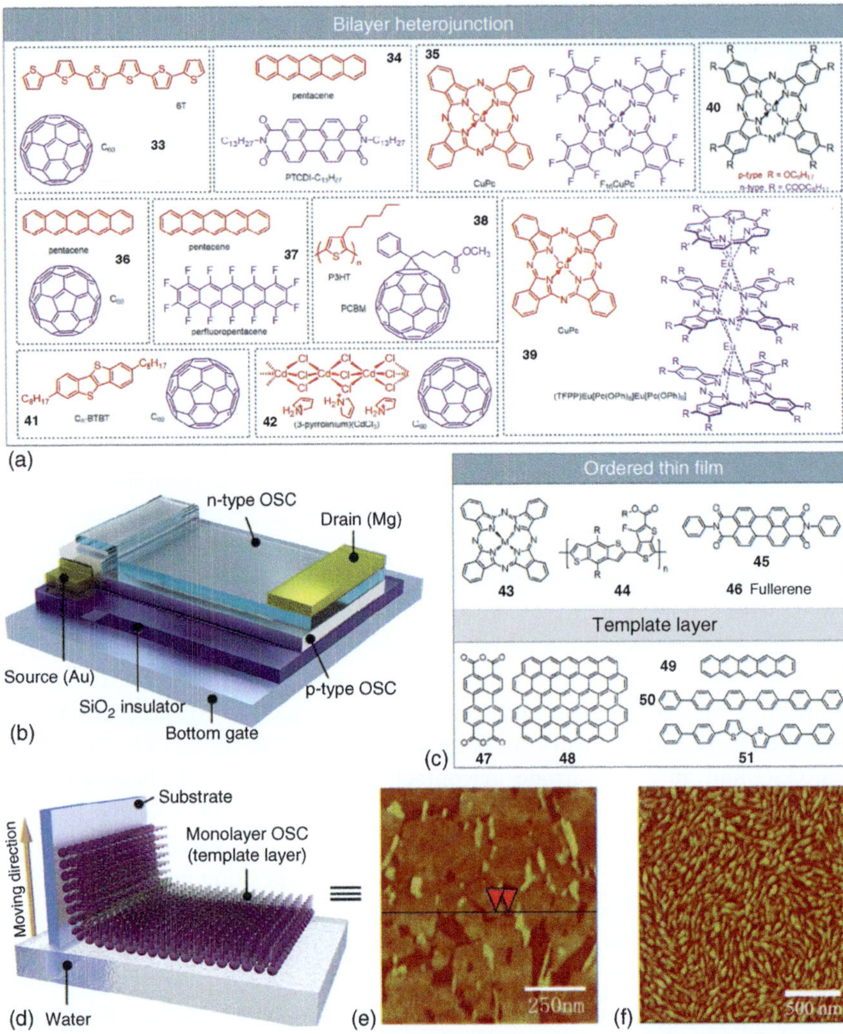

Figure 3.13 Planar bilayer structures. (a) Chemical structures of bilayer heterojunction. P-type is highlighted in red and n-type is highlighted in blue. (b) A bilayer ambipolar OFET with asymmetric contact electrodes. (c) Chemical structures of molecules used in the molecular template growth (MTG) method. (d) LB method to prepare bilayer heterojunction. (e, f) AFM images showing the morphology of a CuPc monolayer (e) and 50-nm-thick $F_{16}CuPc$ films grown on it (f). Source: (b) Adapted from Rost et al. [312]. (d–f) Reproduced with permission from Wei et al. [313], © 2009 American Chemical Society.

even multilayer films [330], through continuous growth on the surface of *p*-6P thin layers, which were later used as channel layers. Improved charge-transport (unipolar or ambipolar) behaviors were observed in the corresponding OFET devices.

In contrast to the gas deposition process, which requires high temperature and atmosphere control (ultrahigh vacuum or inert gas), crystallization of OSCs from solution is sometimes thought to be a more convenient method to produce organic films. Kaneto et al. [331] used PCBM as an n-type semiconductor to form heterojunction P3HT/PCBM (**38**) by spin coating with P3HT as the p-channel material. They fabricated bilayer films with two configurations, PCBM (upper)/P3HT (lower) or PCBM (lower)/P3HT (upper). Although ambipolar transport was observed in the former structure, mobilities decreased dramatically in comparison with the pure materials, which was attributed to the possibility of poor crystallinity.

Langmuir films offer a very useful way to prepare high-quality film from solution. An ultrathin organic monolayer formed at the gas–liquid or liquid–liquid interfaces is called a Langmuir film, which can then be deposited onto a solid substrate surface by means of the Langmuir–Blodgett (LB) or Langmuir–Schaefer (LS) technique (Figure 3.13d). By using the LB technique, Wei et al. [313] were able to fabricate a monolayer of CuPc on substrates (Figure 3.13e), and they found that a single monolayer is enough to achieve the field-effect transport behavior. Then, by using the LB monolayer as a template, an F_{16}CuPc film was thermally deposited on top of the CuPc layer to achieve ambipolar transport. Gao et al. [306] reported the use of a bilayer heterojunction **39** as the semiconductor channel to build the transistor. Eu-complex was grown directly onto a vacuum-deposited CuPc film by the quasi-Langmuir–Shafer (QLS) method. The resulting OFET device showed air-stable carrier mobilities of ~0.16 $cm^2 V^{-1} s^{-1}$ for holes and ~0.30 $cm^2 V^{-1} s^{-1}$ for electrons, which were 10^4 times higher than single component. The improvement was attributed to the template-induced significant improvement over the molecular ordering and change of aggregation mode of the top film. The authors also showed that through the QLS solution-processed method, soluble CuPc derivatives with different functional groups showing p- and n-type behaviors, such as p-type $CuPc(OC_8H_{17})_8$ and n-type $CuPc(COOC_8H_{17})_8$, were integrated into a p/n or n/p bilayer **40**, resulting in quite different electric performances and sensor responses [58].

However, the LB method is far from perfect. Most of the LB films are polycrystalline with pinholes and defects. As a result, the top layer that grows on it represents a less ordered film or, even worse, undesired aggregates (Figure 3.13f), which greatly hampers the charge transport. It is thus of great importance to grow long-range ordered single-crystal films stacked p-n heterojunctions. By carefully selecting proper solvents and evaporation processes, Fan et al. [332] achieved this goal through a droplet-pinned crystallization (DPC) method (Figure 3.14a). They have prepared single-crystalline p–n junction **41** from a mixed solution of p-type C_8–BTBT and n-type C_{60} in their work (Figure 3.14b). The contacting p–n single crystals reported in this work promised ideal interfaces at the junctions for exciton generation/dissociation. The best values obtained were ~0.16 $cm^2 V^{-1} s^{-1}$ for holes and 0.17 $cm^2 V^{-1} s^{-1}$ for electrons (Figure 3.14c), which were among the highest mobilities reported in single-crystal-based ambipolar organic transistors.

There are two important requirements when growing crystalline p–n junctions from a mixed solution (Figure 3.14a). First, the two crystalline OSCs should have different growth rates; that is, one molecule should crystallize more easily and grow first. Subsequently, the other nucleates on the former heterogeneously. Second, the pinner of the DPC method is important as it provides a steady growth condition

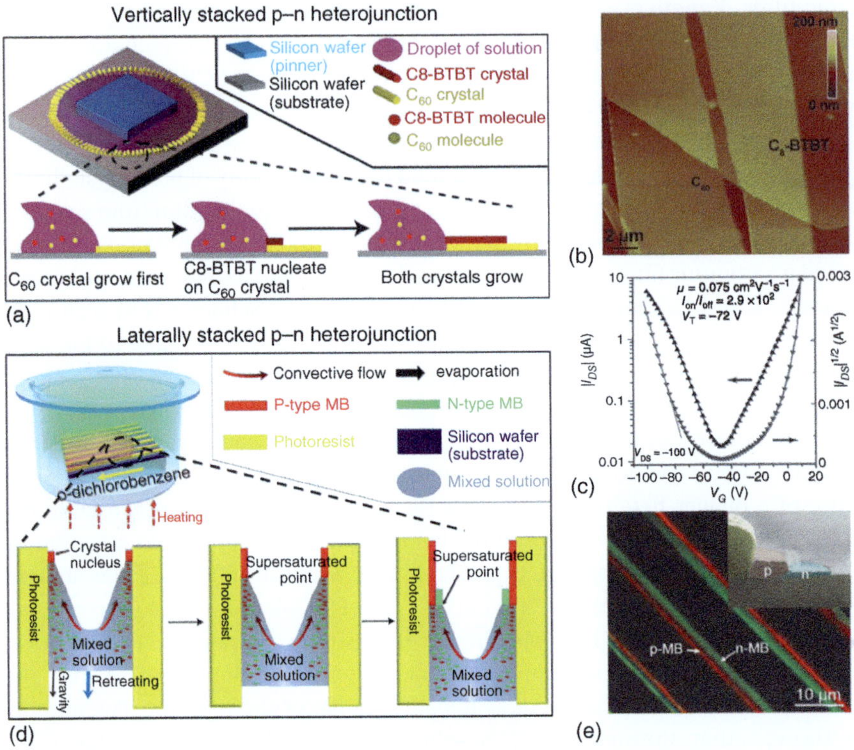

Figure 3.14 Long-range ordered single-crystal film. (a) Schematic illustration showing a droplet-pinned crystallization (DPC) method to prepare the vertical bilayer structure from solution. (b) AFM image showing the vertically stacked p-n junction, showing C_8–BTBT ribbon growing atop C_{60} ribbon. (c) Typical transfer curve of C_8–BTBT/C_{60} heterojunction showing an ambipolar transport behavior. (d) Schematic illustration showing a method to prepare the lateral p–n junction from solution. (e) Polarized OM image and cross-sectional SEM image (inset) showing the structure of lateral p–n junction. Source: (a–c) Reproduced with permission from Wang et al. [332], © 2013 John Wiley & Sons. (d, e) Reproduced with permission from Song et al. [59], © 2018 John Wiley & Sons.

by avoiding the sliding of the solution droplets. Without the pinner, crystals grow randomly and even discontinuously. Based on the same method, Wu et al. [333] prepared well-aligned ferroelectric heterojunction **42** of (3-pyrrolinium)($CdCl_3$) from solution on top of well-aligned C_{60} single crystals, by using an orthogonal solvent to build memory devices. The memory window of these devices was as large as 66 ± 7 V. In combination with a relatively high electron mobility of $1.28 \pm 0.41\ cm^2\ V^{-1}\ s^{-1}$, these bilayer single crystals were potentially useful for high-performance OFET-based memory devices with high operation speeds.

Zhang et al. [59] have demonstrated a modified strategy to prepare lateral p–n junction arrays from mixed solution. In their work, the pinner is prepatterened photoresist, which provides high-surface-energy sites to grow p-type crystal, which has higher growth rate. Then the p-type crystal provided new high-surface-energy positions to induce the growth of n-type molecules at their sides (Figure 3.14d).

This method led to the formation of patterned p-n junction arrays (Figure 3.14e) for potential application in integrated circuits.

3.2.2 Molecular-Level Heterojunction

Organic charge-transfer complexes (CT complexes), which are composed of two or more molecules, with one acting as an electron donor (D) and the other as an electron acceptor (A), are formed together by the weak association of electrostatic attraction, where transfer of a fraction of electronic charge occurs from D to A [334–337]. They are regarded as molecular-level D–A heterojunctions with typical mixed- or segregated-stack motifs within the crystal structure according to the molecular arrangement (Figure 3.15a). Segregated-stack systems, formed from D and A packing into separate columns (-A-A-A- and -D-D-D-), usually show high electrical conductivities in the stacking direction and even metallic behaviors at high temperature [338–341]. In the meantime, mixed-stack systems, with D and A alternating along the stacking direction (-D-A-D-A-), are generally less conductive, being semiconductors or insulators under ambient conditions [342–345].

Apart from the properties of individual D or A in the co-crystal, promising ferro- and opto-electronic features might originate from the combination and/or intermolecular interactions between D and A [347, 348]. The construction of such organic CT complexes with p–n heterojunctions at the molecular level provides an efficient way to realize high-performance ambipolar OFETs and multifunctional devices based on the choices of different organic molecules as D–A pairs and without complicated molecular design, thus exhibiting unique electrical, magnetic, and optoelectronic properties. Although many breakthroughs have been made in the study of CT complex systems since the discovery of the CT crystal tetrathiafulvalene–tetracyanoquinodimethane (TTF–TCNQ) (**52**) [337, 349], it is still challenging to construct such a CT complex heterojunction at the molecular level, partly because of the rather rare D–A pairs suitable enough to satisfy the demand when forming the crystalline structure (Figure 3.15b). Therefore, it is necessary to consider the complicated weak interactions between donor and acceptor components beyond charge-transfer (CT) interactions, such as π–π interactions, H-bond interactions, halogen–halogen interactions, in the formation of organic co-crystals [61, 344, 350]. In fact, the term "organic co-crystal" is currently used as a broad concept to describe complicated D–A systems in general. That is to say, the organic CT complex is actually one member of the organic co-crystal family [351–353].

In general, owing to the composition diversity, the CT complex can be a semiconductor, a metal, or even a superconductor. TTF and TCNQ are typical strong donor and acceptor, respectively, which can form TTF-TCNQ co-crystals (**52**) with a segregated-stack structure [354]. It was first prepared by Ferraris et al. [355] by co-crystallization from their acetonitrile solutions (Figure 3.15c). The ambipolar transport behavior in such CT complexes was first explored by Hasegawa's group with BEDT–TTF/F2-TCNQ (**53**) in organic Mott insulators at low temperature

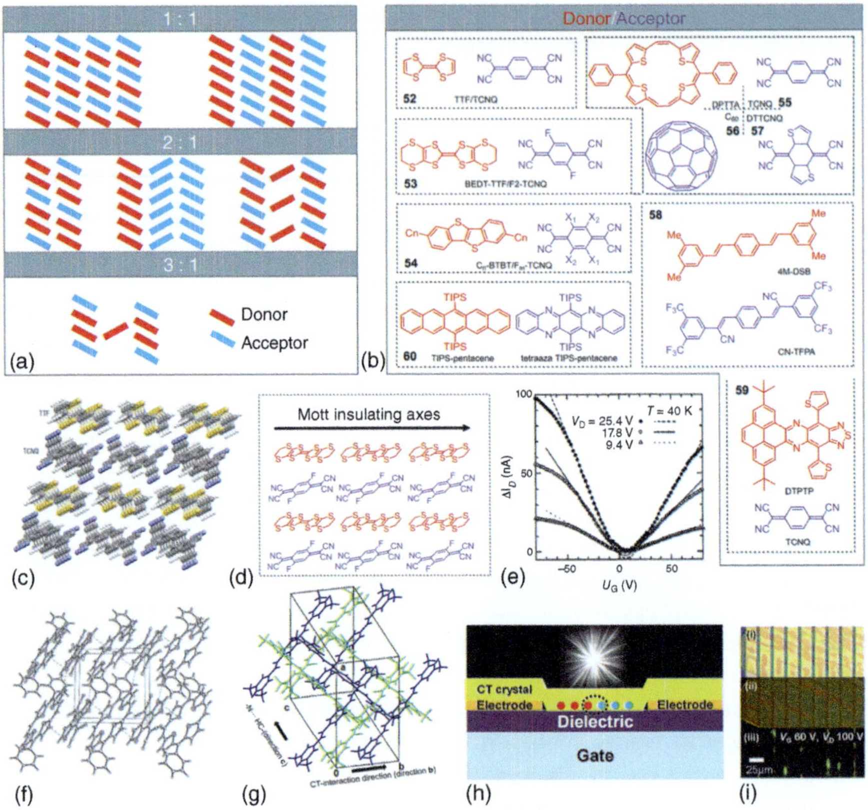

Figure 3.15 Molecular-level heterojunctions. (a) Schematic illustration showing mixed donor–acceptor stacking heterojunctions with various ratios. (b) Chemical structures of charge-transfer complex. (c) Crystal structure of TTF-TCNQ co-crystals. (d) Crystal structure and (e) ambipolar transport behavior of BEDT-TTF/TCNQ co-crystals. (f) Crystal structure of DPTTA/DTTCNQ co-crystal. (g) Crystal structure of 4M-DSB/CN-TFPA co-crystal. (h) Schematic illustration showing the bottom-contact OLET device. (i) OM images of the OLET device under (i) white light irradiation; (ii) UV irradiation; and (iii) operation at $V_G = 60\,V$, $V_D = 100\,V$, respectively. Source: (c) Reproduced with permission from Alves et al. [338], © 2008 Springer Nature. (d, e) Reproduced with permission from Hasegawa et al. [62], © 2004 American Physical Society. (f) Reproduced with permission from Qin et al. [60], © 2014 John Wiley & Sons, Inc. (g–i) Reproduced with permission from Park et al. [363], © 2017 John Wiley & Sons, Inc.

(Figure 3.15d) [62]. Alves et al. [338] studied the electronic properties at the interfaces between TTF and TCNQ molecules in a single crystal. Both are large-gap undoped semiconductors, essentially insulators. Their results showed that highly conducting interfaces could be realized with resistivities ranging from 1 to 30 kΩ per square and the conductivity exhibiting metallic-like temperature dependence. The conduction of the electron–hole metallic conductor originating from a large CT took place at the interface on the molecular scale, illustrating the importance of the CT interaction at the interface [63, 64].

In addition to the traditional donor TTF and its derivatives, other p-type OSCs such as C_n–BTBT, and perylene, have been used to form CT complexes with TCNQ derivatives as well. Tsutsumi et al. [356] reported a series of C_n–BTBT–F_mTCNQ (n = 4, 8, 12; m = 0, 2, 4) CT semiconducting compounds (**54**) with layered crystalline structures that showed high FET mobilities (~0.4 $cm^2\ V^{-1}\ s^{-1}$). The p- and/or n-type operations were controlled by changing the fluorine substitution of TCNQ. Qin et al. [60] have developed a novel sulfur-bridged donor molecules, meso-diphenyl tetrathia[22]annulene[2,1,2,1] (DPTTA), which was capable of forming such D–A complexes with TCNQ [346] (**55**) and fullerenes [340] (**56**), all exhibiting efficient ambipolar behaviors with both hole and electron mobilities exceeding 0.1 $cm^2\ V^{-1}\ s^{-1}$. When mixing the electron-rich DPTTA with an extended π-conjugated acceptor DTTCNQ [60] (**57**) (a TCNQ derivative with a π extension), the namely DPTTA–DTTCNQ complex with 1 : 1 stoichiometry, they obtained quasi-2D crystal structure, within which there existed various intermolecular interactions such as sulfur–sulfur interactions and π–π interactions, as well as hydrogen bonds (Figure 3.15f). Single-crystal FETs based on drop-casting microcrystals of the CT complex presented remarkable ambipolar properties, with a high hole mobility of ~0.77 $cm^2\ V^{-1}\ s^{-1}$ and an electron mobility of ~0.24 $cm^2\ V^{-1}\ s^{-1}$ in ambient atmosphere, owing to the direct coupling and super-exchange interactions within the quasi-2D charge-transport network. The ambipolar properties in such an organic D–A system were thought to arise from the super-exchange nature as well as from the alternate stacking direction [342, 357, 358].

In addition to the traditional acceptor TCNQ and its derivatives, other acceptors such as 1,2,4,5-tetraacyanobenzene (TCNB) and C_{60}, as well as naphthalenediimide (NDI) and perylenediimide (PDI) derivatives, have been used to form CT complexes with different donors as well. By selecting different D/A molecules and/or further modifying their structure with chemical groups as a tool to tune their donor/acceptor strengths, as well as solution processability, the packing structure of CT systems can be changed/tuned, bringing about various optoelectronic behaviors [359–361]. Also, researchers have made every effort to discover new CT pairs with high mobilities and/or luminescence. Park et al. [362] designed a densely packed CT co-crystal with a quasi-1D mixed stack containing two isometric distyrylbenzene- and dicyanodistyrylbenzene-based molecules, 4M-DSB/CN-TFPA (**58**). The formation of this CT co-crystal was driven by CT and N···HC interactions between the two components coming from their tailor-made electronic and geometric properties (Figure 3.15g), leading to very satisfying electronic and optical properties. In the solution-processed D–A (1 : 1) single-crystal FET, anisotropic and ambipolar charge-transport properties were observed, showing a hole mobility up to ~6.7×10^{-3} $cm^2\ V^{-1}\ s^{-1}$ and an electron mobility up to ~6.7×10^{-2} $cm^2\ V^{-1}\ s^{-1}$. Owing to the intermolecular CT state, the crystals exhibited a strong bright-red photoluminescence with high efficiency (Φ_{PL} = 31%). By modifying the common parent molecular backbones, the same group further reported a novel 2D-assembled 2 : 1 D:A co-crystal with mixed stacking, which was thought to have been induced by the dipole interaction of side-by-side CN groups on donors and acceptors [363]. The CT complex showed bright luminescence, with a high photoluminescence quantum yield ($\Phi_F \approx 60\%$) and balanced ambipolar transport with hole and electron mobilities of 10^{-1} cm^2 V^{-1} s^{-1}. This also allowed the authors to demonstrate the first CT complex-based organic light-emitting transistors (OLET) with an external quantum efficiency (EQE) as high as 1.5% (Figure 3.15h,i). Their work sheds light on a way to tailor the supramolecular/optoelectronic properties of CT systems.

Changing the stoichiometry ratios of D:A in the CT complex gave different supramolecular arrangement and electronic characteristic (Figure 3.15a), as reported by Vermeulen et al. [364] who crystallized the perylene-TCNQ system with a new ratio of 2 : 1 (P2T1) and compared its charge-transport behavior with those of previously reported CT complexes with stoichiometries of 1 : 1 (P1T1) and 3 : 1 (P3T1) [365], and they observed ambipolar transport (P2T1) and unipolar electron transport (P1T1) or unipolar hole transport (P3T1) in the corresponding systems. In another work, Zhang et al. [366] took TCNQ as a molecular dopant, which could be inserted into the center of two DTPTP planes. They obtained a 2 : 1 organic crystal of $DTPTP_2$–TCNQ CT complexes (**59**), which exhibited an electron mobility of 0.003 cm^2 V^{-1} s^{-1}, in contrast to pure DTPTP, which showed the p-type character (0.3 cm^2 V^{-1} s^{-1}). These results might provide a new strategy, to switch the charge-transport characteristics of an original system by appropriate molecular co-crystal engineering.

Recently, Miao et al. [367] reported "solid solution crystal" (**60**) of p-type (donor, TIPS–pentacene) and n-type (acceptor, tetraaza TIPS–pentacene) OSCs as a new type of p–n blend for solution-processed ambipolar OTFTs. They found that the hole and electron mobilities of the $(p)_x(n)_{1-x}$ crystals are tunable by varying the ratio of p- and n-type OSC in the solid solution. Besides, their response to light differs from that of microphase-separated D–A heterojunctions or stoichiometric co-crystals because the solid solutions involve mixing of donor and acceptor at the molecular level with π–π stacking.

3.2.3 Supramolecular Arrangement of the Heterojunctions

As one of the means to develop new molecular materials with novel properties, supramolecular systems, which make use of chemistry beyond the molecular scale, have drawn wide attention in many fields [274–278]. The supramolecular arrangement of OSCs in the solid state is as critical as the molecular structure to achieve good electrical properties, and the assembly process has a significant effect on the growth and stacking of supramolecular π-conjugated nanostructures [65, 368, 369]. To control their supramolecular organization in the solid state, weak interactions should be taken into consideration. We have listed below some of the interactions to build supramolecular heterojunctions, to name a few, halogen bond, CT interaction, radical–radical interaction, H-bond, concave–convex interaction, as well as covalently or coordinately bonded metal-organic framework (MOF) and covalent organic framework (COF) systems. Organic conductors, semiconductors, insulators, and even superconductors could then be obtained [370–373].

Yamamoto et al. [374, 375] made use of halide anions, as well as iodine-containing neutral molecules, as the building blocks to construct molecular conductors with supramolecular assemblies. The resulting cation-radical salts presented novel donor (TTF derivatives) arrangements and interesting transport properties, ranging from those of metals to those of semiconductors (Figure 3.16a). The same group [376] further reported the formation of supramolecular insulating networks that sheathed conducting 1D nanowires by organizing and matching the packing demands (Figure 3.16b), providing new insights into designing unique structures and a procedure to achieve sophisticated devices, as the authors mentioned the demand for high OFF resistance in improving the performance of OFETs.

We have mentioned in the previous part that CT complex can co-crystallize into mixed- or segregated-stack motifs within the structure. In the meantime, by tailoring their chemical structures and interactions, CT interactions are capable of driving alternate stacking of aromatic donor and acceptor chromophores, thus generating 1D supramolecular structures in solution [379]. Rao et al. [377] reported cylindrical micelles and hydrogels of a donor–acceptor CT complex that resulted from the self-assembly of coronene–viologen-based donor–acceptor pairs (Figure 3.16c), CS–dimethyl viologen (DMV) (**61**) in water (Figure 3.16d). The unsymmetric amphiphile DMV was used as the acceptor molecule, which would not only allow CT interactions with coronene but also trigger the self-assembly of the resultant D–A pair through hydrophobic interactions. Based on the same system, Sagade et al. [380] reported that in this two-component hierarchical self-assembly, the potassium salt of CS and the dodecyl substituted DMV form non-covalent amphiphiles via CT interactions, which subsequently self-stack into cylindrical micelles following a surfactant-like assembly (Figure 3.16e). This results in micrometer-long nanofibers with bilayers of CT-amphiphiles arranged radially along the width, and the D–A molecules stacked face-to-face along the length of the fiber, making it promising to build supramolecular CT nanofiber-based OFET (Figure 3.16f). These devices showed a high mobility of ~4.4 $cm^2 V^{-1} s^{-1}$ (Figure 3.16g).

Fahrenbach et al. [378] reported the self-assembly and solid-state semiconducting properties of a trisradical tricationic complex (**62**) composed of the diradical dicationic cyclobis(paraquat-*p*-phenylene) ($CBPQT^{2(\cdot+)}$) ring and methyl viologen radical cation ($MV^{\cdot+}$) (Figure 3.16h). They showed that the complex **62** self-assembled into supramolecular microwires as millimeter-long crystalline needles. The single-crystal structure of this complex revealed a continuous π-stack of radical cations in the solid-state exhibiting p-type conductivity with a mobility of 0.05 $cm^2 V^{-1} s^{-1}$. The nature of the recognition processes driving this self-assembly is a radical–radical interaction between bipyridinium radical cations ($BIPY^{\cdot+}$). This strategy can be employed in the design of multifunctional molecular-level heterojunction based on mechanically interlocked molecules (MIMs).

Black et al. [381–383] introduced a dipyrrolopyridine derivative as a new donor capable of interacting with NDI acceptors through complementary H bonding without disrupting charge transport. They found that the arrangement of mixed or segregated stacks was influenced by the substitutions of N atoms in NDI (Figure 3.17a). All new co-assembled organic materials showed the electron mobility in single-crystal FETs (Figure 3.17b), and ambipolar transport behaviors were observed in DP-P2P/NDI-81 crystal (**63**), exhibiting a relatively balanced transport

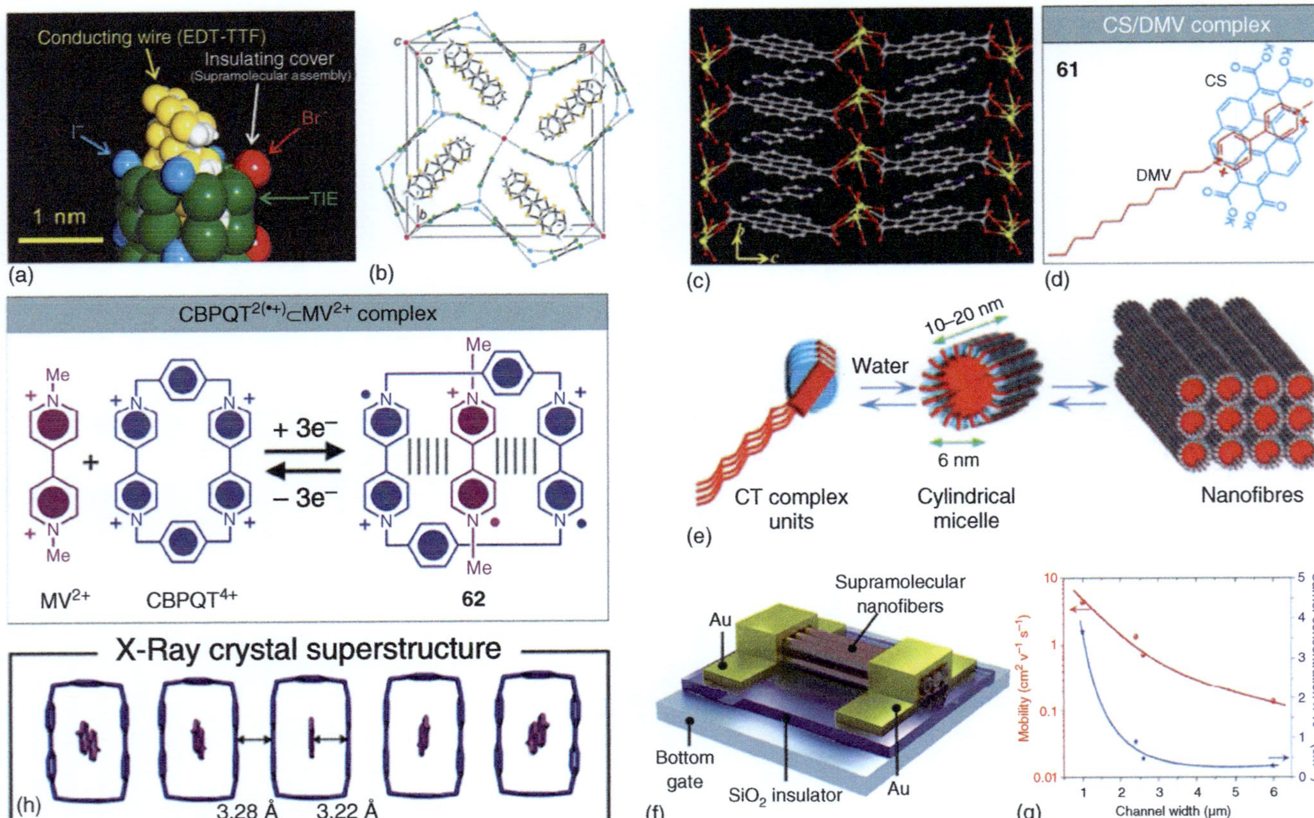

Figure 3.16 Supramolecular arrangements of different heterojunctions. (a) Schematic illustration showing the conducting wire embedded in insulating sheath of halogen ions. (b) Crystal structure showing the interaction of halogen bonds. (c) Crystal structure showing the extended π-columnar stacks of CS–MV cocrystals bridged by potassium ions. (d) Chemical structure of the co-facial CT complex formed from coronene tetracarboxylate (CS) as the donor and viologen derivative (DMV) as the acceptor. (e) Schematic illustration showing the stacking of the complex units to form a cylindrical micelle and eventually nanofibers. (f) Schematic illustration showing the device configuration of a nanofiber OFET and (g) the device performance. (h) Formation of CBPQT$^{2(\cdot+)}$⊂MV^{2+} complex driven by radical–radical interactions. Source: (a, b) Reproduced from Yamamoto et al. [376], © 2008 American Chemistry Society. (c, e, and g) Reproduced and (f) Adapted from Rao et al. [377]. (h) Adapted from Fahrenbach et al. [378].

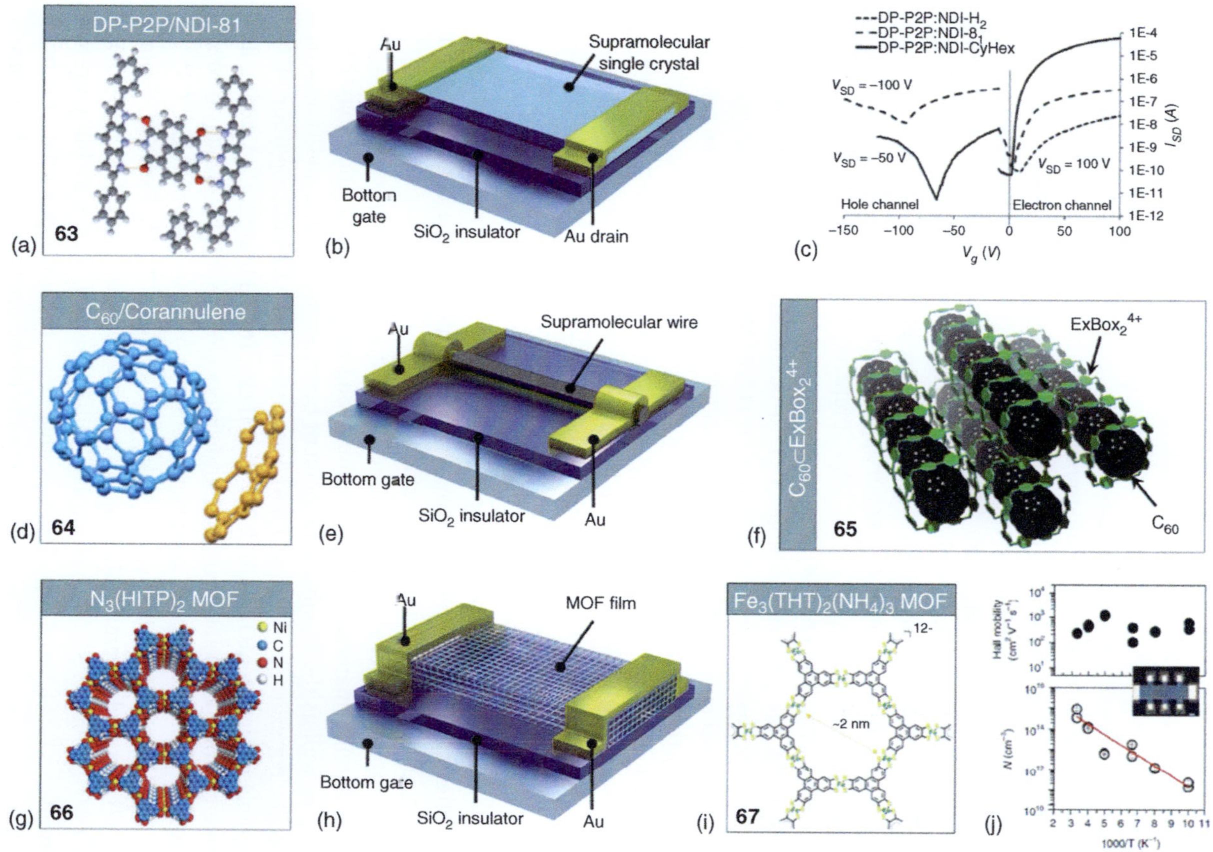

Figure 3.17 Supramolecular arrangements of different heterojunctions. (a) Crystal structure of the organic supramolecular semiconductor arrangement through H-bond. (b) Schematic representation of H-bonded supramolecular single-crystal OFET, which shows ambipolar charge-transport behavior in (c). (d) Crystal structure of corannulene/C_{60} showing the existence of $\pi-\pi$ concave–convex interaction. (e) Schematic representation of the supramolecular microwire OFET. (f) Crystal structure of viologen-based cyclophane, $ExBox_2^{4+}$, as a molecular receptor for C_{60}. (g) Crystal structure of a semiconductive MOF, $Ni_3(HITP)_2$. (h) Schematic representation of a MOF film-based porous FEF. (i) Chemical structure of 2D MOF, $Fe_3(THT)_2(NH_4)_3$ (left) and its high-mobility band-like charge transport (j). Source: (a–c) Adapted from Black and Perepichka [381]. (d, e) Adapted with Wang et al. [385]. (f) Adapted from Barnes et al. [66]. (g, h) Reproduced with permission from Wu et al. [386]. Copyright 2006 American Chemistry Society. (i, j) Reproduced with permission from Dong et al. [387], © 2018 Springer Nature.

with hole and electron mobilities of ~0.043 and ~0.089 $cm^2\ V^{-1}\ s^{-1}$, respectively (Figure 3.17c). Just going beyond the role of supramolecular interactions in the bulk of semiconductor films, it is worth noting that recently Bao's group [384] introduced dynamic on-covalent crosslinking moieties, which could form complementary H-bonds via inter- and intramolecular interactions, into the backbone of DPP-based polymers. Self-healing and stretchable properties were incorporated into the polymer without degrading its mobility (it showed a mobility as high as 1.3 $cm^2\ V^{-1}\ s^{-1}$). Moreover, the damaged devices could be almost fully recovered after a solvent and thermal healing treatment.

Fullerene derivatives are one of the most famous benchmark n-type materials, with unusual shapes (approximately spherical) and chemical natures (unpolarized polyenes), which can form curved π-conjugated complexes when the concave aromatic surfaces match the convex surfaces of the fullerenes. The term "concave–convex interactions" was coined by Kawase et al. [388] to describe such non-covalent π–π interactions between curved aromatic hosts and guests [67, 385]. Following this strategy, Wang et al. [385] prepared large-scale 1D nanowires composed of C_{60}/corannulene (**64**) (Figure 3.17d). Based on strong concave–convex π–π interactions, the cocrystal exhibits a 1D structure as expected and exhibited high electron transport characteristics up to 0.11 $cm^2\ V^{-1}\ s^{-1}$ (Figure 3.17e). The results suggest that co-crystal engineering provides a novel strategy for the rational design of new carbon-based crystalline nanomaterials. In this regard, the Stoddart's group utilized host-guest chemistry to bind C_{60} one-on-one and prepared 1D microwires successfully [66]. They created an extended viologen-based cyclophane, $ExBox_2^{4+}$, as a molecular receptor [389] which was capable of bonding with C_{60} one-on-one, resulting in the columnar self-assembly of a 1 : 1 inclusion complex (**65**) under ambient conditions, and 1D arrays of fullerenes stacked along the long axis of needle-like single crystals (Figure 3.17f) [66]. The crystals showed electrical conductivities of approximately 10^{-7} $S\,cm^{-1}$, which matched the conductivity of high-quality, unfunctionalized C_{60}-based materials.

Recently, there has been a growing interest in studying MOFs as a new type of semiconducting material because of their long-range ordered analogous like graphene, as well as the tunable arrangement between the organic and inorganic constituents. For example, Wu et al. [386] demonstrated the first microporous FETs with a semiconducting MOF, built with $Ni_3(HITP)_2$ (**66**) (Figure 3.17g), acting as the active channel material in organic transistors (Figure 3.17h), which formed free-standing membranes with compact and smooth surfaces. The MOF-OFET showed excellent hole mobilities as high as 48.6 $cm^2\ V^{-1}\ s^{-1}$. Recently, Dong et al. [387] reported a semiconducting π–d conjugated porous $Fe_3(THT)_2(NH_4)_3$ (**67**) 2-D MOF (Figure 3.17i) showing band-like transport with a mobility up to ~220 $cm^2\ V^{-1}\ s^{-1}$ (Figure 3.17j). These results show the great potential of semiconducting MOFs as active materials in optoelectronic applications in the future. The structure of MOFs is highly tailorable. Through appropriate chemical design, we can anticipate controllable conductivity, mobility, and tunable bandgap. For example, in-plane engineering can be addressed via doping the films with metals or small molecules, functioning the ligands during synthesis, as well as filling the pores with host–guest strategy. Out-of-plane engineering can be realized by stacking them with other 2D materials to form novel heterostructures with unique physical and chemical properties.

3.3 Integrating Molecular Functionalities into Electrical Circuits

In general, it is believed that doping semiconductors to integrate functionalities has a negative effect on device performance because it will affect the packing of OSCs. However, recent studies show that by carefully optimizing the processing methods such as phase separation, dopant types, and doping concentrations, a balance can be achieved between performance and functionality. We will discuss some of the progress in the following part.

3.3.1 Charge-Trapping-Induced Memory Effect

Hysteresis is the dependence of the state of a system on its history, also known as the "memory effect." In transistors, hysteresis may be caused by charge storage and/or slow polarization of the gate dielectric, which will be discussed in Section 5.3.3, as well as by charge storage in the semiconductor. Although undesirable for classical transistors, the reversible hysteresis effect can be useful in memory applications [390–392]. Herein, we will briefly discuss two strategies to create interfacial charge-trapping-induced memory effect within semiconductors, namely CT doping on surface (Figure 3.18a) and in bulk (Figure 3.18b).

A brief introduction to the charge-trapping-induced memory effect in semiconductors is given here. Generally, OFETs operate after the accumulation of charge carriers (holes or electrons) through applying a gate (V_G), which is called electrical accumulation [396]. Alternatively, the operation of OFETs can also be manipulated by doping opposite-polarity dopants onto the host OSC, which will act as a chemical accumulation thus creating additional mobile charge carriers [397, 398]. The related mechanism is speculated to be that the existence of the interfacial CT doping preferentially fills charge traps, reduces activation energy (E_a) for the hopping transport process of the injected charge carriers, and accumulates mobile charge carriers at the active channels (Figure 3.18a). In this regard, Park et al. [399] presented a strategy to create directional and unipolar charge transfer doping by utilizing discontinuous nanopatches of a dopant layer to overlay on the surface of the semiconductor film, which in turn increase the threshold voltage (V_{th}) shift upon exposure to photo-irradiation owing to the electron trapping in the gap state of pentacene.

Although the carrier density and V_{th} were effectively controlled, these methodologies could affect the nanomorphology and resistivity of OSCs, on which the carrier mobility strongly depends. To solve this problem, Heeger and coworkers [393] applied chemical vapor treatments by using ammonium (NH_3) and iodine (I_2) (~0.2% doping concentration) on aligned PCDTPT (**68**) thin films (Figure 3.18c). The devices exposed to ammonia vapor (for charge compensation) showed a

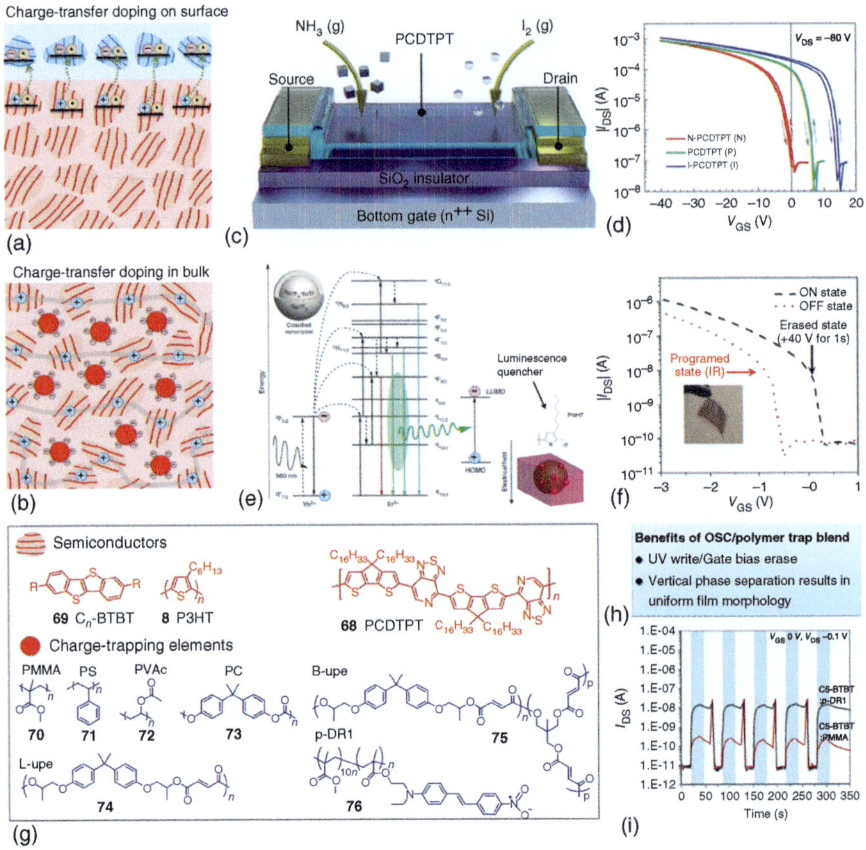

Figure 3.18 Charge-trapping-induced memory effect. (a) Charge-transfer doping on the surface. (b) Charge-transfer doping in bulk. (c) Schematic device architecture and illustration of chemical vapor treatments with NH_3 and I_2 and (d) transfer curves of the devices with N-PCDTPT (NH_3-doped), PCDTPT (pristine), and I-PCDTPT (I_2-doped). (e) Upconversion NPs serving as charge-trapping sites and (f) transfer curves showing the memory characteristics. (g) Chemical structures of representative materials used as semiconductor layer and charge-trapping elements in OFETs. (h) Benefits of OSC/polymer trap blend. (i) I_{ds} as a function of time showing repeatable UV write/erase cycles. Source: (c) Adapted and (d) Reproduced with permission from Lee et al. [393], © 2016 John Wiley & Sons, Inc. (e, f) Reproduced with permission from Zhou et al. [394], © 2014 Springer Nature. (i) Reproduced with permission from Smithson et al. [395], © 2015 John Wiley & Sons, Inc.

decreased carrier density (at zero gate bias) and shifted V_{th} to ~0 V, while the devices treated with I_2 vapor (for doping) produced an increased carrier density with positively shifted V_{th} (~14 V) (Figure 3.18d). Whereas the hole mobility increased up to as high as ~58.6 $cm^2\ V^{-1}\ s^{-1}$ with a higher and higher doping concentration. They attributed the device performances of PCDTPT to oxidation and reduction of nitrogen in the pyridyl unit through facile chemical vapor treatments. The dramatic shift of V_{th} sets the foundation for memory device operation.

Another strategy to manipulate the operation of OFETs is realized by introducing opposite-polarity dopants onto the bulk of OSC (Figure 3.18b). The related mechanism is that the existence of interfacial CT doping preferentially fills charge traps, reduces the activation energy for the hopping transport process of the injected charge carriers, and accumulates mobile charge carriers at the active channels [397, 398]. Furthermore, the presence of charge-trapping/charge-releasing moieties (i.e. electron-withdrawing and electron-donating groups) in the conducting channel has also been demonstrated to optically modulate the carrier density and tune the electrical properties of OFETs through charging/discharging of the trapping sites (Figure 3.18b) [400–402]. Zhou et al. [394] demonstrated a multilevel photonic flash memory cell by utilizing upconversion (UC) materials, $NaYF_4$:Yb/Er@$NaYF_4$, as charge-trapping element when blended with P3HT. The working principle of the tunable memory effect is shown in Figure 3.18e. OFF state (programmed state) was obtained by applying a gate bias of −40 V for 1 second (Figure 3.18f, red line). The holes produced in P3HT are transferred and trapped in the nanocrystals and/or interfaces between nanocrystals and polymer. The trapped holes in turn provided a built-in potential across the channel and reduced the effective electrical field, leading to a shift in the threshold voltage of the two states. Whereas in ON state (erased state) the stored holes are removed by applying a gate bias of +40 V for 1 second (Figure 3.18f, black line). Under IR illumination, the UC nanocrystals emit green light, which is simultaneously reabsorbed by P3HT. The photoinduced electron–hole pairs are separated with the application of a negative gate bias, and more holes are trapped in the nanocrystals during the programming process under IR illumination.

Zhu's group [395] has done a series of systematic works using insulating polymer binder as charge-trapping elements when spin coated with C_n–BTBT (**69**) in a mixture solution (Figure 3.18g). There are two benefits of such OSC/polymer trap blends (Figure 3.18h). First, polar groups in the polymer binder, such as PMMA (**70**), PS (**71**), PVAc (**72**), and PC (**73**), may function as electron-trapping sites to help photo-induced electron and hole pairs. This can be used to build the UV write/gate bias erase memory device. Second, spin coating of such two component blends usually results in vertical phase separation, providing additional benefits in terms of film uniformity and morphology. This plays an important role in transistor performance. In this regard, Ljubic et al. engineered a UV-responsive blend in the OSC channel by choosing C_5–BTBT as the OSC and linear L-upe (**74**) [403] or branched B-upe (**75**) [57] unsaturated polyester polymers as the dielectric binder. With appropriate electron-withdrawing groups acting as the charge-trapping sites, the devices achieved high responsivity as well as photosensitivity through enhanced separation of the photogenerated charge carriers at the semiconductor/polymer interface, which led to a dramatic memory effect [404]. To further increase the electron-trapping capability, Smithson et al. [395] chose C_5–BTBT as the OSC and synthesized a copolymer, p-DR1 (**76**), with the strong donor–acceptor moiety functioning as a deeper trap in the semiconductor layer, thus enhancing the charge separation efficiency. The devices showed a strong current modulation to ultraviolet (UV) light illumination and maintained the high current state after UV light was removed, demonstrating slow relaxation (Figure 3.18i). It was proved that the current could return to its low-current ground state by applying a short (1–2 seconds) gate bias (−60 V) pulse.

3.3.2 Photochromism-Induced Switching Effect

As mentioned earlier, a convenient strategy for incorporating new functionalities into OFETs is to choose light as a remote control to modulate their electrical properties. In particular, a promising approach developed recently relies on the use of photochromic molecules (PCMs) because they undergo reversible photoisomerization between two isomers upon irradiation with light of different wavelengths (Figure 3.19a, left). These conformational changes have resulted in remarkable differences in their electronic structures and their distinct photophysical, chemical, and electrical properties, thus forming the basis for new types of multifunctional devices such as memories, photodetectors, and some other optoelectronic devices [405, 406].

The most common families of PCMs are depicted in Figure 3.19a [407]. In particular, they are azobenzenes (ABs, **77**), spiropyrans (SPs, **78**), and diarylethenes (DAEs, **79**). Very recently, some leading groups explored the multimode photoresponsive behaviors of the traditional system (**80**) (Figure 3.19a, right), which is promising for constructing multifunctional optoelectronic devices [407]. All of these families feature two or more independently addressable states that lead to reversible structural changes such as geometry/steric hindrance, causing electronic changes, including dipole moments, π-conjugation, HOMO–lowest unoccupied molecular orbitals (LUMO) gaps, and redox potential. These molecular-level changes affect macroscopic properties such as shape, aggregation behavior, conductance of the resulting materials, or interfaces containing these photoswitches. Until now, only a few works dealt with PCMs as a single semiconducting component in OFETs [408, 409]. However, these devices suffer from rather poor charge-transport properties. In this regard, blending the OSC material with a suitable PCM should be a possible and efficient strategy for achieving dual functionality in OFETs. A schematic of the energy-level engineering in the bicomponent blending films is depicted in Figure 3.19b, where two isomers in State I or State II that feature different energy levels are supposed to act differently on the hole transport by generating deeper (State I) or shallower (State II) energy-trapping levels in the band gap of semiconductors [68].

The performance of OFETs consisting of a bi-component film with a blend of solution-processed SP and a polymer semiconductor has been investigated (Figure 3.20a) [410–413]. Light-induced conformational changes of SPs in the polymer semiconductor matrix modulated the channel conductance in OFETs. Charge transfer between the OSC P3HT and the pentoxide unit of SPs occurred as the SP open form with photogenerated phenoxide groups acted as charge traps. The conductance state of the OFET channel was changed gradually from low conductance to higher conductance with UV irradiation for approximately three minutes, thus resulting in increased output current. Upon exposure to white light, the back reaction took place, which required approximately eight minutes for the device to return to the initial drain current value (Figure 3.20b). Noteworthy, the device displayed long-term stability, which was tested in air for up to three hours with no degradation.

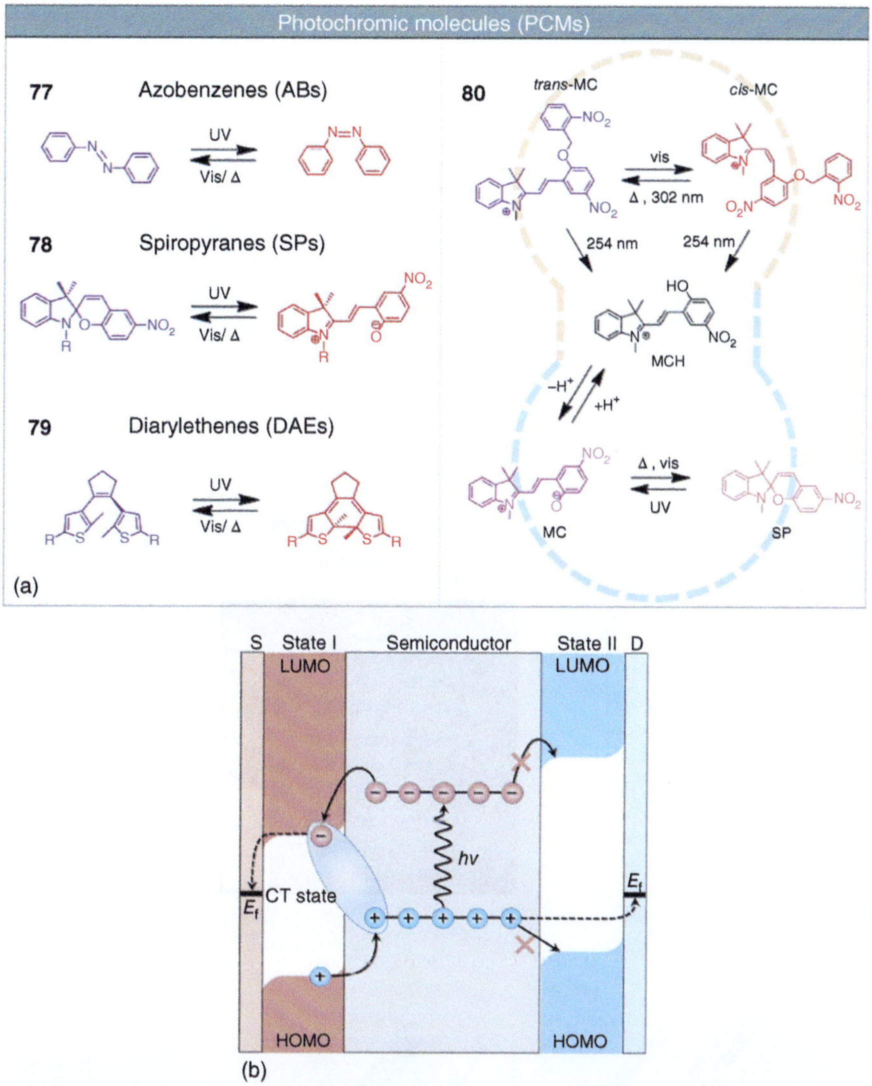

Figure 3.19 Interfacial energy diagram of photochromic blends. (a) Left: The sketch depicts typical photochromic molecules: azobenzenes (ABs), spiropyrans (SPs), and diarylethenes (DAEs). Right: Multimode photoresponsive behaviors of SPs: cis–trans isomerization, photodecaging, and spiro-merocyanine isomerization. (b) Energy diagram showing the HOMO levels of the employed components and depicting the photomodulation mechanism occurring in the device. Source: (a, right) Adapted from Fleming et al. [407]. (b) Reproduced with permission from Orgiu et al. [68], © 2012 Springer Nature.

A similar approach has been reported that relies on the introduction of phototunable energetic levels in the bandgap of P3HT by using DAEs. DAEs were chosen as PCMs because both switching states feature radically different electronic properties. In addition, both switching states offer significant advantages in terms of high fatigue resistance and high thermal stability. By blending two molecular components, the photochromic DAE derivative and a P3HT polymer semiconductor,

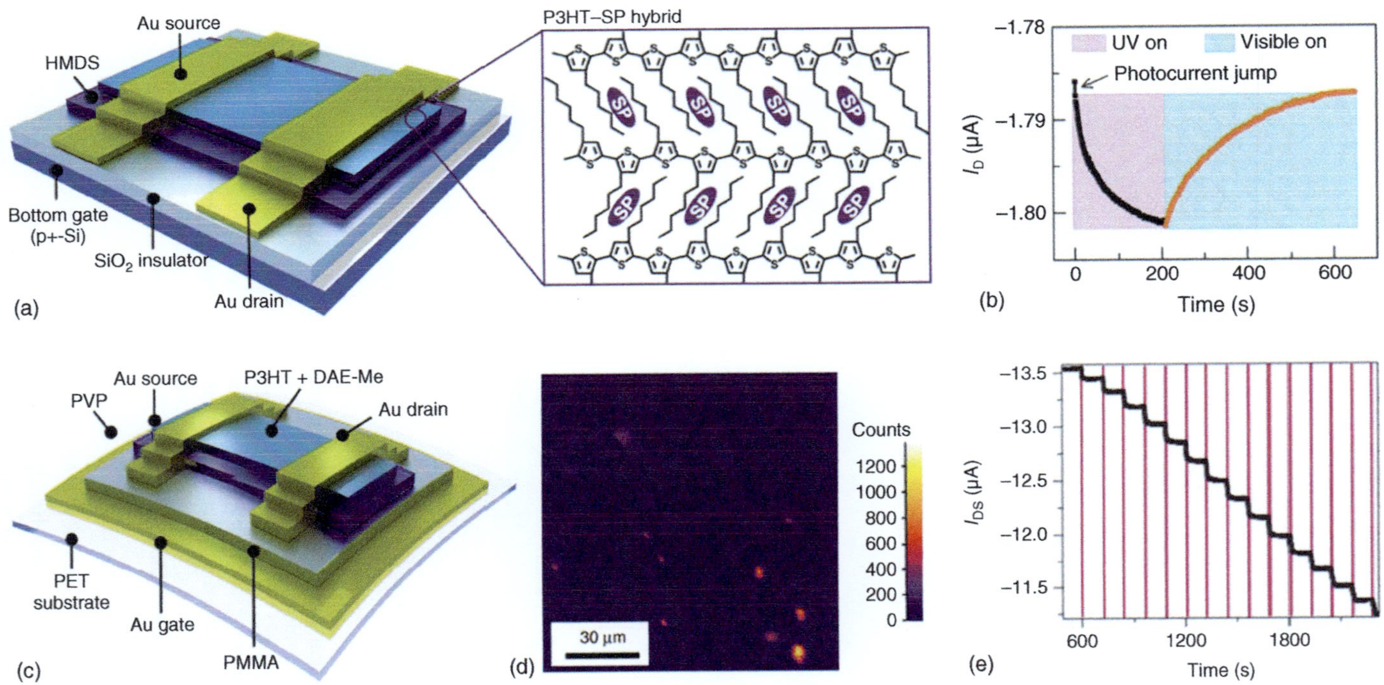

Figure 3.20 Photochromism-induced switching and memory effect. (a) Schematic representation of a P3HT/SP blend-based OFET. Inset shows the bicomponent semiconducting film is a blend of P3HT and SP, within which the SP molecules are thought of being randomly dispersed between the interdigitated P3HT side chains and close to the polythiophene backbone. (b) One full switching cycle of the time-dependent behavior of the photoresponsive memory device. (c) Schematic representation of the integration of the memory unit onto a flexible PET substrate. (d) Scanning confocal fluorescence microscopy image showing the uniformity of the PEHT/DAE-Me blend film. (e) Multilevel memory behavior of the device. Source: (a) Adapted and (b) Reproduced with permission from Li et al. [410], © 2012 The Royal Society of Chemistry. (c) Adapted and (d, e) Reproduced with permission from Orgiu et al. [68], © 2012 Springer Nature.

Orgiu et al. [68] attained phototunable and bi-stable energy levels for P3HT hole transport and reported the photocurrent modulation of OFETs (Figure 3.20c). The devices were tested with light irradiation over second-scale cycles while switching the light source between two different wavelengths. This work showed that the output current of the devices could be dynamically controlled by light in a reversible fashion on a timescale of seconds without obvious fatigue. However, these blended films suffered from poor FET mobilities ranging from 5×10^{-4} to 2.2×10^{-4} cm^2 V^{-1} s^{-1}, probably caused by large-scale phase segregation. To solve this problem, tuning the interactions between the photochromic diarylethene system and the OSC is greatly needed. Gemayel et al. [414] achieved this goal via *ad-hoc* side functionalization of DAEs. Thereby, phase segregation was avoided (Figure 3.20d) and superior miscibility was provided while retaining optimal π–π stacking to warrant efficient charge transport as well as to attenuate the effect of photoinduced switching on the extent of current modulation, demonstrating a photoresponsive bi-functional OFET with enhanced mobilities exceeding $0.2\,cm^2$ V^{-1} s^{-1}. By blending P3HT and a photochromic diarylethene, they also realized a non-volatile flexible optical memory thin-film transistor (Figure 3.20c) with 256 (8-bit storage) distinct current levels, that can be switched with ultraviolet and green light irradiation (Figure 3.20e). The memory states could be switched with 3 ns laser pulses, showing good non-volatility [415].

In summary, multifunctional devices such as charge-trapping memories and photoswitchable transistors can be realized via blending OSCs with the external components of functional materials. However, numerous challenges must be overcome toward future practical applications, including (i) how to control phase segregation; (ii) how the crystallization of the semiconductor is affected by PCMs in the mixture; (iii) how to minimize the mobility decrease caused by carrier scattering in the presence of PCMs; and (iv) how to speed the switching rate since the conversion of one isomer to another is a slow process usually requiring a long irradiation time. It is a promising strategy to integrate switching and memory effects into OFET, however; there is still a long way to go before actual application.

4

Semiconductor/Electrode Interface

Although there has been tremendous progress with organic field-effect transistors (OFETs) in recent years, they still suffer from poor charge-injection/collection, which limits their performance for numerous potential applications. The organic semiconductor (OSC)/electrode interface usually determines the carrier injection characteristics. Conventional single-crystal Si MOSFETs exhibit excellent contact properties because their charge injection is through tunneling from a metal contact into the bulk of the heavily doped silicon semiconducting area, that is, ohmic contact occurs [416]. Ohmic contact, the ideal contact between the electrode and the organic layer without the energetic barrier, is always desired but usually hardly reachable for OFETs [9, 417, 418].

In general, the fact is that in most cases, non-ohmic contact occurs because the work function (WF) of the electrode is not equal to the HOMO (for p-type) or the LUMO (for n-type) of the OSCs, called Schottky contact, thus creating a carrier injection barrier between the metal electrode and the semiconductor layer (Figure 4.1). The phenomenon of the Schottky barrier can be understood by the energetic alignment of a metal/organic interface, with interface dipole and band bending occurring simultaneously [419–427].

Inspired by this, an injection barrier is capable of reduction by carefully tuning the work function of source/drain electrodes to better match the HOMO/LUMO of an OSC, which is usually a fixed value. After establishment of the thermal equilibrium of the surface charge with the same Fermi energy level of the two contact materials, a space-charge region is formed by the redistribution of mobile carriers near the heterojunction, and energy band edges in the semiconductor are continuously shifted owing to the electric field, inducing a phenomenon called band bending [428]. In other words, the basic reason for band bending in the metal/semiconductor interface is the difference in work function for both organic and inorganic samples. The potential distribution, which is governed by the Poisson equation, leads to a built-in potential in the organic layer and results in the same Fermi energy-level alignment of the metal electrode and OSC. Band bending in OSCs with low concentrations of carriers, although it happens over very short lengths, a few to tens of nanometers from the interface resulting from the redistribution of mobile charge carriers, can act as an important factor influencing the device performance. In conclusion, many physical and chemical processes are involved when discussing

Interface Engineering in Organic Field-Effect Transistors, First Edition. Xuefeng Guo and Hongliang Chen.
© 2023 WILEY-VCH GmbH. Published 2023 by WILEY-VCH GmbH.

the energy-level alignment at the interface. In addition to the two most important processes related to the device performance (interface dipole and band bending), possible chemical reactions, diffusion, and charge transfer are also crucial and effective for tuning the optoelectronic behaviors [418, 423, 429–434].

Learning from the relevant processes in conventional crystalline OSC counterparts, charge injection from the source electrode to the OSC channel layer can be classified in principle into two processes, either Fowler–Nordheim (F–N) tunneling (Figure 4.1a) or Richardson–Schottky emission (Figure 4.1b). When classic F–N tunneling happens (Figure 4.1a), it is a direct process where an electron at the Fermi level tunnels through a triangular potential barrier set by the interfacial energy barrier and the applied potential (sometimes at the large interfacial electric field strength), which effectively lowers the potential barrier. As for the Schottky emission case (Figure 4.1b), it means that a thermally excited electron from the Fermi level overcomes the energy barrier caused by the contact of two different

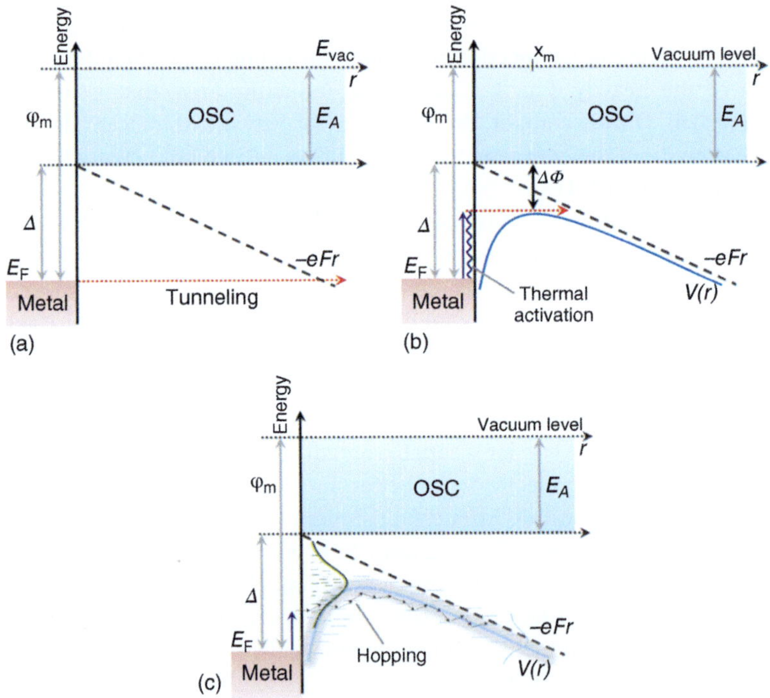

Figure 4.1 Schematic representation of electron injection from a metal electrode into a semiconductor via (a) Fowler–Nordheim tunneling, (b) Schottky emission, and (c) hopping in a disordered organic solid. [Various relevant energies: vacuum energy level (E_{VAC}), electrode Fermi level (E_F), electrode work function (φ_m), energy gap of OSCs (E_A), potential barrier from electrode to OSC (Δ), and lowered potential barrier ($\Delta\Phi$). r denotes the distance from the electrode. The dashed line indicates the potential energy due to the applied (external) electric field, F. The potential in the device, $V(r)$, is shown as a gray solid line. Note that the activation energy and tunneling distance are reduced in (c) in comparison with (b)]. Source: Adapted from Bassler and Kohler [205].

materials, traveling across the maximum of the electrostatic potential, as mentioned above. In consideration of the distinct properties of an OSC/metal interface, such as the existence of an image charge, hopping-type transport, and the presence of disorder in non-crystalline films, another possible injection mechanism has been proposed involving defect-assisted transport, in which carriers bypass the barrier by hopping through mid-gap states into the disordered organic solid (Figure 4.1c). In this mechanism, a thermally activated jump raises an electron from the Fermi level of the electrode to a tail state of the density-of-state distribution of transport sites in the dielectric medium, subject to the condition that there exists at least one hopping neighbor at equal or lower energy. Therefore, the injected carrier is able to continue its motion away from the interface rather than recombine with its image charge in the electrode [9, 205, 435, 436].

4.1 Work Function Tuning for Better Contact

By making good use of the energy alignment diagram at the interface, researchers have come up with a variety of methods to solve the poor charge-injection issue in OFETs by mainly focusing on how to realize the lower barrier. It is obvious that for a given OSC whose HOMO, LUMO, and band gap can be derived by characteristic experiments, an ohmic contact interface can be theoretically fabricated by simply choosing a metal as source/drain electrode with a suitable work function close enough to the HOMO or LUMO of the organic molecule [69]. To be specific, noble metals such as Au, Pt, and Ag (especially Au), whose high work functions are close to the deep HOMOs of OSCs, are used to fabricate p-type transistors; transition metals can also be used occasionally. Alkaline earth metals and alkali metals are used to achieve n-type operation because their low work functions are more suitable for LUMOs of n-type organic molecules, but their atoms tend to diffuse into organic layers, making the device unstable [437]. Also, for top-contact (TC) OFETs, the thermally deposited metal can diffuse into the organic layer occasionally, forming an interaction whose strength is related to the choice of metal and organic materials [438, 439].

In fact, there is far more need to take into consideration the interaction between metal electrodes and organic molecules, for example, interface dipole and band bending, as discussed earlier, as well as the possibility of chemical reactions when active metals are used. For example, if the Au electrode is polycrystalline or its surface is modified by other organic molecules (including contamination from ambient air exposure), the growth of OSCs on the surface would be influenced by different orientations; the interface energy-level alignment would be dramatically affected, too [440–442]. In general, the work function of Au is often cited as 5.1 eV, but measurements on Au surfaces exposed to air yield work functions of approximately 4.5 eV because of the contamination, which could decrease the energy barrier for electrons [443]. In another study, the work function of polycrystalline Au after ambient exposure prior to making contact with a semiconductor resulted in a reduction of 0.4–0.6 eV, which was measured as 4.7 eV for contaminated polycrystalline Au in comparison with clean polycrystalline Au of 5.4 eV [444]. In conclusion, the work function of metal can be influenced by the Fermi level inside the metal and the electrostatic potential near the surface. The energy diagrams of the interfaces can be obtained by ultraviolet photoelectron spectroscopy (UPS), providing an enriched understanding of how the work function of metal is affected by the surface absorption of organic molecules [69, 425, 443].

Therefore, it is extremely important to use contact engineering to reduce contact resistance and improve the charge-injection efficiency. With deep understanding of the injection process at the OSC/electrode interface, various methods have been developed. Previous works have proven that by fine-tuning the work function of the electrodes, the injection barrier can be changed and contact resistance can be significantly reduced that SAM and CIL can tune the work function over a large range. With different SAMs, the work function and contact resistance range widely as a result of the complicated influence of the SAM molecules on the metal electrode surfaces, usually Au, whose work function can be finely controlled. This effect will be discussed in the semiconductor/dielectric section in detail. By introducing a CIL to OFETs, it will affect the interfacial dipole formed at the metal/organic contact and be capable of leading to an increase in charge-injection efficiency. There is an approximate rule where the work function and contact resistance can both be reduced when a thinner layer with a work function matching the energy alignment of the semiconductor is employed. Furthermore, the work function and contact resistance can be finely tuned by using a mixture of SAMs or CIL compounds with different ratios, which is used as a strategy in ambipolar transistors.

Another choice is to use carbon-based electrodes. In consideration of the diversity of the carbon-based materials, we subdivide this major principle into polymer-based electrodes, carbon nanomaterial-based electrodes, and covalent bond formation at the molecular level.

4.1.1 SAM Modification

Self-assembly is a common phenomenon happening everywhere. The metal electrode surface, which is capable of absorbing adventitious organic materials, offers a good platform for studying SAMs formed by chemisorption. SAM modification is not only a universal method to modify the surface properties but also a method for providing more functionalities [109, 110, 445, 446]. Herein, we focus on SAMs on the metal surfaces of source/drain electrodes to fine-tune their work function in OFETs (Figure 4.2).

Commonly used SAM molecules are thiol derivatives (**81–104**), which have a sulfhydryl group for docking to the surfaces of metals such as Au, the most-used element on account of its high stability and conductivity as electrodes in organic electronics. This forms metal–sulfur (M–S; here Au–S) interactions through the sulfur long-pair electrons, called the thiolate—gold bond (RS–Au), whose strength is close to that of the Au—Au bond [38, 114]. Organic compounds with disulfide (S-S) (**105–107**) or sulfide acetyl (–SAc) (**108–110**) bonds can also form SAMs on gold surfaces. Among the highest work function reductions, recently Alt et al. [115]

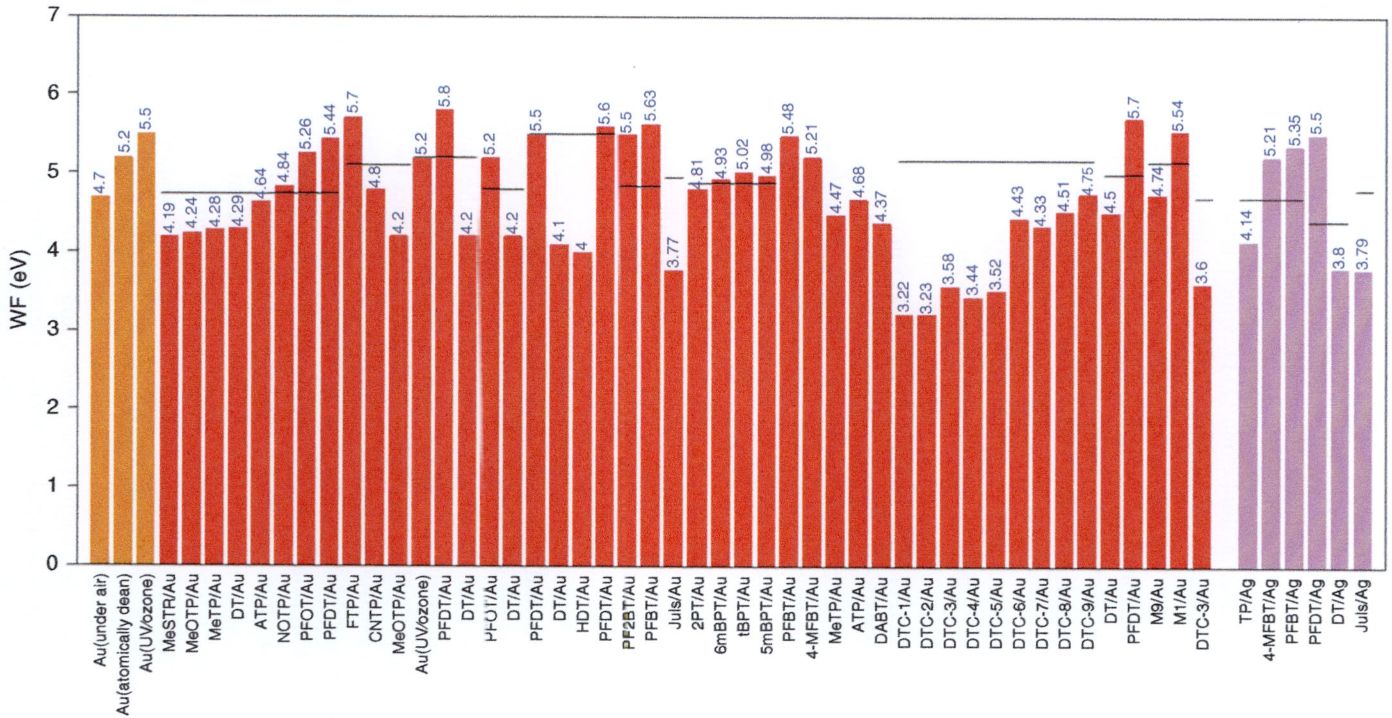

Figure 4.2 Modified work function by the SAM methods. Materials are classified by color as: bare Au (orange); SAM-modified Au (red); SAM-modified Ag (pink); black lines show the work functions of common metal contacts. Materials are classified by color as: bare Au (orange); SAM-modified Au (red); SAM-modified Ag (pink); each black line represents the values of work functions of Au or Ag in the corresponding original literatures.

presented a novel SAM molecule (**105**), referred to as Juls, which has a rigid S—S single bond linker connecting an amino moiety to an aromatic core. Casted via immersing Au or Ag electrodes into a Juls ethanol solution, the formation of SAMs lowered the work function by ~1.2 eV. Meanwhile, with this SAM layer in an n-type OFET, the contact resistance decreased by two orders of magnitude, thus improving the device performance.

Tulevski et al. [116] used a thioketone (**111**), which is structurally similar to pentacene, as a SAM to modify Au electrodes prior to the deposition of pentacene, which induced a complementary surface for the growth of pentacene and reduced the contact resistance, improving the hole mobility. In addition to this, dithiocarbamate (DTC) based molecules [117] have been proven to be stably adsorbed on metals under various types of environmental stresses, which could form robust and low-contact-resistance DTC-based molecular junctions in organic-based electronic devices [423, 447–449]. Ford et al. [118] designed rod-shaped molecular structures with a DTC-piperazine/piperidine-R motif (R represents an aryl or alkyl substituent) (**112–120**) to form robust SAMs on the surfaces of metal electrodes, which could reduce the work function of noble metals (Au and Ag) to an ultralow value (such as 3.1 eV for Au), thus enabling electron injection into the LUMO of organic channels because of the high intrinsic dipole moment of the dipolar monolayers.

All thiolate–SAMs are polar molecules in a sense, and the intrinsic dipole of SAMs toward the out-plane is determined by the functional group within the rest of the molecular structure and by the terminal sulfur-containing group. In other words, a dipole layer arises simultaneously as soon as a SAM forms on the metal surface (Figure 4.3a). When a SAM is deposited onto a clean metal electrode surface in an OFET, two dipole layers result. One is the M—S bond dipole interface ($\mu_{M\text{-}S}$) and the other is the aligned dipole moment of the SAM (μ_{DT}), which together change the work function of metal electrodes (Figure 4.3b). Furthermore, the charge-injection behavior into the semiconductor layer is changed accordingly owing to the final dipole formed at the electrode/semiconductor interface [450, 451]. The molecular structure of SAMs has been shown to have a vital impact on the final dipole, even for the simplest alkanethiols; the carbon number matters, known as the odd–even effect [452–454]. It has also been proven that alkanethiols reduce the electron injection barrier from Au electrodes to the semiconductor layer (φ_m decreases to 4.1 eV), whereas perfluorinated thiols improve hole injection (φ_m increases to 5.8 eV) [455]. For a bifunctional SAM, which contains one more functional group on the other end of a thiol or comprises heteroatom substitutions, its effect on the change in work function is a little more complicated when deposited on the metal surface. More specifically speaking, different tail groups [456, 457] attached by aryl or alkyl linkage units onto the framework of thiol–SAMs with the same or opposite direction of the dipole moment as that of the M–S dipole can give diverse results in tuning the work function [451, 458–462].

Also, the packing behavior, such as the density and orientation of the molecules in the SAMs on the surfaces of metal substrates, is an influential factor (Figure 4.3c) [463, 465]. Because the SAM can form dense packing layers on metals, thus changing the surface energy, the SAM modification of metal electrodes will avoid

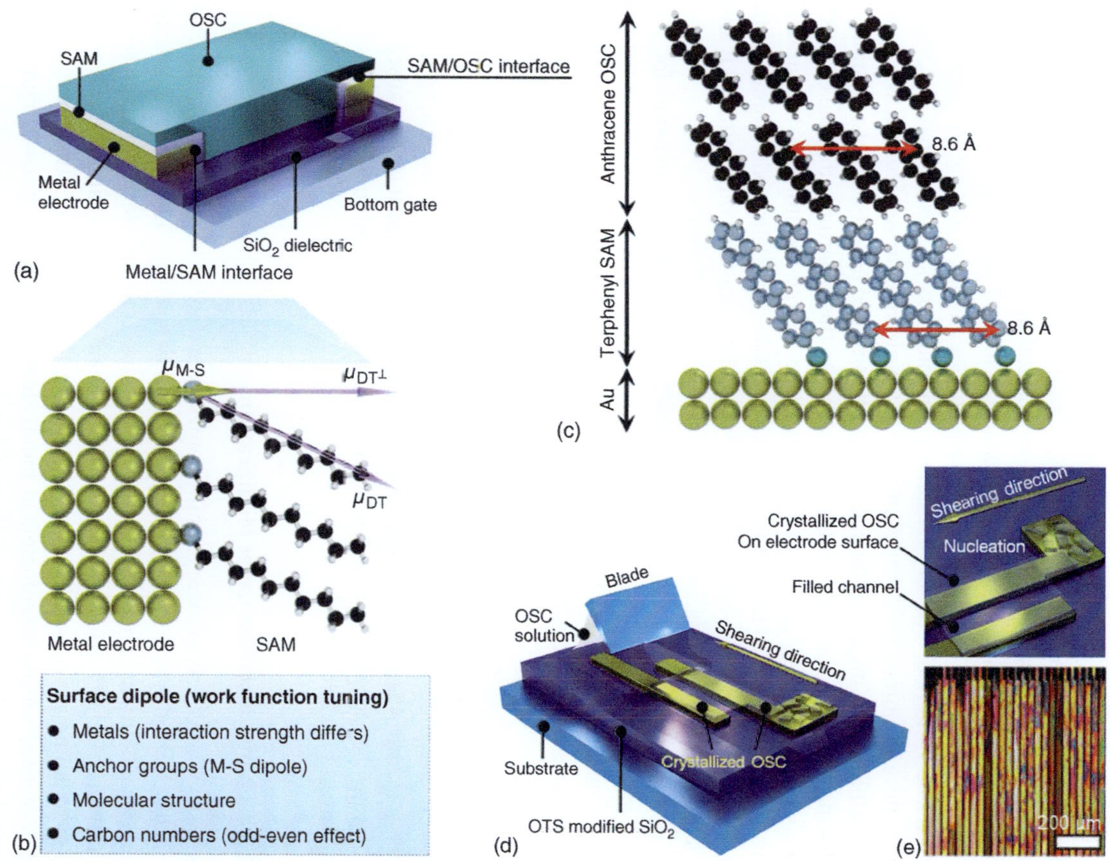

Figure 4.3 SAM modification of metal electrodes in OFETs. (a) Schematic device configuration of an OFET with SAM modification at OSC/Metal interface. (b) Schematic illustration showing the mechanism of SAM tunes the surface dipole of metal electrodes. (c) Schematic illustration showing the mechanism of SAM that affects the film morphology of OSC. (d) Schematic depiction of film growth during solution shearing. Inset shows the electrode design with labeling of different regions. (e) Optical image showing well-aligned patterned crystals on the electrodes and within the channel regions.
Source: (c) Adapted from Briseno et al. [463]. (d, e) Reproduced with permission from Park et al. [464], © 2015 Proceedings of the National Academy of Sciences PNAS.

the undesired dewetting phenomena observed when directly depositing OSCs onto metals; that is, a highly ordered crystalline OSC film with a reduced density of grain boundaries might be grown on the electrodes, thus improving the electron/hole injection [466] as reported in the pioneer works by Katz et al. [467] and Kymissis et al. [112]. Moreover, OSCs can grow into large, oriented single crystals directly on SAM-modified Au substrates [463]. Gundlach et al. [468] reported the fabrication of high-performance acene-based OFETs by using contact-induced crystallinity. Fluorinated diF-TESADT (**25**) was deposited on PFBT (**86**)-treated Au source/drain electrodes by spin coating. A highly ordered, crystal-like oriented film with a structure very similar to that of the bulk crystal was formed with edge-on orientation, as confirmed by GIXD data, whereas semiconductor molecules on an untreated Au surface exhibited face-on orientation. The authors suggested that the weak sulfur–fluorine interactions between adjacent molecules in semiconductors were responsible for the improved tendency of the fluorinated diF-TESADT to crystallize in comparison with the unfluorinated TESADT. In the meantime, the role of PFBT was crucial beyond simply "tuning" the surface energy, resulting in the growth of 10 μm or more of di-TESADT into the channel from the electrode boundary. In contrast, untreated Au, even benzenethiol-treated Au, induced only minimal if any crystallization along contact edges.

Park et al. [464] presented controlled OSC nucleation and extension for the circuit technique, termed "CONNECT", to grow self-aligned crystalline domains of TIPS-pentacene (Figure 4.3d). Au electrodes were patterned on SiO_2 substrates as bottom contacts (BCs). After being treated with OTS and thiophenol successively, SAMs were formed on both dielectric and electrode surfaces. Based on the surface energy difference, the nucleation of OSCs during solution shearing could be controlled, thus allowing the channel to be bridged by the forming of crystals between electrodes with low device-to-device variability when fabricating OTFT arrays (Figure 4.3e). Contact-induced nucleation of TIPS-pentacene caused by fluorinated benzenethiol-SAMs with various numbers and positions of F atoms has been investigated, too [469].

4.1.2 Charge Insertion Layer Modification

Because the injection barrier generally comes from the gap between the Fermi level of metal electrodes and the HOMO/LUMO energy levels of OSCs according to the energetic alignment, a reasonable solution is to insert a suitable additional layer to modify the electrode and function as a springboard for charge carriers, thus possibly overcoming the poor injection [70]. Initially, this method of introducing a thin buffer layer of several nanometers at metal electrode/OSC interfaces (Figure 4.4), referred to as a charge insertion layer (CIL), had already been applied to increase the efficiency of OLEDs by improving the contact [71, 72, 470].

Besides the application in OLED, the introduction of a CIL when fabricating the electrodes in OFETs has also received attention. A preliminary study was carried out by Hajlaoui et al. [73] who inserted a thin film of TCNQ (**122**), working as molecular dopant, between Au electrodes and α-conjugated quaterthiophene (4T) active layers

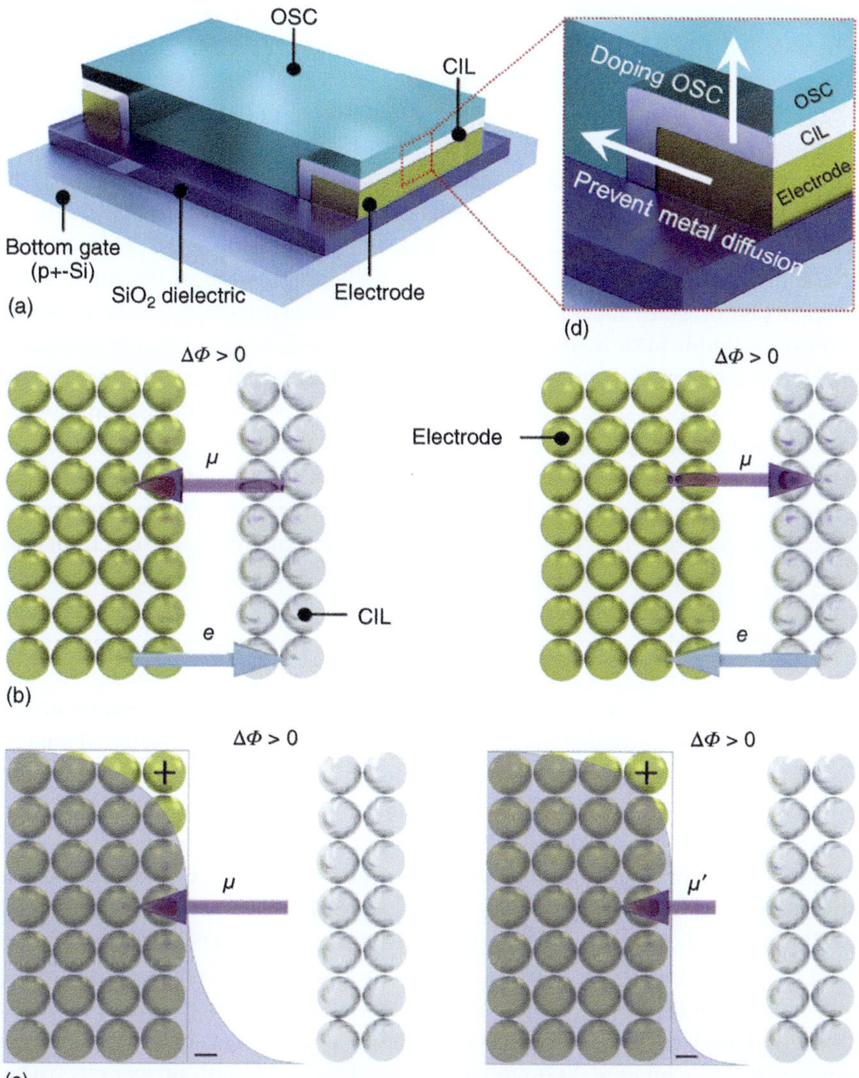

Figure 4.4 CIL modification mechanism of metal electrodes in OFETs. (a) Schematic device configuration of a CIL-modified OFET. Inset (d) shows the OSC/CIL/Electrode interface. CIL can dope the OSC as well as prevent metal diffusion into the OSC. (b, c) Two major contributions to the change in work function Φ induced by an ultrathin dielectric layer on a metal: (b) Charge transfer at the interface; (c) compressive electrostatic effect.

in OFETs to achieve increased hole mobility. By using an ultrathin LiF interlayer as CIL between Al electrodes and $F_{16}CuPc$ active layers, enhanced n-channel performance was achieved [33]. Since then, this method has attracted more and more attention in terms of enhancing the hole/electron injection efficiency. Various materials, such as inorganic compounds (metal oxides and inorganic salts) and organic compounds (small molecules and polymers), have been used as CILs, as shown in Figure 4.4a.

The presence of CILs at the electrode interface induces a charge transfer across the interface, creating a dipole (μ), which shifts the position of the metal Fermi energy (Φ) (Figure 4.4b,c), which basically depends on the materials used. In general, if an inorganic compound is deposited onto a metal electrode, the induced change of the work function comes from charge transfer owing to the metal-to-oxide surface dipole for most transition metal oxides (such as MoO_3, TiO_2, and WO_3) (Figure 4.4b), or it could be electrostatic in origin owing to the compression of the metal electron density enforced by the ionic materials (MgO, NaCl) (Figure 4.4c) when charge transfer is negligible [471–473]. Another factor that needs to be taken into consideration is doping effect that if the CIL materials are capable of diffusing into the bulk of semiconductors or if strong charge transfer happens between the CIL and semiconductor materials, especially when the CIL is capable of setting off a chemical interaction with the semiconductor layer (Figure 4.4d) [22, 266]. The CIL-caused doping can play a role in tuning the charge-transport property. When introducing organic molecules as CILs, they can serve as surface modifiers for electrodes or dopants in semiconductor layers, depending not only on the chemical properties but also on the fabrication process [474].

Metal oxides provide a variety of candidates for CILs because of their wide range of work functions (Figure 4.4a) [74]. Transition metal oxides, in particular, whose work functions range from very low ($\varphi \approx 3$ eV for ZrO_2) to extremely high ($\varphi \approx 7$ eV for V_2O_5) [75, 76], are suitable for modifying different metals with matched energy levels, for example, a low-work-function metal such as Ca (2.8 eV) [475], or a high one such as Au (5.4 eV) [444]. In addition, the interface dipole formed at the metal/metal oxide interface can increase or decrease the work function of metals, depending on the two major contributions, charge transfer at the interface and the compressive electrostatic effect [473]. Thus, they can be applied to meet the need for well tuning the charge injection into different p- or n-type OSCs to decrease the contact resistance, increase the current level of the channel, and further improve the carrier mobility of the device. Metal oxides can be easily deposited on the electrodes by solution-based coating, conventional vacuum evaporation, or oxygen plasma treatment [9, 78]. Chu et al. [476] demonstrated TC OFETs with bilayer electrodes consisting of a metal layer and a metal oxide CIL (MoO_3, WO_3, or V_2O_5). They succeeded in presenting a high-performance OFET with Al, a low-work-function metal, as source/drain electrodes and pentacene as the channel by introducing a metal oxide CIL between them. By taking MoO_3 as an example, its work function of 5.3 eV is close to the HOMO energy level of pentacene, 5.0 eV. MoO_3 was thermally evaporated onto the channel layer, followed by the depositing of metal electrodes such as Al. In this case, additional impurities such as Mo species with different oxidation states (from free Mo to MoO_3) would cause an increase in the doping concentration in the CIL, thus reducing the depletion region width between the CIL and the metal layers and increasing the tunneling probability. An ohmic contact was resulted to ensure a high performance [77]. The CIL layer also acted as a protection layer against the diffusion of metal into the active layer. The devices with MoO_3/Al electrodes, as well as those with WO_3/Al and V_2O_5/Al electrodes, gave much higher performance than those with Al or even with bare Au. Also, the MoO_3 worked well as CIL for modifying different metals, such as Au and Ag. Moreover, the MoO_x CIL led to an increased performance in a BC OFET by promoting effective connection between Au and pentacene, with one order of magnitude reduction in the contact resistance. The best performance was achieved with 2-nm-thick MoO_x with the lowest threshold voltage (approximately −0.2 V) and the highest hole mobility (~0.42 $cm^2\ V^{-1}\ s^{-1}$) [477]. Cho et al. [78] introduced a CIL in $PC_{61}BM$- and $PC_{71}BM$-based TC OFETs by spin casting a TiO_x solution onto a fullerene film and then thermally depositing Al or some other metal as the electrodes. The TiO_x electron injection layer between metal electrodes and the semiconductor layer induced a strong interfacial dipole, leading to a decrease in the electron injection barrier, as well as a decrease in the contact resistance. As a result, this interfacial dipole improved charge injection and device stability when exposed to air because the coverage of the TiO_x prevented the permeation of O_2 or H_2O into the n-channel active layer.

Different metal oxides have been utilized to obtain high-performance OFETs with a variety of OSC materials, illustrating the universality of this CIL method [78, 478–485]. As mentioned above, after introducing the CIL, the thickness of the depletion layer between the metal and metal oxide junctions decreases to a few nanometers, producing a reachable distance for tunneling injection, and the contact resistance could be significantly reduced once the tunneling injection is dominant. The factors that influence the behavior of the CIL include the thickness and roughness of the insertion layer [70, 74]. In conclusion, as for the mechanism of how the metal oxide CIL works, a widely applicable explanation is the increase of the direct tunneling path of charge injection from the electrode to the active layer [486, 487].

Metal salts are another big family of inorganic materials working as CILs (Figure 4.4b) [79, 80]. They experience a similar mechanism as metal oxides. Minari et al. [488] reported that the insertion of an $FeCl_3$ layer between Au electrodes and the C_8-BTBT layer resulted in a significant reduction of the contact resistance from 200 to 8.8 kΩ cm and a high mobility of ~7.0 $cm^2\ V^{-1}\ s^{-1}$. The doping effect caused by $FeCl_3$ could drive the charges generated at the contact to occupy the trap states in the access region, as well as to narrow the depletion layer, enhancing the hole injection by tunneling. Baeg et al. [489] reported that inserting a layer of salt, Cs_2CO_3 or CsF, via a simple spray coating method to tune the work function of the metal, significantly enhanced the electron injection in ambipolar OFETs at the Au/OSC (different conjugated polymers were used) interface, without significant degradation of the p-channel performance. Kim et al. [490] reported that using solution-processed ultrathin $Ba(OH)_2$ or $BaCl_2$ CIL to tune the work function of Au down to 3.5 eV could significantly improve the electron carrier mobilities in $PC_{61}BM$ and P(NDI_2OD-T_2)-based OFETs. The thickness of the CIL affects electron transport. For instance, a thicker layer of CsF (more than 0.6 nm) decreased the mobility owing to its insulating nature [83]. In a further step, by mixing two different materials to form a CIL, the injection barriers for different types of carriers were finely tuned according to the molecular ratios of the two compounds, as reported by Noh's group [84]. With Mo as source/drain electrodes, they used the metal oxide

V_2O_5 as a hole injection layer and the metal salt Cs_2CO_3 as an electron injection layer. The work function of Mo was regulated approximately in the range of −4.2 to −4.8 eV. Thus, the modified diketopyrrolopyrrolethieno[3,2-*b*]thiophene copolymer (DPPT-TT)-based OFETs showed reduced contact resistance and high performance, with hole and electron mobilities as high as ~2.98 and ~2.6 $cm^2 V^{-1} s^{-1}$, respectively. When depositing a CIL into ambipolar OFETs only using a single component, the outcome performance of the p- and n-channels cannot be enhanced simultaneously. Determined by the electrical properties of the metal oxides or metal salts, once the injection of the p-channel is enhanced, the n-channel will decrease.

Recently, CuSCN was used as a novel CIL to treat electrodes with an intrinsic electron-blocking property and matching of work function with the polymer semiconductors. Zhang et al. [81] used CuSCN to enhance the charge injection from Au to the novel D–A polymer pIDTTBT and obtained hole mobilities to 8.7 $cm^2 V^{-1} s^{-1}$. In another work, Fei et al. [82] used spin-casted CuSCN as Au electrode modifier, thus their newly designed low band gap polymer of PDPPCp could show mobility up to 3.1 $cm^2 V^{-1} s^{-1}$.

When it comes to organic small molecules as an insertion layer (Figure 4.4c), sometimes direct charge transfer occurs at the interface between the CIL and the semiconductor, which helps to reduce the contact resistance by the additional interface dipole, yet a problem that still exists is the possibility of matrix–dopant hybridization [90, 91, 491]. On many occasions, small-molecule CILs function based on the molecular contact doping effect [22, 492–495]. With a CT complex of TTF-TCNQ (**42**) acting as the buffer layer [85] the single-crystal OFETs based on DNTT with a Cytop-modified SiO_2 gate dielectric and TC Au/TTF-TCNQ electrodes resulted in hysteresis-free characteristics showing a mobility of ~8.3 $cm^2 V^{-1} s^{-1}$ and an on/off ratio over 10^8. Metal complex compounds (**128–130**) can also be used as CILs [86, 87, 496]. Recently, Choi et al. [87] reported the reduction of the contact resistance in OFETs by using different contact-modification strategies to modify the Au source/drain electrodes by using a SAM of PFBT (**86**), an evaporated thin layer of the metal–organic molecular dopant molybdenum $Mo(tfd)_3$ (**129**), or a CIL of MoO_3. In conclusion, the improved performance observed in the devices modified with $Mo(tfd)_3$ or MoO_3 demonstrated the usefulness of the surface-doping effect. Several metal tetraphenylporphyrin and phthalocyanine compounds (**131–135**) have been demonstrated as the buffer layers functioning doping effect at organic/metal contact regions. For example, Yan et al. [88] reported the improved performance of n-type F_{16}CuPc-based organic transistors by introducing a CuPc layer (**136–140**) under source/drain electrodes, which formed a conductive organic heterojunction (a F_{16}CuPc/CuPc buffer layer) to decrease the contact resistance. The role of several N-heterocyclic compounds (126) as buffer layers that enhanced device performance in OFETs have also been attributed to a reduction in contact resistance at the organic/metal interface [497, 498].

The organic polymers can also act as surface modifiers to influence the dipole of the interface. Zhou et al. [92] introduced polymers containing aliphatic amine groups, PEIE (**141**), and branched PEI (**142**), to achieve electrodes with a low work function. Owing to the dipolar amine groups in PEIE and PEI, physisorption onto the electrodes' surfaces occurred when the polymer was used as an insertion layer, helping to decrease the work function. So, these polymers are suitable for a variety of conductor electrodes such as metals, metal oxides, polymers, and graphene. The authors fabricated different optoelectronic devices such as organic solar cells, OFETs, and OLEDs, and all showed good results. The improved n-type OFET performance was obtained because of an enhanced electron injection. On the basis of these satisfactory facts, the authors believed that using these simple aliphatic amines as polymer "surface modifiers" could serve as a universal method to produce low-cost, air-stable, low-work-function electrodes.

If a polymer comprises groups capable of being charged with cations or anions such as the CIL (**143–153**), it will act as a dopant to the OSC layer [92, 499–502]. Seo et al. [503] demonstrated an effective strategy for improving carrier injection in n-channel C_{60} OFETs through the insertion of a conjugated polyelectrolyte (**143**) layer beneath top-contact Au source/drain electrodes, by which the interfacial dipoles induced at the metal/OSC interface modified the energy-level alignment. Tang et al. [23] recently reported a general strategy to achieve solution-processed doped films over a wide range of work functions (3.0–5.8 eV) through first charge doping of conjugated polyelectrolytes (**146–153**) and then the occurrence of internal ion exchange to give self-compensated heavily doped polymers. These polymers could be used to modify the charge-injection electrodes to realize ohmic contact for high-performance organic electronic devices. Seah et al. [93] showed that self-aligned polyelectrolyte counterion monolayers on metal electrodes could induce stable carrier doping at the semiconductor/electrode interface and provide ohmic contacts for both p- and n-type OFETs.

Other materials (Figure 4.4e), such as DNA and graphene oxides (GOs) have also been used as CILs. Nguyen et al. [94] inserted a thin DNA layer between the OSC layer and Au source/drain electrodes, and observed improvements in the electron mobility of $PC_{71}BM$ and ambipolar transport in BTDPP2 OFET devices. The presence of phosphate groups in the DNA leads to a dipole layer, which could change orientation in response to applied gate voltage and enhance the charge injection of both types of carriers. Moreover, atomically thin aryl-functionalized GOs were applied to act as molecular insertion layers on the surfaces of metal electrodes in OFETs [95–97, 504]. The improvement in device performance could be attributed to the good contact nature of carbon-based materials with semiconductors, which will be discussed in Sections 4.1.3–4.1.5.

4.1.3 Polymer-Based Electrodes

Assuming that the conducting polymers exhibit good compatibility with the OSCs, as well as provide opportunities to achieve low-cost, all-organic devices, the application of conducting polymer-based electrodes (Figure 4.5) in OFETs has been reported already, especially as printed electrodes [505–512]. For example, poly(3,4-ethlenedioxythiophene) (PEDOT)/PSS (**154**), a water-based conjugated π-electron bonding system with good conductivity over 500 $S\,cm^{-1}$, is a promising conductive material that enables the fabrication of cost-effective and flexible devices

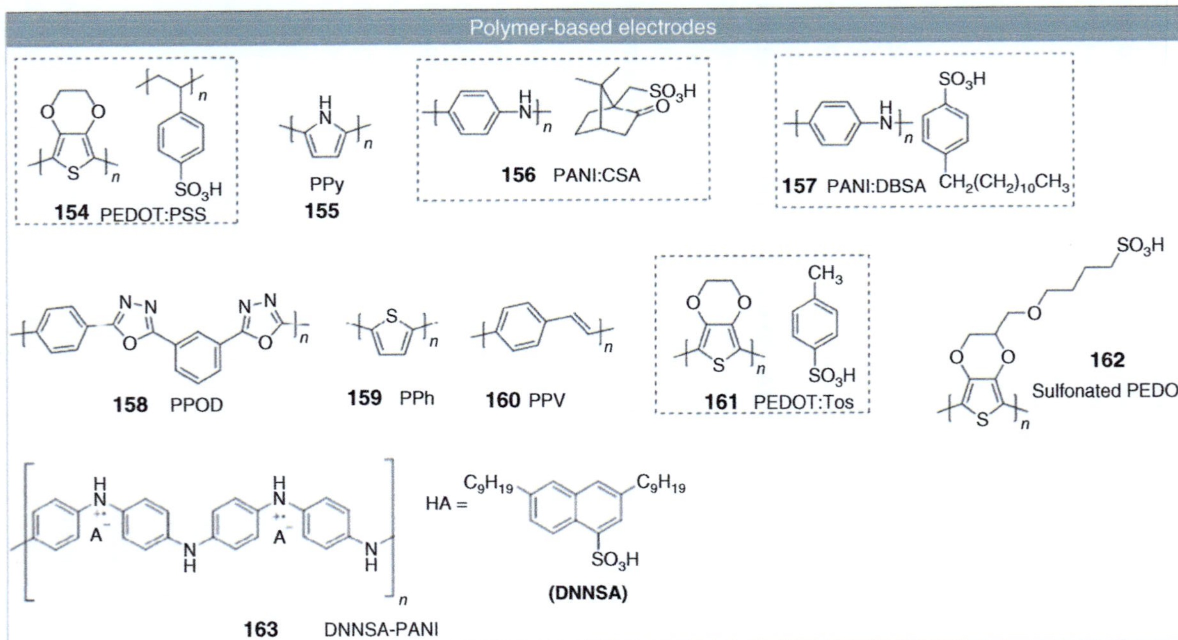

Figure 4.5 Chemical structures of polymer-based electrodes.

as well as large-area printing-based electronics [513–515]. Sirringhaus et al. [25] reported the use of PEDOT/PSS ink to act as electrodes for the construction of high-resolution inkjet printing of all-polymer transistor circuits. They made use of substrate surface energy patterning to direct the flow of conducting polymer inkjet droplets, enabling high resolution of the channel lengths down to 5 μm (limited by the resolution of the photolithography setup). Cosseddu et al. [516, 517] reported the realization of flexible and transparent n-type and ambipolar all-organic C_{60}-based OFETs by using PEDOT/PSS polymer electrodes with a pentacene buffer layer whose thickness affected the operation of the transistors. Schmidt et al. [514] presented the post-press treatment of printed PEDOT/PSS source/drain contacts with ethylene glycol, which improved the performance of fully printed OFETs because of the increased conductivity and reduced work function of the composite electrodes.

Other conducting polymers have also been widely evaluated as transparent electrodes. Koezuka et al. [518] fabricated polythiophene FETs with polypyrrole (PPy) (**155**) layers acting as a source/drain electrode. The Chi's group [511, 519] described a simple method to fabricate well-defined PPy on rigid and flexible substrates without affecting the conductivity by *in situ* area-selective surface-grafting polymerization. They reported highly stable, high-performance p-/n-type OFETs and complementary inverter circuits with patterned PPy as the pure electrodes owing to the excellent compatibility and low contact resistance at the contact interface, i.e. the well-matched interface energy level, as well as the quite uniform morphology of semiconductors in the channel and across the electrode/channel area [433].

Drury et al. [520], Matters et al. [521], and Gelinck et al. [522] used photochemical patterning to form doped polyaniline (PANI) (**156**, **157**) conducting films to serve as interconnects and electrodes and high-mobility semiconductor materials such as pentacene or P3HT films as channels of low-cost, all-polymer (organic) FETs. They successfully built a 15-bit mechanically programmable code generator combining over 300 FETs. Some other conducting polymers, such as polyoxadiazole (PPOD) (**158**), polythiophene (PPh) (**159**), poly(para-phenylene vinylene) (PPV) (**160**), PEDOT/Tos (**161**), sulfonated PEDOT (**162**), have also been used as electrodes in OFETs.

Moreover, the work function of a polymer-based contact electrode could be easily tuned to match that of the organic channel layer. Kim et al. [515] reported that the work function of PEDOT/PSS could be tuned to be higher by the addition of Nafion, inducing vertical phase separation in the (PEDOT/PSS)/Nafion electrodes and variation in surface properties. By using this method, significantly increased charge-carrier mobilities (~0.39 $cm^2\ V^{-1}\ s^{-1}$) were obtained in pentacene-based OFETs in comparison with those with neat PEDOT:PSS electrodes (~0.021 $cm^2\ V^{-1}\ s^{-1}$), as well as a reduction of the threshold voltage shift approximately from −1.84 to −0.28 V. By blending polymer electrodes dinonylnaphthalene sulfonic acid (DNNSA)–PANI (**163**) with single-walled carbon nanotubes (SWCNTs), Lefenfeld et al. [505] and Blanchet et al. [523] reported the successful printing of the hybrid electrodes as contacts in organic transistors. Hellstrom et al. [524] employed P3HT/single-wall carbon nanotubes (SWNT) composite electrodes to drive pentacene and C_{60} transistors. Finally, it is worth mentioning that conducting polymers can work as surface modifiers of metal electrodes to enhance charge injection [525–527]. The Park's group reported the improved hole mobility in pentacene-based transistors [528], as well as the improved n-channel operation of PTCDI–C_{13} n-type FETs [529] by coating PEDOT/PSS onto Au source/drain electrodes, which reduced the contact resistance for both p- and n-type BC OFETs.

4.1.4 Carbon Nanomaterial-Based Electrodes

Fabricating OFETs with contacts made from carbon nanotubes (CNTs) or graphene (Gr) is another efficient method to enhance charge injection [9, 530–534]. Early works reported the fabrication of molecular optoelectronic devices with single-walled CNTs as source/drain electrodes. A medium work function (~4.8 eV) [535–537] was obtained by using CNTs as contacts in OFETs [113, 536, 537]. Qi et al. [537] fabricated molecular FETs with cut metallic SWNTs with sub-10 nm gaps as quasi-1D electrodes, where the short channel length was bridged by organic materials (pentacene nanocrystallites and P3HT) (Figure 4.6a,b). The FETs exhibited good electrical performance because of the good contact [9, 531, 533]. Tsukagoshi et al. [536] reported the use of evaporated pentacene on the gaps of metallic multiwalled carbon nanotube (MWCNT) electrodes to build OFETs, where pentacene selectively grew around the MWNT electrodes during fabrication owing to the interaction between the two. The resulting nanotransistor showed controllable transistor operation at room temperature.

Because of their low dimensionality, high aspect ratio, and easy fabrication process, the quasi-1D CNT electrodes were often used in 1D monolayer transistors and single-molecule devices [113, 536, 537]. Guo et al. [170, 539] reported efficacious 1D molecular FETs by using oxidative cutting of individual metallic SWNTs to come into contact with monolayers of polycyclic aromatic hydrocarbons capable of responding to their chemical environment (Figure 4.6c). To evaluate the potential of CNTs in OFETs having larger dimensions, the Martel's group [538] fabricated CNT arrays to make contact with the OSC channel layer (Figure 4.6d) and found that the contact resistance could be reduced to several hundred ohms by using CNT arrays for n-type PCBM OFETs (Figure 4.6e), in comparison with that by using Au electrodes (some tens of millions of ohms) [540]. By fabricating CNT arrays as source/drain electrodes and vacuum annealed CuPc as the semiconductor channels, ohmic contacts were formed for both electron and hole injection [541]. Xie et al. [541] applied thin layers of the CNT network at the charge-injection interface of bis(trifluoromethyl)-dimethyl-rubrene crystals and substantially reduced the contact resistance for both holes and electrons. Consequently, they achieved benchmark ambipolar mobilities, reaching mobilities as high as 4.8 $cm^2\ V^{-1}\ s^{-1}$ for hole transport and 4.2 $cm^2\ V^{-1}\ s^{-1}$ for electron transport.

Sarker et al. [542] reported a new device architecture fabricated with P3HT crystalline nanowires grown directly on solution-processed high-density-aligned SWCNT interdigitated electrodes, which were aligned from their aqueous solutions via dielectrophoresis (an electric field-assisted deposition technique) [543] and

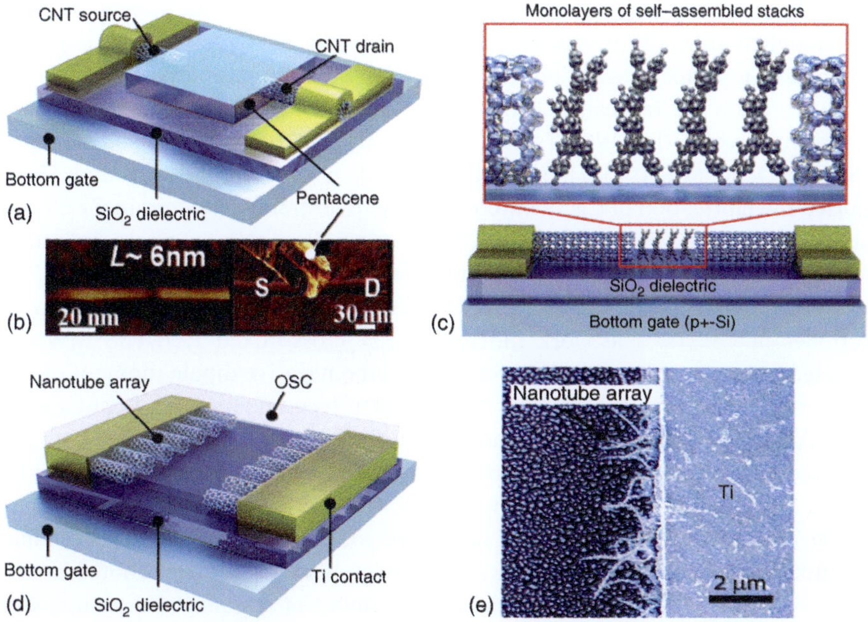

Figure 4.6 CNT as electrodes for OFETs. (a) Schematic drawing of a cut SWCNT with a sub-10 nm gap to be used as source (S) and drain (D) electrodes of an OFET. (b) AFM image of a cut SWCNT (diameter ~2 nm, gap size L measured to be ~6 nm after correction of the tip size effect). Right inset: AFM image of a vapor-deposited pentacene crystallite bridging a cut SWCNT. (c) Monolayers of self-assembled stacks being probed with SWCNT electrodes separated by only a few nanometers. (d) Schematic of a carbon nanotube array electrode in the configuration used to fabricate the pentacene OFET devices. (e) SEM image of the array-contacted OFET showing carbon nanotubes sticking out from titanium electrodes. Source: (a, b) Reproduced with permission from Qi et al. [537], © 2004 American Chemical Society. (c) Adapted from Guo et al. [170]. (d, e) Reproduced with permission from Aguirre et al. [538], © 2009 American Chemical Society.

patterned by electron beam lithography (EBL) and oxygen plasma etching. The performance of the corresponding transistor was enhanced, showing a maximum mobility of ~0.13 $cm^2\ V^{-1}\ s^{-1}$ and an on/off ratio of ~3.1×10^5, respectively. This was attributed to the improved interfacial contact between SWCNT electrodes and the crystalline P3HT nanowires via strong π–π interactions. In addition, the authors built short-channel pentacene-based OFETs with aligned arrays of SWNTs as electrodes by using the same procedure [544]; they also studied the charge-injection mechanism at the SWCNT/pentacene interface [545, 546]. By increasing the density of arrays of the SWCNT electrodes, the best performance was achieved by maximizing the SWCNT/pentacene interfacial area, where a low charge-injection barrier (~0.16 eV) that was much smaller than that of a metal/pentacene device for thermal emission was achieved. A transition to the tunneling mechanism happened at temperatures below 200 K, which was able to transition to F–N tunneling at lower temperature, suggesting the small barrier width.

Wrapping CNTs with small molecules or polymer surfactants, as well as conductive polymers, could overcome the poor solubility of CNTs [547–550]. Zhang et al. [551] reported a solution-processable method for employing SWNTs as BC source/drain electrodes in OFETs with a significant reduction of the contact resistance, in which the SWCNTs used were highly COOH-functionalized owing to H_2SO_4:HNO_3 treatment [547] and dispersed in deionized water. Hong et al. [552] reported the fabrication of CNT composites as source/drain electrodes in OFETs by dispersing polymer/MWCNTs in water. The authors demonstrated photo-patterned composites of conductive MWCNTs with both poly(acrylic acid) (PAA) and PSS (referred to as PAA/PSS–MWNTs, conductivity up to ~210 S cm^{-1}) as contact electrodes in OFETs. The work function of UV-cross-linked PAA/PSS–MWNTs increased from ~4.46 to ~4.76 eV as a result of the negative dipole moment of the polar groups normal to the surface, suitable for p-type OFETs. This led to a nine times higher mobility in TIPS–pentacene transistors than in the case of using Au as source and drain contacts. In another work [553], the authors found that PSS without a photo-crosslinker could be photo-crosslinked by exposure to UV light (255 nm) under a vacuum, and thus they were able to fabricate photo-patternable crosslinked PSS/MWCNT composites as electrodes, with the work function decreasing to ~4.53 eV. Although the cleavage of a number of sulfonic acid groups was observed, the composite electrodes were still appropriate, and the corresponding devices exhibited higher performance (12 times higher mobility) than those with Au electrodes, owing to the much lower contact resistance. In addition, the same group also developed a size-sorted MWCNT/PSS composite ink for electrohydrodynamic jet printing in the preparation of directly patterned conductive lines in organic transistors [554].

Since the discovery of graphene, another form of carbon materials, in 2004, it has attracted considerable attention [555–562]. Graphene is chemically stable, mechanically flexible, optically transparent, and an excellent electrical and thermal conductor, which makes it a promising candidate for an electrode or an insertion layer between metals and OSCs. Under normal conditions, the work function value of graphene is ~4.6 eV [563, 564]. Many studies have been carried out using graphene as electrodes for enhancing contact injection in OFETs because of the tuning capability of the injection barrier at the semiconductor/graphene interfaces [533, 565–567]. For example, in a p-type semiconductor such as pentacene, although the barrier height under vacuum alignment, i.e. the work function difference, of pentacene/graphene is a little higher than that of pentacene/Au, a reduction in the hole injection barrier occurred after the formation of the interfacial dipole and vacuum shifting for the pentacene/graphene interface. The final result of the hole-injection barrier was ~0.29 eV for graphene electrodes, thus forming a better contact in comparison with that based on Au (with a barrier of ~0.54 eV) [556]. Di et al. [568] demonstrated the use of patterned graphene as source/drain electrodes for BC pentacene-based OFETs. The shape of graphene films was controlled by growth on Cu and Ag electrodes patterned on the SiO_2 dielectric layer. With the excellent graphene electrode/organic interface contact with low injection, the corresponding devices showed high performance with saturated mobilities of

$0.47–0.53\,cm^2\,V^{-1}\,s^{-1}$, one of the best values reported for BC pentacene transistors with a bare SiO_2 dielectric. The Guo's group reported high-performance photoresponsive OFETs with single-layer graphene as 2D electrodes directly produced through oxidative cutting of individual planar graphene by EBL and oxygen plasma etching, and P3HT as the semiconductor (Figure 4.7a,b) [565, 566]. The device

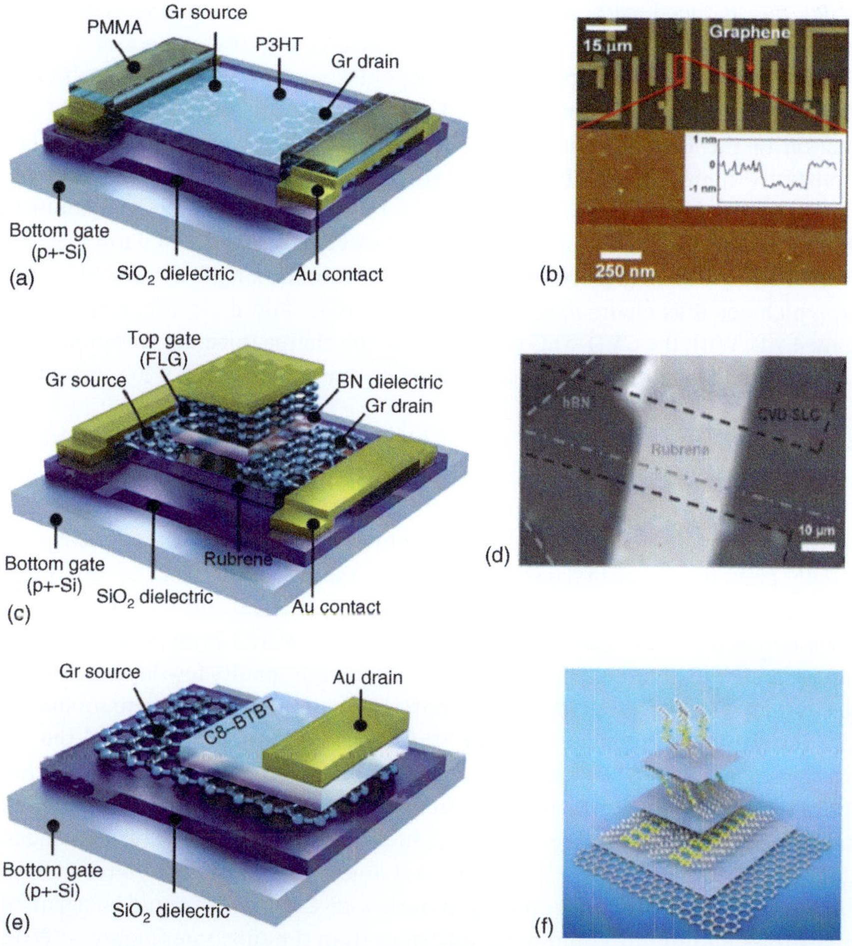

Figure 4.7 Graphene-based carbon electrodes for OFETs. (a) Schematic and (b) AFM image of a pentacene transistor with graphene electrodes (up) and representative device structure in which the graphene–metal junctions were protected by PMMA (bottom). (c) Schematic and (d) SEM image of rubrene single-crystal FETs fabricated by stacking chemical vapor deposition-single-layer graphene (CVD–SLG), hBN, and pristine few-layer graphene (FLG) as electrodes, dielectric, and top-gate electrode, respectively (left bottom), together with circuit diagram of the final device structure (right top). (e) Schematic shows the device architecture of the vertical OFETs based on graphene–C_8–BTBT heterostructures, and (f) schematic illustration of C_8–BTBT packing on graphene. Source: (a, b) Reproduced with permission from Cao et al. [565], © 2009 John Wiley & Sons. (c, d) Reproduced with permission from Kang et al. [570], © 2014 John Wiley & Sons. (e, f) Reproduced from He et al. [571]/with permission from Springer Nature.

showed a high on/off ratio and bulk-like carrier mobility. Lee et al. [567] fabricated graphene-based pentacene OFETs for high-performance electronic devices. They found that OFETs using as-fabricated monolayer graphene electrodes with PMMA residues showed a higher mobility of ~1.2 $cm^2\ V^{-1}\ s^{-1}$ than thermally treated graphene without residues. The pentacene molecular stacking arrangements on graphene electrodes were responsible for the differences in the contact resistance, thus R_c was reduced with a stand-up orientation in π–π stacking. Therefore, thermal annealing can affect both the work function and the contact resistance. Lee et al. [556] improved the performance of pentacene OFETs by using multilayer graphene electrodes instead of traditional Au electrodes, primarily because of the formation of the favorable interfacial dipole. In another work, Lee et al. [569] demonstrated high-quality rubrene films grown on h-BN dielectric layers via vdW epitaxy, which exhibited a high mobility by forming interfacial electrical contacts with graphene electrodes. Kang et al. [570] fabricated heterostructured top gate/top contact (TGTC) OFETs (Figure 4.7c) consisting of rubrene single crystals, SLG, and hBN, which acted as channel, source–drain electrode, and dielectric, respectively (Figure 4.7d). With the CVD–SLG as electrodes, the charge injection was improved, exhibiting high mobilities in the range of 10–15 $cm^2\ V^{-1}\ s^{-1}$ and a low operating voltage of 5 V.

Being an atomic crystal with a 2D flat planar structure, graphene can also be utilized as a good substrate for the epitaxial growth of 2D organic molecules. He et al. [571] reported vdW epitaxial growth of C_8–BTBT on graphene or BN substrates into large-area single crystals with precisely controlled atomic thicknesses that preferentially grew in a layer-by-layer fashion. Also, the thicknesses of the initial two layers (interfacial layer, IL, and the first layer, 1L) were ~0.6 and ~1.7 nm, respectively, whereas the thicknesses of the subsequent layers were ~3 nm as bulk crystals, demonstrating the different packing modes. When high-quality few-layer C_8–BTBT molecular crystals grew on Gr, pristine charge-injection interfaces were produced, where the substrate graphene could be utilized as bottom electrodes and the 2D quasi-freestanding C_8–BTBT films as the semiconductors. Then, after the deposition of Au as top electrodes, a vertical OFET was built, exhibiting the diode-like rectifying behavior with ultralow leakage current (Figure 4.7e). Also, the authors concluded that the IL did not likely contribute to lateral charge transport that might result from small intermolecular π–π interactions and the few-layer C_8–BTBT could be regarded as quasi-freestanding with minimal disturbance from the substrate (Figure 4.7f). In addition, planar OFETs by using 1L C_8–BTBT crystals grown on the insulating BN with Au as contacts showed mobilities up to 10 $cm^2\ V^{-1}\ s^{-1}$, much higher than the reported values with a similar transistor structure. Based on the same strategy, the same group reported some other works, such as band-like transport with an intrinsic hole mobility of over 30 $cm^2\ V^{-1}\ s^{-1}$ [572, 573] and phototransistors with over $10^4\ A\ W^{-1}$ responsivity and over 10^8 photoconductive gain [574].

Reduced graphene oxide (RGO) can also be used as electrodes and is a more cost-effective choice. The work function value of graphene is ~4.6 eV, whereas the work function of RGO is 4.7–5.4 eV, which is suitable for p-type OFETs [575–577]. Pang et al. [578] demonstrated the first highly defined patterned graphene source/drain electrodes fabricated by means of a novel oxygen plasma etching approach from solution-processed GO films. Lee et al. [95] used RGO as contact electrodes in BGBC pentacene transistors and observed a similar performance (mobility of ~0.03 $cm^2 V^{-1} s^{-1}$) to those with Au electrodes (mobility of ~0.045 $cm^2 V^{-1} s^{-1}$), regardless of the far poorer conductivity of the RGO electrodes (five orders of magnitude less) than Au. In addition, if using RGO-coated Au as source/drain electrodes, the highest mobility that could be obtained was ~0.13 $cm^2 V^{-1} s^{-1}$ owing to a lower injection barrier. Becerril et al. [579] demonstrated a solution-processing method to introduce RGO, with a work function of ~5.01 eV, as source/drain electrodes in p-channel and n-channel OFETs with BC configuration to improve charge injection. Because of a lower contact resistance, high mobilities were obtained in comparison to Au electrodes. Kang et al. [580] reported a direct surface functionalization method to tune the work function of RGO by using two different types of alkylsilane compounds as SAMs, (tridecafluoro-1,1,2,2,-tetrahydrooctyl) trichlorosilane (FTS) and aminopropyl triethoxysilane (APTS), to produce p-type or n-type doping of RGO. Through covalent reactions with the oxygenated sites via Si—O—Si bonds on the RGO surface, FTS and APTS were able to change the work function of RGO (~4.94 eV), and FTS–RGO exhibited an increased work function of ~5.51 eV, whereas that of APTS–RGO decreased to ~4.31 eV owing to interfacial doping induced by the different electron-donating and withdrawing properties of alkylsilanes. By using the doped RGO as electrodes in transistors, the operation types could be altered owing to the different injection barriers for hole or electron.

In summary, both graphene and CNTs are suitable materials as electrodes or as insertion layers at metal/OSC interfaces. The difference between a graphene film electrode and a CNT electrode is mainly that the former is planar, whereas the latter shows a large length/diameter ratio, resulting in a much higher electric field at the tip of CNTs than with the planar graphene. It is believed that this strong electric field can assist the tunneling process from CNTs into semiconductor channels, thus improving charge injection [9, 531]. Also benefiting from the excellent mechanical and electronic properties and convenient fabrication methods of carbon-based electrodes, graphene and CNTs can form ohmic contacts with many OSC materials, which makes it feasible to build nanoscale optoelectronic devices [532–534, 581].

4.1.5 Covalent Bond Formation at the Molecular Level

Carbon-based electrodes have also been applied in molecular-scale electronics, acting as a reliable way to wire molecules into electrical circuits toward future inexpensive, functional, and atomically precise electronic devices [532, 582–584]. The chemical structure of CNTs and graphene is composed of sp^2-hybridized carbon, which could be oxidized into carboxylic groups under suitable oxidation conditions, thus offering opportunities to realize a covalent bond connection through amide linkage and improving the stability of electrode–molecule junctions. The Guo's group developed a novel way to build a carbon-based electrode system by using covalent bonding to form point contacts [532]. In 2006, the first example of covalently bonded carbon-based single-molecule devices was successfully fabricated based on SWCNTs as the electrodes through a lithography-defined oxidative cutting method [585]. The gap in SWCNTs was formed by oxygen plasma with precise oxidation, and the terminal carboxyl groups were obtained afterward, which were able to react with different molecules with amino groups on a quasi-1D geometric scale. The size of the gap was controlled by varying the oxidation time. Then, different molecules with diamine groups were able to covalently connect by the robust amido bond, demonstrating the universality of this platform. On the basis of robust amide linkages, SWCNT-based single-molecule devices can be utilized in many research fields [530]. By taking advantage of the ability to investigate the distinct properties at the single-molecule level, Guo et al. [586] were able to explore the electrical conductivity of individual DNAs (Figure 4.8a). The DNA sequences were first modified with amines on the terminal. The amines were then able to assemble into amide linkages between SWCNT electrodes, thus allowing direct measurement of their charge-transport behaviors, which enriched the understanding of the electrical process within DNA-based materials. Zhu et al. [587] reported an example of a solution-processable assembly method to build single-molecule junctions with metallic SWCNT nanoelectrodes to investigate the molecular conductance of a series of oligophenyls at the single-molecule level.

Graphene is another allotrope of carbon with a 2D planar structure, which could be utilized to form electrodes on a large scale with a high yield and a simple fabrication process. The length of the nanogaps of graphene can be tuned by using a method similar to that of SWCNTs through controlling the etching time. The Guo's group developed a new dash-line lithography (DLL) approach to build graphene electrode arrays for fabricating reliable and functional molecular transport junctions in high yield [589]. Cao et al. [590] also reported the installation of molecular functionalities into an electronic device based on the new DLL-generated molecular junctions capable of acting as photon- or proton-gated sensors, with the change of conductance under the illumination of light with different wavelengths or exposure to different pH conditions. Recently, the same group reported the first single-molecule junction (photo-switch) showing stable and reversible photo-switched conductivity with a single diarylethene covalently bonded to graphene electrodes (Figure 4.8b) [588]. The key to the successful realization of this fully reversible, two-mode electrical conductance was the decreased molecule–electrode coupling by the three methylene (CH_2) groups serving as spacers, demonstrating the importance of the electrode interface at the molecular level [534]. The switching devices exhibited remarkable properties in terms of accuracy (on/off ratio of ~100), stability (over a year), and reproducibility (46 devices with more than 100 cycles for photo-switching). The formation of covalent bonds at the single-molecule level presents superior opportunities to build reliable single-molecule devices. Carbon electrode–molecule junctions have been proven to be a very useful and helpful platform, providing a new understanding of physical and chemical phenomena in the field of molecular electronics. Thanks to the robust and reliable graphene-based single-molecule devices with covalent bond linkages, Wen et al. [591] were able to study the complex formation dynamics in solution because host–guest interactions

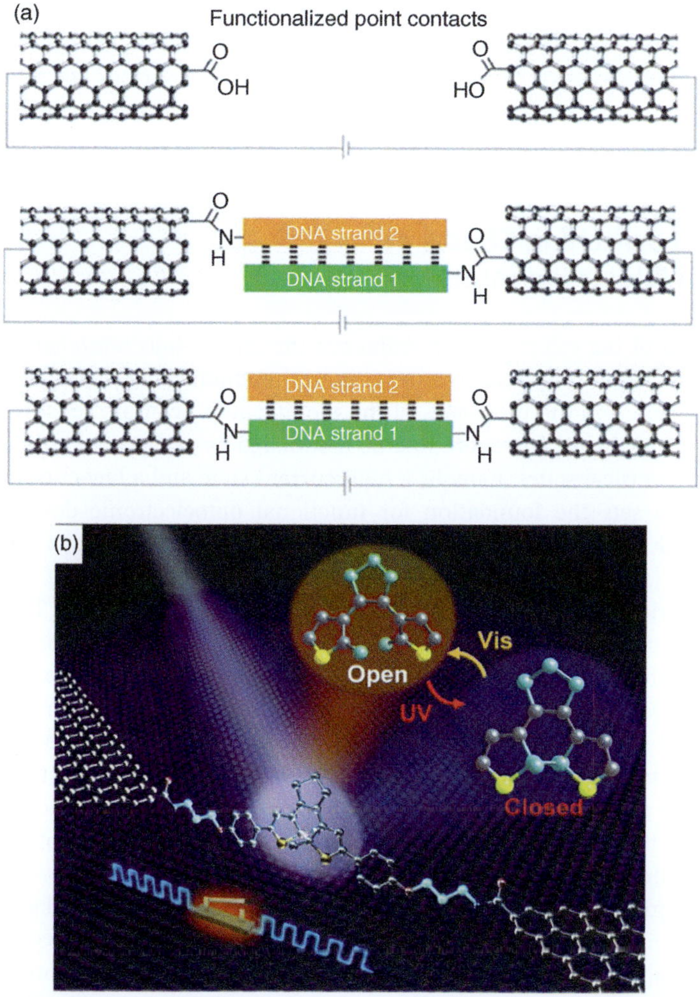

Figure 4.8 Carbon-based molecular junctions. (a) Schematic of using SWCNTs as electrodes to build molecular circuits for measurement of the conductivity of DNA strands. Source: Reproduced with permission from Guo et al. [586], © 2008 Springer Nature. (b) Schematic of a graphene–diarylethene–graphene junction that highlights the expansion of the molecular bridge by methylene groups. Source: Reproduced with permission from Jia et al. [588]. American Association for the Advancement of Science.

affected the conductivity of molecular junctions. By using the covalently bonded graphene-based single-molecule junction devices, stereoelectronic effect-induced conductance switching was also observed in the aromatic chain [592]. Also, Xu et al. [593] built a single-electron transistor with graphene electrodes made by using a feedback-controlled electroburning process. Then, aromatic ring molecules were able to covalently connect to the graphene nanogaps.

In conclusion, the fabrication of single-molecule devices based on the above-mentioned carbon-based electrodes, together with other technologies for developing single-molecule devices based on metal electrodes, not only opens the door for fundamental research by directly observing and studying the physical/chemical/biological events on a molecular scale but also brings out new opportunities for the future practical molecular electronics with multiple functions [581, 594, 595].

4.2 Installing Switching Effects at Semiconductor/Electrode Interface

Surface modification of the electrodes can influence the charge-injection/charge-extraction efficiency at the interface, thus leading to the conductance change of the devices. Furthermore, if the properties of the modification layer are sensitive to the environment, the device currents could be modulated by external stimuli to create two distinct electrical states, namely, a high-current state and a low-current state. This operation sets the foundation for functional optoelectronic devices such as optoelectronic switches, photodetectors, and nonvolatile memories. In this aspect, Nouchi et al. [596, 597] demonstrated electrically stimulated switching of a charge-injection barrier at electrode/semiconductor interfaces (Figure 4.9a). This was investigated by using various benzenethiol (BT) derivatives as a disordered monolayer (DM) on the electrode surfaces. The switching behavior was induced by the structural changes of DM molecules and manifested as a reversible inversion of the polarity of DM-modified Au electrodes (Figure 4.9b). The pushback effect of the thiol anchor group was found to dominantly determine the switching direction, whereas the terminal group modulated the switching strength. All of the tested BT derivatives exhibited the same switching direction, regardless of the direction of the overall dipole of the derivatives.

However, this approach is based on electrical control of the electrical signal and needs an external high voltage to induce the switching. This will cause a disturbance with the reading voltages. Another technique developed recently based on an optical stimulus should overcome this problem because it enables remote control as well as diminishes the disturbance between the electrical and optical signals. For example, Zhang et al. [598] demonstrated an efficient route to construct optically switchable OFETs featuring a photochromic spirothiopyran (SP) SAM-functionalized electrode/semiconductor interface (Figure 4.9c). The photoisomerization of SPs induced a reversible change in the dipole moment of the SP-SAMs (~6.30 D for SP-closed and ~12.7 D for SP-open). The interfacial dipole moment changes affected the work function of Au electrodes. These changes in the electrode work function enabled tuning of the contact resistance between the OSCs and the metal electrodes (Figure 4.9d). Consequently, the channel conductance of these devices was modulated in a nondestructive manner, leading to a new type of cost-effective OFET-based photodetector with high photosensitivity. In another example, The Wee's group [600] reported that the molecular dipole moments within H– and CN–azobenzene SAMs varied during photoisomerization from the *trans* form to the *cis* form and back upon thermal recovery. The change of φ_m during the *cis*-to-*trans* thermal

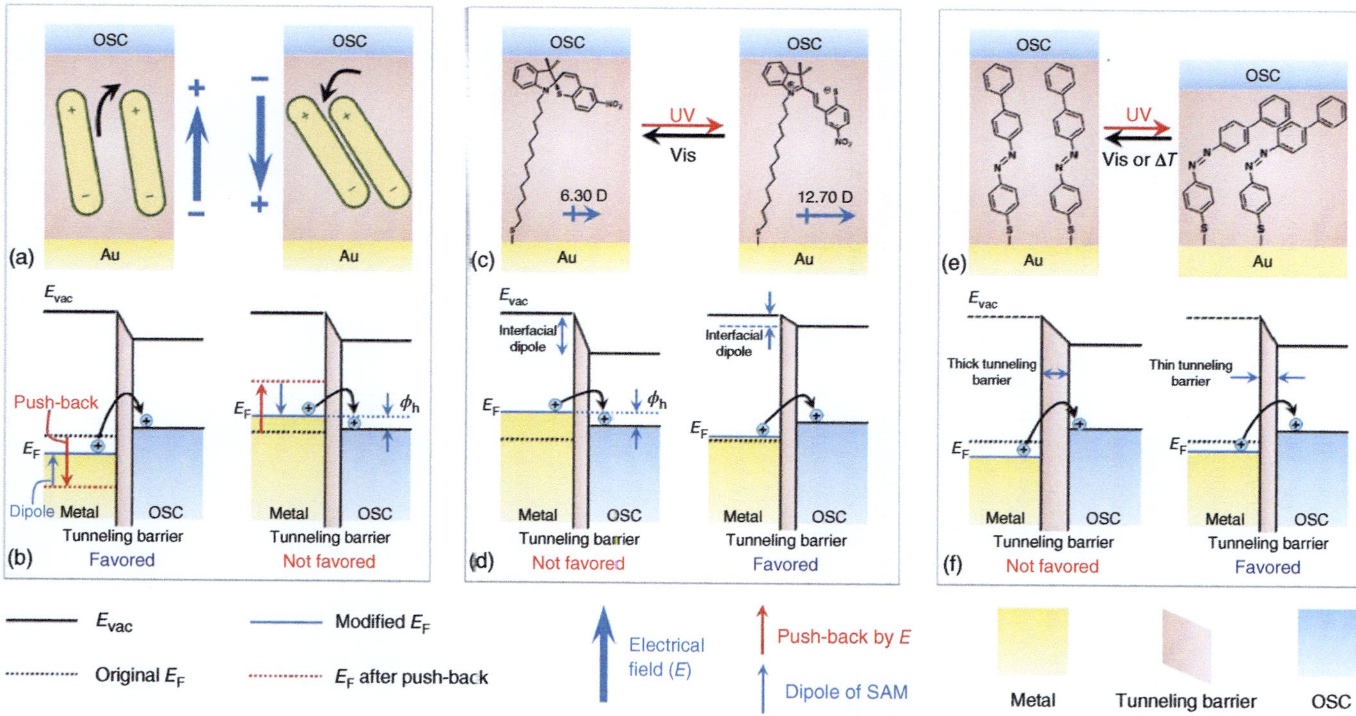

Figure 4.9 Three switching mechanisms. (a, b) Switching behaviors induced by the structural change of disordered monolayers. (c, d) Photo-switching behaviors induced by the dipole moment change of SP–SAMs. (e, f) Photo-switching behaviors induced by the tunneling resistance change of AZO–SAMs. Source: (a, b) Adapted from Nouchi and Tanimoto [597]. (c, d) Adapted from Zhang et al. [598]. (e, f) Adapted from Crivillers et al. [599].

recovery was monitored in real time and showed a reversible tunability up to 0.2 eV.

On the basis of the theoretical prediction, the photochemical F–AZO modification-induced modulation of the Au work function could be as large as 220 meV (Figure 4.9e). However, Crivillers et al. [599] used a Kelvin probe to measure the work function difference between *trans*- and *cis*-AZO SAM. They showed only a subtle difference of approximately 70 meV between the two isomeric states of the films. Therefore, they claimed that the observed differences in the output current could not be ascribed to a substantial work function shift leading to a more favorable energy-level alignment. They proposed that the switching effect of the transport properties could be explained by the smaller thickness of the SAM, i.e. a decrease in the tunneling barrier thickness from *trans* to *cis* (Figure 4.9f) [601]. This is reasonable because the tunneling rate decreases exponentially with the increasing barrier thickness [602]. In addition to the factors mentioned earlier, some other reasons, such as the change in film morphology, may also influence the electrical FET parameters. Rational modification of Au electrodes by stimuli-responsive SAMs could lead to more effective and controllable regulation of the electrical characteristics of the devices, promising a new generation of multifunctional optoelectronic circuits and sensors.

5

Semiconductor/Dielectric Interface

In general, organic field-effect transistors (OFETs) operate by charge accumulation at the interface between the semiconductor and the dielectric when a bias gate voltage is applied. Under enough gate bias (known as the threshold voltage, V_{th}), efficient mobile carriers are induced, leading to the outcome of the source–drain current (I_D) once a source–drain voltage (V_D) is applied [4, 603]. In this regard, the properties of dielectric materials play very important roles on the performance of OFETs, such as the carrier type and mobility, threshold voltage, and on/off ratio, as well as environmental and thermal stability. Generally, it is believed that charge transport takes place mainly in the first few layers within the semiconductor at the semiconductor/dielectric interface in OFETs (Figure 5.1a) [604]. This can be understood from the energy-level alignment at the interfaces.

Although the detail of the carrier distribution perpendicular to the organic semiconductor (OSC)/dielectric interface is still unclear, there is a consensus that, experimentally and theoretically, charge transport predominantly occurs two-dimensionally in the first 1–3 monolayers adjacent to the dielectric interface (Figure 5.1b) [12]. In Figure 5.1c, the band structures of the metal gate, the insulator (gate dielectric), and the OSCs are depicted when no gate bias is imposed. In this case, the OFET does not work. Figure 5.1d shows what happens at the interface when a negative V_G is imposed. After applying a negative gate voltage, the field causes the semiconductor band edges (HOMO and LUMO levels) to bend upward. A potential well will form that allows positive charges to be confined and accumulated at the OSC/dielectric interface, forming a conductive channel flowing through an ideal path perpendicular to the drawing plane. Because of this confinement, the field-induced transport is 2D. That is to say, if a continuous channel layer is deposited, charge transport is theoretically independent of the thickness of the OSC (Figure 5.1b). For an n-type OSC, negative charges are accumulated at the interface by applying a positive V_G bias and current flows for a positive V_D (Figure 5.1e). Regarding the transport mechanism that the generation and transport of charge carriers generally occurs at the semiconductor/dielectric interface, the choice of the dielectric materials and any modification of the interface will be ready to influence the device performance and even incorporate new functions.

Interface Engineering in Organic Field-Effect Transistors, First Edition. Xuefeng Guo and Hongliang Chen.
© 2023 WILEY-VCH GmbH. Published 2023 by WILEY-VCH GmbH.

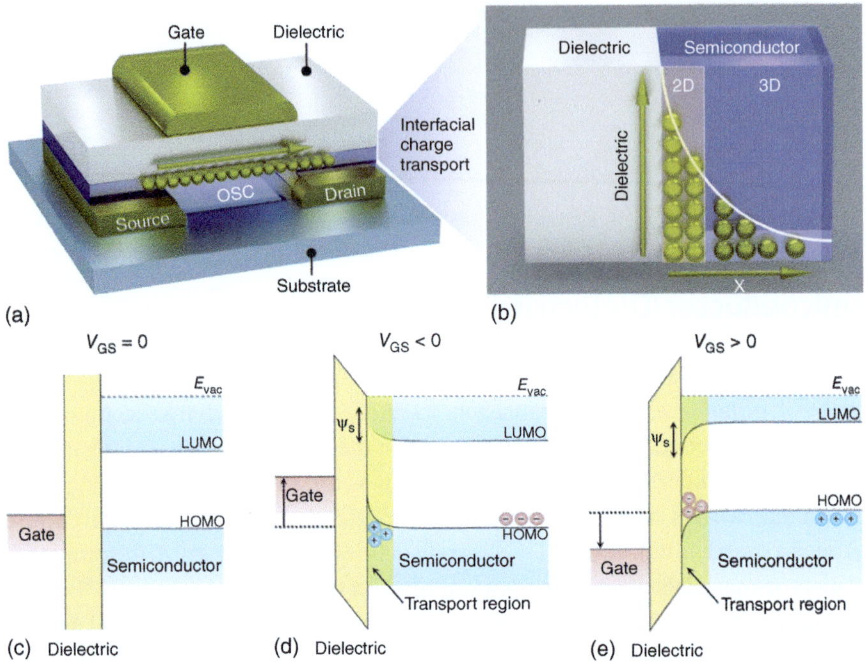

Figure 5.1 Charge-transport physics at the semiconductor/dielectric interfaces. Schematic representation of (a) the OFET geometry and (b) the 2D/3D carrier distribution profile in the accumulation layer of the transistors. (c–e) Energy-level alignments at the semiconductor/dielectric interfaces. (c) $V_{GS} = 0$; (d) $V_{GS} < 0$; (e) $V_{GS} > 0$. Source: (a, b) Adapted from Kronemeijer et al. [605]. (c–e) Adapted from Torsi et al. [12].

Up to now, a variety of materials have been applied as dielectric layers in typical OFETs based on single components or combinations of hybrid multilayers/nanocomposites, such as the most classical silicon oxide, inorganic oxides, and insulating small molecules/polymers [27, 81, 105, 146, 148, 606–612], even biomolecules (such as gelatin [613] keratin [614], and DNA [615, 616]), as well as the newly used self-assembled monolayers (SAMs) as an additional layer deposited onto the surfaces of dielectrics [37, 150, 466, 617, 618]. In general, a gate dielectric material should show good environmental, thermal, and electrical stability. The term permittivity (k) is used to indicate the ability of a dielectric to modulate the channel current, an intrinsic characteristic of the materials (Figure 5.2).

A general fact is that the dielectric capacitance (C) in a transistor is given by:

$$C = \varepsilon_0 \frac{k}{d}$$

where capacitance (C) is decided by the permittivity (k) and thickness (d) of the given dielectric layer(s). ε_0 is vacuum permittivity. It is desired to obtain a device where a rather high current is generated with a low voltage to reduce power consumption, which could be achieved by making the dielectric layer very thin or by using high-k gate dielectrics. SiO_2 is a classical material used as the gate dielectric layer in metal–oxide–semiconductor (MOS)-based integrated circuits, with a low k

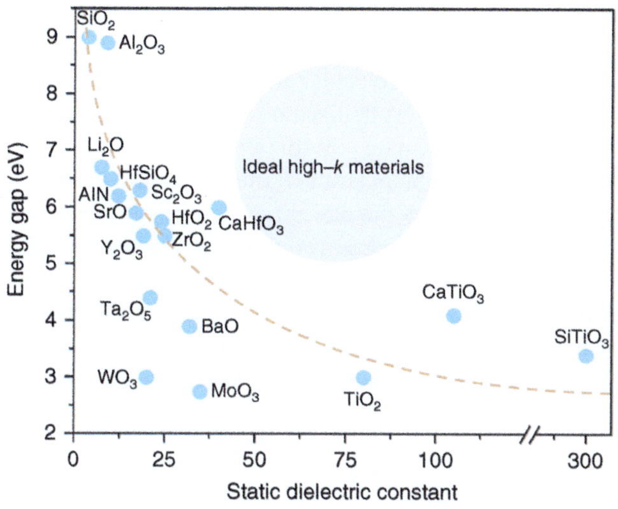

Figure 5.2 Energy gap and static dielectric constant of representative dielectric materials.

(3.9). These developments in traditional silicon-based electronics have also benefitted research on organic optoelectronics. However, its use in low-voltage devices is limited because when it gets so thin (under 2 nm), gate leakage due to direct tunneling becomes very high and cannot be ignored [27]. Then, high-*k* dielectrics, for example, HfO_2-based materials ($k = \sim 23$), were developed to overcome this current leakage problem to achieve stable and reliable gate dielectrics without the need for very thin layers [619–621]. In fact, high-*k* dielectrics have already been introduced to organic transistors to enable a lower gate voltage for sufficient charge accumulation in the active layer [26, 144].

Besides, when choosing a gate dielectric, some other requirements need to be taken into consideration, such as a trap-free interface as well as a smooth surface [27, 466]. Both are important issues governing charge-transport behaviors within the channel layer in an actual OFET. This calls for the engineering of semiconductor/dielectric interfaces, which shall benefit from the items listed in the following:

(1) Control the growth and packing behavior as well as morphology of OSC;
(2) Eliminate trap sites on dielectric surface;
(3) Manipulate the polarity of dielectric surface to improve the mobility;
(4) Alleviate the bias stress to improve the stability;
(5) Install new functionalities at the interface by using stimuli-responsive dielectrics.

5.1 Dielectric Modification to Tune Semiconductor Morphology

For OFETs with the back gate (BG) configuration, the semiconductor layer is usually directly deposited onto the gate insulator through thermal evaporation or a solution process. The surface properties, that is, the roughness and surface energy, of the gate dielectric have a significant effect on the morphology and microstructure of the films grown on the gate dielectric. For molecular materials that are capable of exhibiting new polymorphic forms when crystallized on a solid surface, usually called substrate-induced phases, taking advantage of this property helps realize a well-stacked structure order [622, 623]. In the meantime, the nucleation density of OSCs, depending on the surface properties of the dielectric, changes the grain size of organic films. A larger grain size means fewer grain boundaries, leading to higher mobility (as discussed in Section 5.2). Thus, by controlling the surface energy, roughness, and/or microstructure of the dielectric surface, the semiconductor morphology (and thus the tuning performance of the corresponding organic transistors) will be significantly influenced. We will discuss the strategies developed based on these factors to improve the quality of OSC thin film in the following sections.

5.1.1 Dielectric Surface Energy Control

Generally, dielectrics with a low total surface energy (γ^{tot}) tend to result in semiconductor tin film with smaller grain size; however, some researchers have reported higher mobility for smaller grains [624, 625]. Whereas another study found higher mobility for larger grains grown on higher γ^{tot} dielectric surface [626]. Other researchers have concluded that a matched γ^{tot} between dielectric and OSC is more preferred for larger grains and higher mobility [101, 627–629]. However, the fundamental understanding is still unclear. We do not wean to go into details of these empirical findings. Instead, we organized this section from the point of different methods used to manipulate the dielectric surface energy, which include but are not limited to SAM modification and polymer encapsulation, and specially highlighted the recent development of modification of polymer dielectric surface.

5.1.1.1 Modify with SAM

The polarity of the dielectric interface has been widely shown to have a significant impact on the quality of the semiconductor layer by affecting local morphology, as first demonstrated by the deposition of pentacene on SiO_2 under different growth conditions [246, 630–633]. SAM modification is the most common method to tune the properties of the dielectric surface. For example, the surface energy of SiO_2 could decrease after the absorption of SAMs, such as octadecylsilanes (OTSs) (**154**) [138, 634], as a result of the hydrophobic nature of SAMs. In this section, we mainly focused on the powerful ability of SAMs to control the surface energy of inorganic dielectrics and the dramatic influence on the nucleation and growth of OSCs thin films.

Surface energy is usually obtained by measuring the contact angle at different areas on a surface, and thus it does not provide enough fundamental understanding to explain its impact on device performance. In fact, SAM modification affects the surface energy of dielectric in two ways (Figure 5.3a). One is the crystallinity of the SAM, which will in turn affect the nucleation and growth of OSC. This is determined by processing method of the SAM. The other is the chemical structure

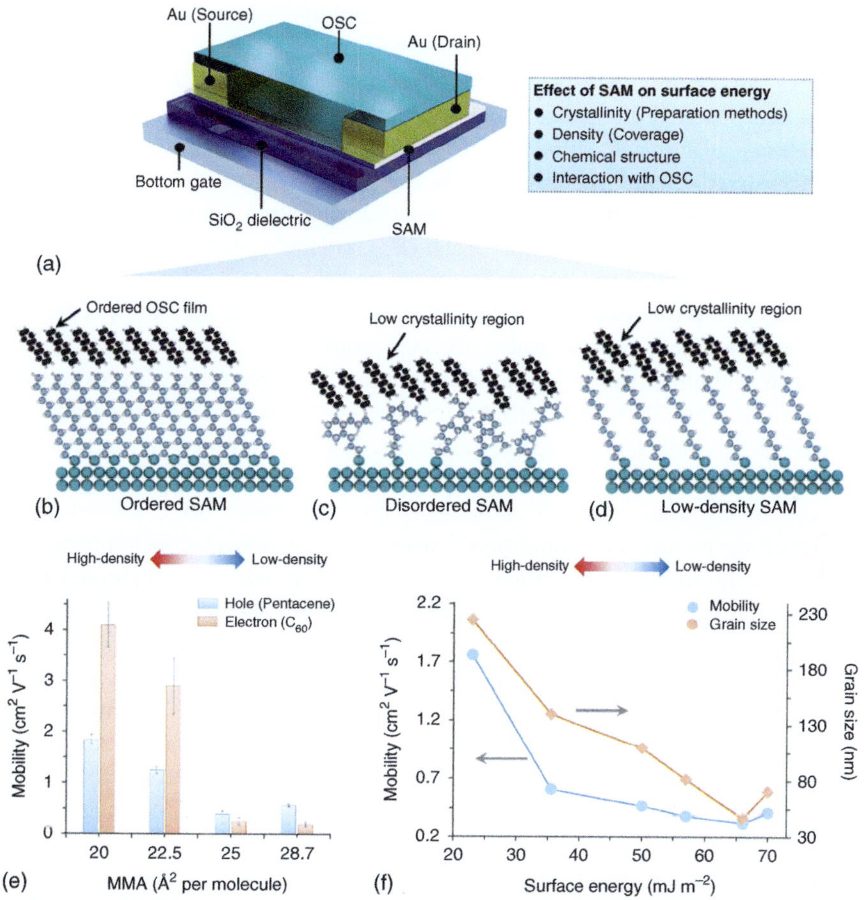

Figure 5.3 Semiconductor morphology control by SAMs at the semiconductor/dielectric interface. (a) Schematic representation showing the device configuration of a SAM-modified OFET. (b–d) Schematic illustration showing the nucleation of OSC is influenced by (a) ordered SAM, (b) disordered SAM, and (c) low-density SAM. (e) Mobility as a function of SAM coverage. (f) The correlation between mobility and grain size as a function of surface energy. Source: (e) Adapted from Virkar et al. [136]. (f) Adapted from Prisawong et al. [624].

of SAM, which will result in different intermolecular interactions with the OSC. In the following, we will discuss both effects in detail.

An earlier attempt reported that the presence of an ordered substrate caused by a SAM-treated dielectric increased the mobility and reduced the threshold voltage of pentacene-based OFETs [635]. From then on, many studies have been carried out to control the surface properties of dielectric using SAMs [37, 150, 636–638]. A crystalline and densely packed SAM through surface modification is preferred as it promotes 2D semiconductor growth, that is, quasi-epitaxy growth of organic semiconducting molecules with higher crystallinity and better interconnectivity between domains, which could result in a higher mobility than that on disordered SAMs [129, 136]. In general, the crystallinity of the SAM is determined by the preparation conditions, that is, temperature, humidity, and substrate roughness, although the surface energy of the prepared SAMs is the same [129, 639, 640]. In the case of the ordered SAMs, the alkyl chains are tightly packed because they are strongly held together by means of van der Waals interactions. This will help to grow high-crystallinity OSC films on top (Figure 5.3b). In contrast, the alkyl chains of the disordered monolayers are flexible due to the weak interactions between neighboring chains (Figure 5.3c) or the low chain packing density (Figure 5.3d). This may assist the surface diffusion of the OSC molecules, which results in a low crystallinity of the OSC films.

The Bao's group described a simple spin-coating technique that allowed for the deposition of a smooth, crystalline OTS SAM onto Si/SiO_2 substrates [138]. For both p- and n-channel vapor-deposited OSCs upon the crystalline OTS SAM, they grew in a favorable 2D layered growth manner, which is generally the preferred morphology for high charge-carrier transport. The crystallinity of the SAMs is particularly influenced by the preparation temperature [641, 642]. For example, octadecyltrichlorosilane (ODTS) monolayers prepared below 30 °C are relatively rigid, resulting in ordered monolayers. In contrast, at preparation temperatures above 30 °C, the monolayers exhibit a disordered state, which is originated from the loose packing and higher mobility of the alkyl chains. Kang et al. [142] reported that the grain structure of OTS SAMs could be tailored at much lower temperatures (20, −5, and −30 °C), which influences the growth of pentacene on them. Although the surface energy was measured to be very similar, the −30 °C OTS showed the best homogeneity over a wide area, with large crystalline grains, thus enabling the quasi-epitaxial growth of pentacene films, which led to the improved performance.

In addition, it has been shown that the density of the SAM correlates with effects on the semiconductor's nucleation mechanism and grain size. Virkar et al. [136] utilized the Langmuir–Blodgett (LB) technique to systematically vary the density of the OTS monolayers. The charge-carrier mobility of p-type pentacene and n-type C_{60} is plotted in Figure 5.3e as a function of the density [inverse mean molecular area (MMA)] of the OTS monolayer. They found that the hole mobility of pentacene measured on each type of OTS increased with increasing OTS density from 0.4 $cm^2\ V^{-1}\ s^{-1}$ for the least compressed LB film to 2.1 $cm^2\ V^{-1}\ s^{-1}$ on the most compressed (highest order) LB film, despite only small differences in surface energy. This suggests that the density of OTS is a critical factor affecting performance. N-type C_{60} showed a similar trend.

Recently, vacuum-UV (VUV) irradiation of SAMs has emerged as an accessible and robust method to controllably and reproducibly vary the density and coverage of SAMs and, finally, the surface energy of the dielectrics over a wide range [643]. It is realized by partially eliminating the SAMs through highly reactive oxygen atoms created by light. Using this method, Prisawong et al. [624] have observed a one-to-one correlation between the grain size of the surface and the saturated field-effect mobility (Figure 5.3f). They found that low surface energies (high-density SAM) result in rough films with large grains, which in turn give the best field-effect mobility. For high-surface-energy (low-density SAM) substrates and thin dinaphtho[2,3-b:2′,3′-f]thieno[3,2-b]thiophene (DNTT) films, the resulting films were smoother, with the DNTT molecules effectively wetting the substrate. However, the early-stage growth resulted in small grain sizes of OSC films and thus lower field-effect mobilities.

The chemical structure of SAMs also matters, and if the chain length and terminal group are changed, morphological differences will occur, leading to different device performance [644]. In general, the SAM consisting of fluoroalkyl groups may provide a rather low surface energy (~10 mN m^{-1}). The alkyl chain substituted SAMs locate in the medium. While SAMs with polar terminal groups – i.e. hydroxyl and carboxyl groups – usually result in high surface energy (~50 mN m^{-1}). Umeda et al. [134] reported surface-energy-dependent field-effect mobilities for polymer thin-film transistors by using four kinds of silane-based SAMs (**164–167**) (Figure 5.4a). When using 1h,1h,2h,2h-perfluorooctyltrichlorosilane (FOTS) (**165**), a fluoroalkyl-containing molecule, as a SAM on SiO_2, with a rather low surface energy, the mobility was up to 1 cm^2 V^{-1} s^{-1} for a highly crystalline conjugated polymer, PB16TTT (**169**).

The molecular orientation of OSCs critically depends on the relative strength of molecule–substrate interactions versus intermolecular interactions. Regarding this, SAMs with high surface energies tend to induce face-on growth of

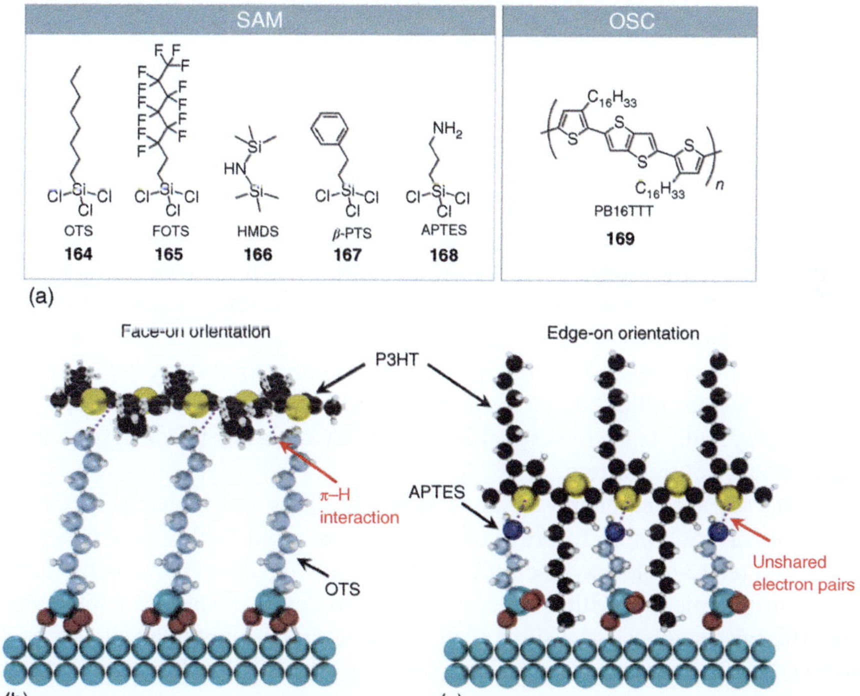

Figure 5.4 Molecular structure effect of SAMs in surface energy. (a) Chemical structure of SAM and OSCs used. (b) Schematic representation showing different orientations (b) face-on and (c) edge-on according to interfacial characteristics. Source: (b, c) Adapted from Kim et al. [645].

the semiconductor, whereas for SAMs with low surface energies, edge-on growth is favorable [162]. By using various functionalized SAMs with $-NH_2$ ((3-Aminopropyl)trichlorosilane [APTES], **168**) and $-CH_3$ (OTS, **164**) as terminal groups, Kim et al. [645] found that the orientation of P3HT could be transformed from face-on (Figure 5.4b) to (Figure 5.4c), which was attributed to the difference in intermolecular interactions between P3HT and the terminal groups of SAM. For the P3HT and OTS-SAM systems, no unshared electron pairs existed, and the π–H interaction between them led to parallel orientation to the insulator substrate (Figure 5.4b). However, the perpendicular orientation of P3HT was induced by the combination of the repulsive force between the unshared electron pairs on APTES–SAM with the π electron clouds of thienyl units and the π–H interaction (Figure 5.4c). Similar dependence of morphology and orientation control of liquid-crystalline polythiophene for high-performance OFETs through substrate surface chemistry by using different SAMs was also reported by others [134, 624, 646, 647].

As mentioned earlier, with different terminal groups on SAMs, a wide range of surface-energy tuning could be realized [121, 165, 648]. Utilizing this strategy, the Miao's group recently demonstrated an interface engineering approach to modify high-*k* AlO_y/TiO_x dielectric in n-channel OFETs by designing and synthesizing novel SAMs of phosphonic acids [40]. Chemical structures of the SAMs, which have one or more polar oxygen atoms at different positions along the long alkyl chain, marked as PA1–5 (**170–174**), is shown in Figure 5.5a. All of the SAMs exhibited very smooth surfaces, and they had greater surface energies than octadecylphosphonic acid (OPA) (**177**) (Figure 5.5b). 6,13-Bis((triisopropylsilyl)ethynyl)-5,7,12,14-tetraazapentacene (TIPS–TAP) [649] crystallized from its solution onto the high-surface-energy organic phosphonic acid SAMs (PA1, PA2, PA4, and PA5, on which the oxygen is located at most two carbon atoms away from the surface) and exhibited continuous films containing highly ordered and well-connected fibers (Figure 5.5c), thus affording n-channel OFETs showing high mobilities of up to $2.5\,cm^2\,V^{-1}\,s^{-1}$. In contrast, the PA3 and OPA SAMs had relatively low surface energies, leading to unfavorable crystallization of OSCs. These results indicated that the design of SAMs to reach a sufficient substrate surface energy suitable for the crystallization of OSCs is a key factor in governing the performance of transistors.

However, the surface energy of SAM is not the only factor that governs the morphology of OSC films. To prove this, Miao's group put forth a new design of SAMs based on 12-cyclohexyldodecylphosphonic acid (CDPA) (**175**) (Figure 5.5a) [167], a phosphonic acid with a cyclohexyl terminal group attached to a long alkyl chain. They also synthesized a short alkyl chain 4-cyclohexylbutylphosphonic acid (CBPA) (**176**) for understanding the structure–property relationship (Figure 5.5a). The bulky cyclohexyl terminal group in CDPA and CBPA is larger than the methyl terminal group in well-studied long alkyl chain SAMs, which can be fitted into a highly ordered monolayer on the surface of AlO_y/TiO_x dielectric surface. The results showed that the CDPA-modified dielectric exhibited a smooth surface and much better wettability than OPA with a larger surface energy by increasing the dispersion

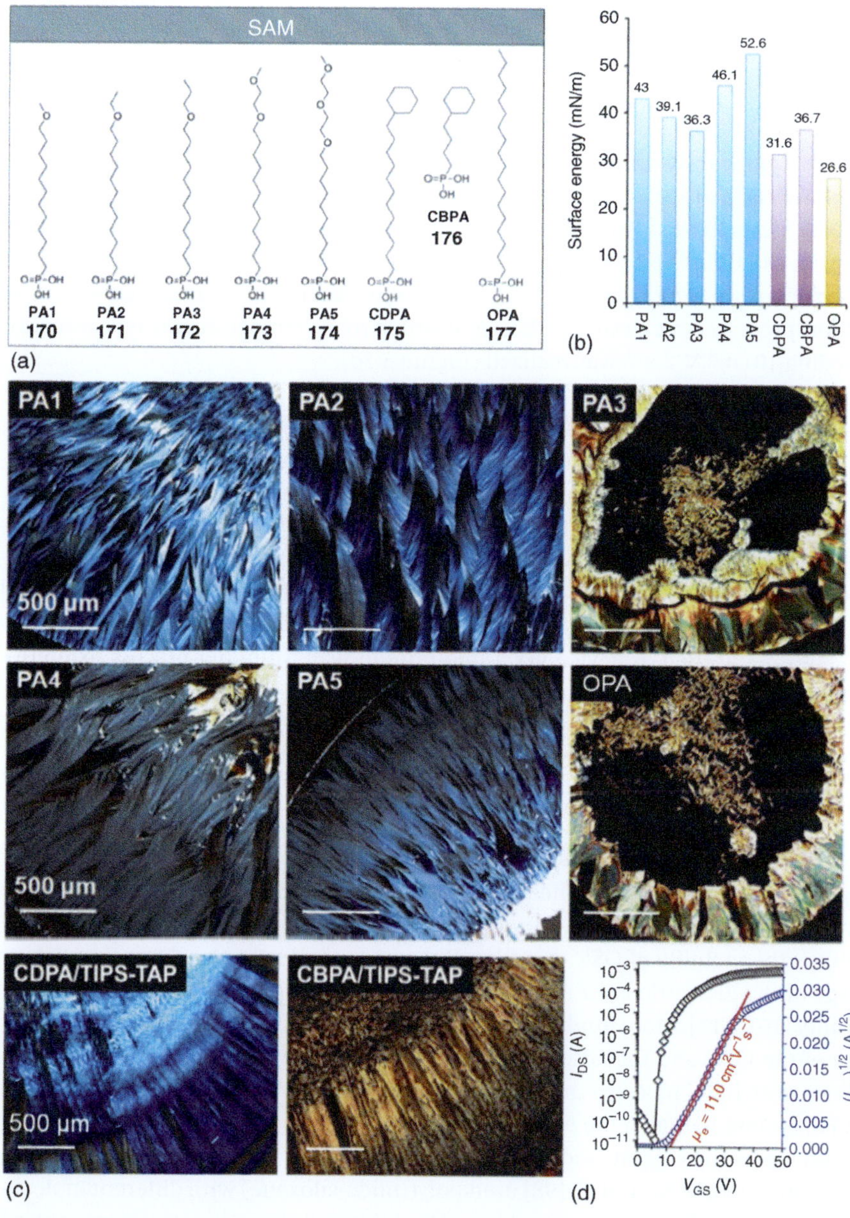

Figure 5.5 Semiconductor morphology control by SAMs at the semiconductor/dielectric interface. (a) Chemical structures and (b) surface energies of SAMs mentioned. (c) Reflected polarized-light micrographs for TIPS–TAP films that were cast from a solution onto an AlO_y/TiO_x dielectric surface modified with different SAMs. (d) Transfer curve showing an electron mobility of 11 $cm^2\ V^{-1}\ s^{-1}$ for TIPS–TAP film on CDPA–SAMs modified dielectric surface. Source: (c) Reproduced with permission from Liu et al. [40], © 2013 John Wiley & Sons and Liu et al. [167], © 2014 John Wiley & Sons. (d) Reproduced with permission from Xu et al. [41], © 2016 John Wiley & Sons, Inc.

component with an almost unchanged dipolar component. In the meantime, CBPA showed the largest surface energy, but it formed a disordered monolayer because of the short chain [167]. In addition, the grazing incidence x-ray diffraction (GIXD) data confirmed that an ordered monolayer of CDPA was indeed crystalline, which was important because semiconductor films grown on crystalline SAMs always showed significantly improved mobilities over films deposited on amorphous, disordered SAMs [129]. On the basis of these results, they demonstrated the ability of CDPA SAMs to modify the dielectric surface to study the temperature-dependent transport of TIPS–TAP by using an AlO_x/SiO_2 dielectric with fewer defects instead of the previously employed AlO_x/TiO_y [41]. Band-like transport was observed in the solution-processed n-channel OFETs, and a high room-temperature n-type mobility exceeding $10\,cm^2\,V^{-1}\,s^{-1}$ was realized (Figure 5.5d).

In another example, Das et al. [650] introduced unique bioinspired zwitterionic molecules, Z–Cat–Ben and Z–Cat–Ben–F [651], which could form molecularly smooth and uniform thin SAMs on various dielectric surfaces. With a strong adsorption ($W_{ad} > 20\,mJ\,m^{-2}$), the "monomolecular" SAM layers showed uniform thicknesses (~0.5 or ~1 nm) and ordered orientations (all catechol head groups facing the oxide surface). As a result, for poly[4-(4,4-dihexadecyl-4H-cyclopenta[1,2-b:5,4-b′]-dithiophen-2-yl)- alt-[1,2,5]thiadiazolo-[3,4- c]pyridine] (PCDTPT)-based (**68**) transistors with different modifications on nano-grooved SiO_2, the Z–Cat–Ben–F device via a one-minute water-dipping process exhibited an enhanced average saturation hole mobility of $\mu = 45 \pm 4\,cm^2\,V^{-1}\,s^{-1}$, reaching the highest value for OFETs reported to date [393, 652, 653] with no hysteresis, in comparison with the devices with conventional n-DTS SAMs, which showed mobilities of $27 \pm 3\,cm^2\,V^{-1}\,s^{-1}$.

5.1.1.2 Surface Modification with Polymers

Polymer encapsulation is another choice for modifying the dielectric surface. There is still a debate on how the polymer modifier surface energy influences the grain size and thus the mobility of OSC thin film. But as a general rule, the closer the surface energy of the dielectric layer is to that of the semiconductor, the higher the carrier mobility [101, 617]. More fundamentally, it is a competition between the aggregation force of the OSC and the adsorption of the OSC on the substrate. Experimentally, Yoshida et al. [654] reported that when the surface energy value of substrate (PMMA and PS) is close to that of the pentacene film surface, nuclear density is the smallest, domain size is the biggest, and mobility is the highest.

In another work, Yang et al. [98] used poly(imide-siloxane) with different molecular weights as surface modifiers and found that the growth mode of pentacene varied with the surface energy change, affecting the coverage of organic molecules. On the substrate with a lower surface energy of $30\,mJ\,m^{-2}$, the growth of pentacene was dominated by a 3D island growth mode, where organic molecules moved freely on the surface and formed large, separated islands of crystals after adsorption, whereas with a higher surface energy of $48\,mJ\,m^{-2}$, the pentacene exhibited a "layer + island" growth mode, that is, after the formation of an initial monolayer, the subsequent island growth occurred [423] The difference between the two growth modes was attributed to the surface energy differences between pentacene and the dielectric. As mentioned earlier, the relative magnitude of the interactions determined whether the deposited molecules covered the surface or aggregated together to minimize the total surface energy of the system [625, 627].

In addition to this, there are some exceptions. Kim et al. [99] employed two surface-functionalized SiO_2 substrates with poly(4-vinylpyridine) (PVPyr) or PS to control the semiconductor–dielectric affinity by surface hydrophobicity. They pointed out that besides the effect of surface energy mismatch between pentacene and PVPyr, the pyridine groups could act as electron-donating components, therefore pulling out the holes produced in the channel, degrading the mobility, and shifting the threshold voltage; that is, the electrochemical activity of a dielectric is strongly correlated with the device performance.

Another exception is reported by Sun et al. [100]. They studied the interfacial heterogeneity of the surface energy by using PS with varying molecular weights, for which the surface energy could be calculated. Interestingly, they found that the lower the surface energy, the poorer the mobility. This observation was quite contrary to the observation that an enhanced connection between grains is usually observed in low-surface-energy dielectrics. The authors then supposed that for low-molecular-weight polymers, the high chain-end density led to a low statistical average surface energy, resulting in the varied heterogeneities of the surface energy on the PS dielectrics. This heterogeneous surface energy strongly affected the morphology of the semiconductors, causing deep grain boundaries, decreased grain size, and a disordered arrangement of molecules. As a general rule, the larger the grain size, the fewer the number of grain boundaries and the higher the mobility [655–659].

Polymer modification of dielectrics has also been applied to the surface of high-k gate insulators [660, 661]. For example, Wang et al. [662] reported a low-voltage pentacene-based OFET with PS thin layer (17 nm) modification on HfO_x. They found that after annealing the PS at 120 °C, the quality of the surface of the hybrid dielectrics was improved owing to the induced flatter phenyl group orientation, better matching the surface energy of pentacene. The device with PS–120/HfO_x showed mobilities up to ~1 $cm^2\ V^{-1}\ s^{-1}$, whereas the mobilities of bare HfO_x and PS–RT/HfO_x were ~0.2 and ~0.4 $cm^2\ V^{-1}\ s^{-1}$, respectively.

5.1.2 Dielectric Microstructure Design

Beyond surface energy, the microstructure of the dielectric surface, including its isotropic roughness and anisotropic structure, can also guide the growth of OSC thin films on top. In general, by increasing the roughness, the grains are likely to be smaller and less oriented, with more valleys produced in the electrical channel, which is harmful for efficient charge transport between source and drain electrodes [663]. However, rough surfaces have proven to be preferential nucleation areas for depositing molecules because of the increased coordination between them, which could be used to obtain patterned semiconductors by selective growth, such as by fabricating large-area arrays of organic crystals and conjugated polymers [664–666].

In addition, anisotropically ordered semiconductor films can also be formed during the fabrication process via the methods including mechanical rubbing [667, 668], directional solidification [669], and surface treatment techniques such as aligning polymers on rubbed/grooved substrates [670, 671]. In this section, we discuss the microstructure design of the dielectric surface to guide the growth of OSCs into well-ordered morphologies in order to achieve high-performance OFETs.

5.1.2.1 Roughness Effect

The roughness of the dielectric surface has a distinct effect on the performance of OFETs by altering the morphology and the structural properties of OSCs. For the morphology of pentacene films deposited on different substrates, it was found that on smooth thermal oxide or polymer dielectrics with low root mean square (RMS) roughness at the sub-nanometer/Å scale, the obtained films were generally highly ordered, exhibiting high performance [663, 672–676]. Because the growth of large and uniform crystal domains was inhibited by the increased roughness, leading to the formation of the valleys in the channel region, which acted as carrier traps and scattering sites, the efficient charge transport was limited (Figure 5.6a). Similar results were reported for solution-processable polymer semiconductors, where the charge-carrier mobility was reduced atop rougher dielectrics [283, 677, 678]. In order to establish the impact of dielectric roughness on film microstructures for highly crystalline, high-mobility polymer semiconductors systematically, Jung et al. [679] evaluated the effect of dielectric roughness on the microstructure and hole mobility of thin films of poly(2,5-bis(3-tetradecylthiophen-2-yl)thieno[3,2-*b*]thiophene) (pBTTT), which formed highly oriented terraces on flat surfaces (Figure 5.6b). The rough dielectrics altered the microstructure of the channel layer by reducing the domain size and long-range layer order, whereas the molecular conformation and the orientations of domains remained unchanged. The hole mobility of as-cast films was similar for all the different dielectric roughness values, but a clear dependence on dielectric roughness was observed after annealing, where the films exhibited substantial reductions in the hole mobility with increasing roughness (Figure 5.6c), which was due to the increased number of lateral grain boundaries and packing defects.

However, Li et al. [680] demonstrated that the interfacial microstructure played only a minor role in the charge-transport behaviors of OFETs with ordered polymer multilayers (Figure 5.6d). They proposed an effective one-step solution method to precisely tune the RMS value of the SiO_2 surface on a sub-nanometer scale within a narrow range from 0.15 to 0.39 nm. This in turn allowed precise control of the microstructure of the semiconductor in the interfacial layer without affecting the morphology in the next layers (Figure 5.6e). In the case of a monolayer, a higher surface roughness led to the reduction of the domain size within the entire difluorobenzothiadiazole (FBT)–Th4 film, which was the only possible pathway for the migration of charge carriers, and thus the poorly aligned monolayer yielded poor performance (Figure 5.6f, red). In the case of multilayers, the contribution of the interfacial layer was negligible (Figure 5.6f, blue) because the charge carriers could bypass the structural defects at the interface, which means that the

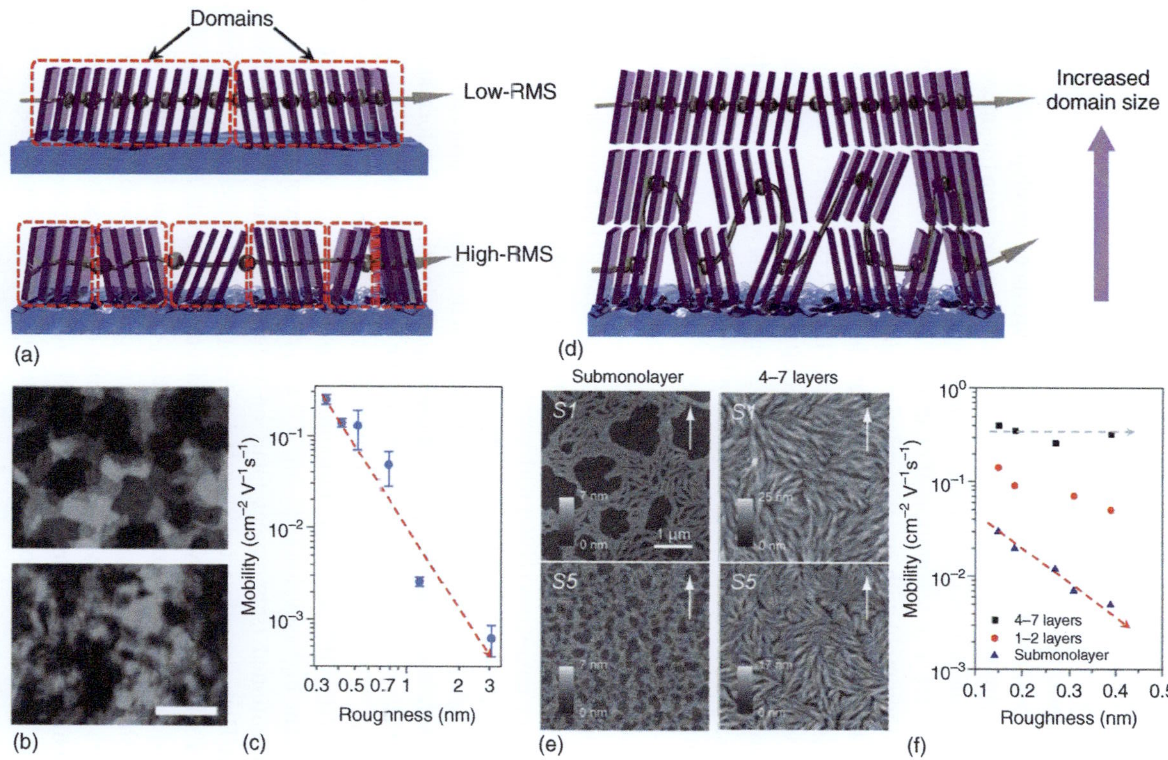

Figure 5.6 Roughness effects. (a) Schematic illustration showing the difference in domain size with respect to dielectric surface roughness. (b) AFM images of pBTTT atop various rough dielectric surfaces. (c) Dependence of FET mobility for pBTTT thin film on substrates of various dielectric roughnesses. (d) Schematic illustration showing the mechanism by which charge carriers could bypass the structural defects at the interface. (e) AFM images showing the morphology difference between submonolayer and multilayer OSC deposited on S1 (RMS = 0.149 nm) and S5 (RMS = 0.390 nm). (f) Dependence of FET mobility on dielectric roughness. Source: (b, c) Reproduced with permission from Jung et al. [679], © 2008 John Wiley & Sons. (e, f) Reproduced with permission from Li et al. [680], © 2016 John Wiley & Sons.

ordered layers on top of the interfacial monolayer mainly dominated the charge transport (Figure 5.6d). This work provided additional insight into the details of charge transport at the semiconductor/dielectric interface. In the meantime, the charge-carrier transport of monolayer transistors still exhibits a strong dependence on the dielectric roughness, which was attributed to the higher densities of grain boundaries, charge-trapping sites, and surface scattering induced by the rougher surface, reported by the same group [681, 682].

5.1.2.2 Nano-fabrication Created Microstructure

A variety of micro- and nano-patterning techniques have been developed for organic (opto-)electronics [683–685]. Moreover, from a broader viewpoint on the microstructure design of dielectric surfaces, selected patterning is a widely applied technology by template transfer. The orientation of channel films is often realized by putting a well-designed top mold with suitably patterned nanochannels atop the semiconductor films and then depositing insulating materials serving as dielectric layers. The Huck's group [686] demonstrated the formation of liquid-crystalline (LC) polymers (F8BT) into a well-aligned uniaxial structure through the use of nanoconfinement caused by a Si mold modified with nanochannel arrays during nanoimprinting [687]. The well-aligned F8BT films were used as active layers in OFETs, resulting in comparable hole and electron mobilities on the order of 10^{-4} $cm^2 V^{-1} s^{-1}$.

In addition, nanoimprinting lithography (NIL) technology has been proven to be another effective method and is widely used in inducing molecular orientation of semiconducting polymers into ordered structures [688–690]. For example, recently Kim et al. [691] developed a nanopattern-transfer technique to induce uniaxial growth of organic nanowires (NWs) with single crystals. After the TIPS–pentacene solution was dropped onto the PI/SiO_2 surface, a nano-grated polyurethane acrylate (PUA) template was placed with conformal contact on the top of the droplet to guide the crystal growth in the geometrically confined space, leading to the formation of single-crystal NWs with a lattice-strained packing motif, which was efficient for charge transport. With the TIPS–pentacene NWs as the channel, the bottom-gate top contact (BGTC) OFETs exhibited mobilities up to ~9.71 $cm^2 V^{-1} s^{-1}$, which is comparable to the highest value (~11 $cm^2 V^{-1} s^{-1}$) reported previously [219].

The aforementioned technologies focus on inducing the growth of OSC with external tools. However, microstructure design technologies for dielectric substrates that focus on guiding the growth of semiconducting materials *in situ* may also achieve high-quality domains, thus leading to high carrier mobilities. The Heeger's group [692] showed that the anisotropic order and macroscopic alignment of polymer fibers could be achieved by introducing directional solvent evaporation (Figure 5.7a) of high-molecular-weight PCDTPT (with number average molecular weight, M_n = 300 kDa) solutions on nanoscale grooved substrates (Figure 5.7b) [693]. Surface grooves were introduced on SiO_2 gate dielectrics by scratching the surface with diamond lapping films with nanoparticle diameters ranging from 100 to 500 nm (Figure 5.7c). The long-range orientation/alignment of PCDTPT was obtained on the nanostructured dielectric surface (Figure 5.7d) after a combination of directional solvent evaporation through a tunnel-like configuration and slow

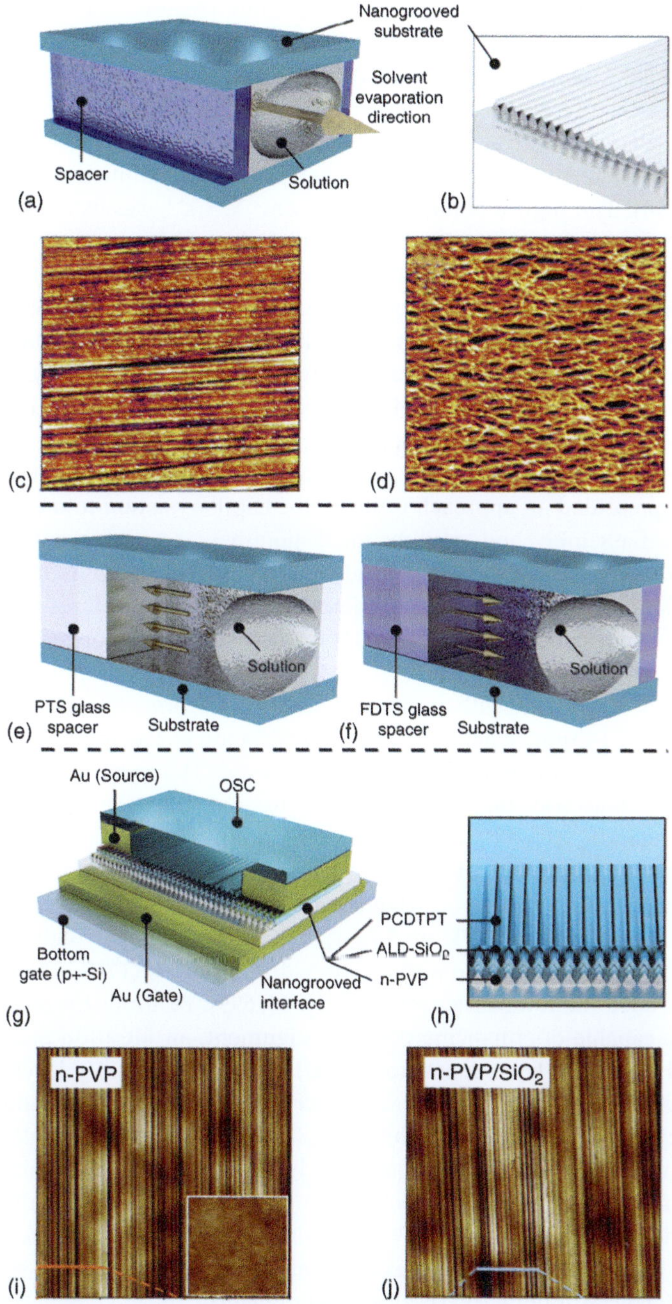

Figure 5.7 Effect of the nanostructured dielectric surfaces on the device performance. Schematic illustration of (a) the tunnel-like configuration setup and (b) the nanoscale grooved substrate. (c, d) AFM images of dielectric substrate surface (c) and related polymer fiber morphology with nanostructures (d). (e) Schematic diagram showing the opposite actions to the polymer solution exerted by the spacers with two different surface treatments, PTS (e) and FDTS (f), respectively. Schematic architecture of (g) the nanogrooved flexible devices and (h) the nanoscale grooved flexible substrate. (i, j) AFM topographic images of the nanogrooved PVP (i) and nanogrooved PVP/SiO_2 (j) substrate. Source: (a–d) Reproduced from Tseng et al. [692], © 2012 American Chemical Society. (e, f) Adapted from Luo et al. [652]. (g–j) Reproduced from Lee et al. [698], © 2016 American Chemical Society.

drying, which led to large changes in charge-carrier mobility (Figure 5.7a). After thermal annealing at 200 °C, a hole mobility for transport along the conjugated chain of ~6.7 $cm^2 V^{-1} s^{-1}$ was achieved for PCDTPT at 100-nm structured substrates. In a subsequent study designed to obtain chain alignment by using the nano-grooved substrate strategy, the authors used PCDTPT with molecular weights (M_W) ranging from 30 k to 300 kDa, and they obtained hole mobilities up to ~23.7 $cm^2 V^{-1} s^{-1}$ [694]. The high mobility of over 20 $cm^2 V^{-1} s^{-1}$ was extracted from aligned PCDTPTs with M_W = 50 kDa. The mobility showed insensitivity to M_W, and it was quite different from the M_W-dependent mobility under non-oriented conditions. Yet M_W-dependent transport happened to be a commonly observed phenomenon, with higher M_W values resulting in higher mobilities. These higher mobilities were affected by various factors, ranging from the crystallinity degree of the conjugated polymers to the morphologies with different chain lengths [221, 248, 695–697].

In order to develop a general strategy for achieving such high mobilities, the Heeger's group [652] presented an approach utilizing the capillary action to mediate polymer chain self-assembly and unidirectional alignment on nanogrooved substrates (Figure 5.7e,f). They designed a sandwich tunnel system separated by functionalized glass spacers to induce capillary action for controlling the polymer crystallinity and nanostructure, and then successfully demonstrated saturation mobilities with average values of over 20 $cm^2 V^{-1} s^{-1}$ in semiconducting polymer-based transistors with a channel length of 80 μm. By using a longer channel length of 140 μm, where the contact resistance is less important, a hole mobility as high as ~36.3 $cm^2 V^{-1} s^{-1}$ was achieved. However, because the nanogrooved SiO_2 dielectric, the key component for inducing chain alignment, is a brittle material, it is challenging to achieve high-mobility flexible OFETs based on this nanogrooved dielectric strategy. To solve this problem, the authors [698] fabricated OFETs, a chemically and mechanically stable nanogrooved polymer dielectric layer (Figure 5.7g,h) made of polyvinylpyrrolidone (PVP):(4,4′-(hexafluoroisopropylidene)-diphthalic anhydride) (HDA) (**176**), which has high solvent resistance and low surface roughness [274, 699]. To mimic the SiO_2 substrate, which was capable of enhancing polymer alignment, an ultrathin SiO_2 layer was introduced by atomic layer deposition (ALD) on top of the n-PVP layers (Figure 5.7h). No obvious morphology change was observed before (Figure 5.7i) and after (Figure 5.7j) the deposition of the 2-nm-thick SiO_2 layer. However, significantly improved mobilities were observed on the nanogrooved bilayer dielectrics, showing a highest mobility of ~19.3 $cm^2 V^{-1} s^{-1}$, which was much larger than that of the f-PVP/ SiO_2 dielectrics (~1.2 $cm^2 V^{-1} s^{-1}$). With polyimide as flexible substrates, the devices showed excellent mechanical flexibility and no obvious decrease in mobility occurred after a continuous 1000 cycles of bending tests.

5.1.2.3 Self-assembled Morphology of Dielectric

Without nanofabrication, some dielectric materials can self-structure into highly ordered structures that could guide the growth of OSCs. Yang et al. [700] synthesized a crystalline material, hexagonal aluminum nitride (AlN), as a gate dielectric

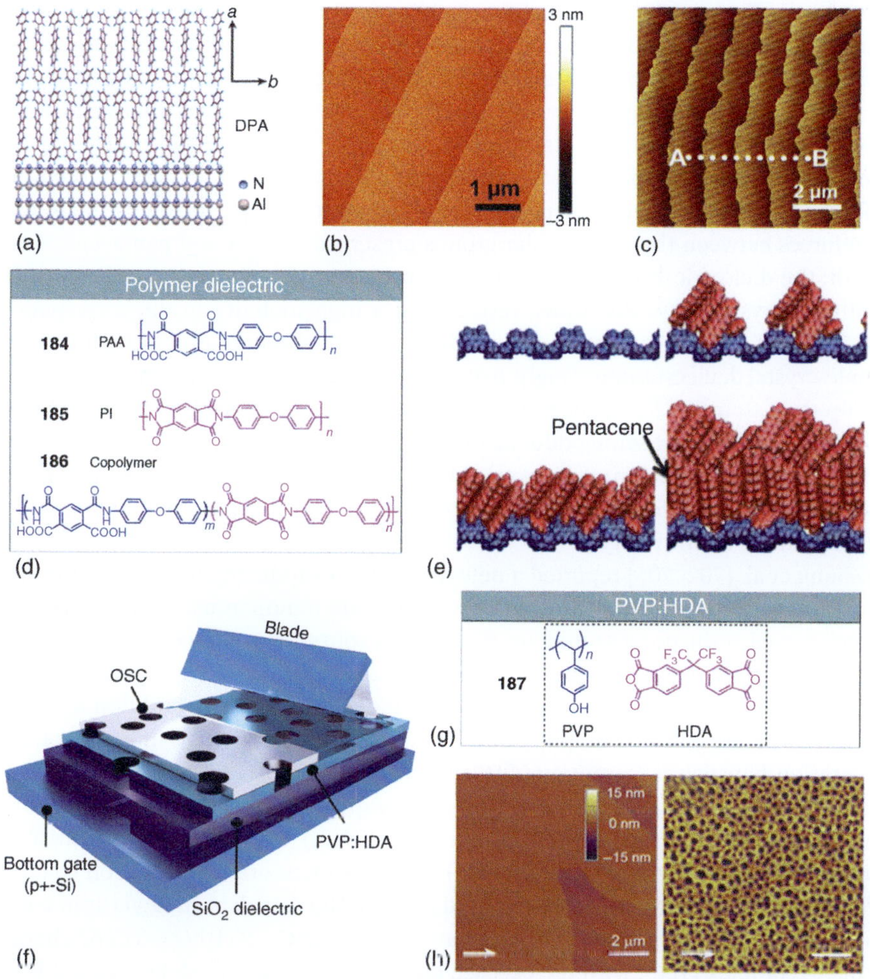

Figure 5.8 Self-structured dielectrics. (a) Schematic illustration showing the packing of ultrathin DPA molecules on the AlN crystal surface. (b, c) AF images showing atomic terracing on AlN crystal dielectric surface (b) and the molecular terracing on the DPA crystal surface (c). (d) Molecular structures of PAA. (e) Molecular orientation of pentacene molecules on PAA with different coverages of pentacene. (f) Schematic illustration showing the meniscus-guided coating on porous templates (g) Chemical structures of the porous dielectric PVP:HDA. (h) AFM images of nonporous (left) and porous (right) C_8–BTBT films. Source: (a–c) Reproduced with permission from Yang et al. [700], © 2017 John Wiley & Sons, Inc. (e) Reproduced with permission from Ji et al. [701], © 2017 American Chemical Society. (f, h) Reproduced with permission from Zhang et al. [702], © 2017 John Wiley & Sons.

(Figure 5.8a). The atomic flatness (Figure 5.8b) makes it a suitable substrate to prepare ultrathin 2,6-diphenylanthracene (DPA) molecular crystals atop via a van der Waals epitaxial strategy (Figure 5.8c), which exhibits promising properties including a saturation mobility of 6.8 $cm^2\ V^{-1}\ s^{-1}$ and a high on/off current ratio of 10^8.

Recently, Ji et al. [703] reported that poly(amic acid) (PAA, **184**) dielectric layer can self-structure into nanogrooves that could guide the growth of OSCs by taking advantage of its surface polarity collaboratively (Figure 5.8d,e). It was realized owing to the existence of OH groups at the surface and the amide bond linkages, which allowed the adjacent polymer strands to interact via H-bonds, forming a self-rippled surface perpendicular to the corrugation formed compared with PI (**185**). Moreover, a favored vertical packing of pentacene on the dielectric was realized by the repulsive forces between the strong polar groups presented in PAA and pentacene. As a result, the dielectric surface was capable of promoting the formation of large-area and high-crystallinity OSC films, resulting in a high mobility in the corresponding OFET of up to ~30.6 $cm^2 V^{-1} s^{-1}$, which is comparable to that of high-quality single-crystal devices. Interestingly, pentacene film as thin as 5 nm exhibited mobility up to 11.0 $cm^2 V^{-1} s^{-1}$, suggesting an efficient charge transport in the first few molecular layers at the semiconductor/dielectric interface. In further work, the same group reported a copolymer dielectric (**186**) with balanced chain-packing density and surface polarity for high-performance flexible organic electronics, demonstrating its general applicability [704].

Zhang et al. [702, 705] reported a new method of introducing nanopores into the semiconductor channels by directly depositing onto nanoporous insulating layers, which served as the porous template, prepared by spin coating a PVP with HDA (**187**) as the cross-linking agent onto SiO_2 in order to achieve effective doping while maintaining the structural integrity of the OSCs (Figure 5.8f–h). The addition of HDA led to a decrease in the miscibility of PVP with THF (Figure 5.8g), inducing phase separation between a continuous PVP-rich phase and a discrete THF-rich phase. After the evaporation of THF, nanopores were prepared in the films with typical film thicknesses and pore depths of 20–30 nm (Figure 5.8h). These nanopores were able to transfer into the semiconductor layer by a subsequent one-step solution coating with almost the same depths (Figure 5.8f). F_4–TCNQ was deposited via spin coating from a water/acetone solution and was able to form C_8–BTBT/F_4–TCNQ charge transfer complex [706], with the exposed conjugated cores of C_8–BTBT on the wall in the edge-on stacking thin films. The introduction of nanopores and dopant deposition did not alter the macroscopic morphology of the semiconductors. But, owing to the effective doping effect, the average apparent mobility of doped nanoporous devices increased from ~2.8 to ~18.2 $cm^2 V^{-1} s^{-1}$.

5.2 Eliminating Interfacial Traps

Besides the surface roughness and surface energy that influence the device performance by altering the growth of OSCs, another issue that needs to be taken into consideration is the existence of interfacial traps, i.e. electrically active defects, located at the interface between dielectrics and semiconductors. These detects block charge transport by acting as scattering centers and/or by trapping the hole/electron carriers, thus degrading the performance of organic transistors. In fact, the trapping/detrapping process at the organic/dielectric interface or within the semiconductor layer might cause high threshold voltages and also induce hysteresis in OFETs [707–710]. In this section, we mainly focus the discussion on the control of the device performance by changing the interfacial conditions at the semiconductor/dielectric interface.

The investigation of the chemical nature of gate electrics is helpful to understand how the interfacial traps affect charge transport, especially the chemical components present on the surface, thus enabling the elimination of the interfacial traps by applying suitable technologies. The most traditionally used dielectric material in OFETs is silicon oxide. Some other high-*k* inorganic oxides were also mentioned, drawing lessons from inorganic silicon-based electronics. It is believed that the surface chemical groups on the gate dielectric may act as charge-carrier traps. Charge-trapping sites typically are the product of (–OH) groups on the surface of inorganic oxides, which greatly impede charge transport through the channel layer. For instance, the residual valences of the initially hydrophobic siloxane (Si–O–Si) surface, which can react with water to form silanol groups (Si–OH), and the physisorbed O_2/H_2O on the silicon dioxide surface, can both quench the FET activity [711–715]. Todescato et al. [128] reported the systematic research on the role of the semiconductor/dielectric interface in determining the electrical parameters of OFETs by employing thermally grown SiO_2 layers in the pristine state that had been chemically treated with OTS and hexamethyldisilane (HMDS) or covered with polymer dielectric layers such as BCB, PVA, and PMMA. Their research showed that the presence of silanol groups strongly affected the hole transport of MEH–PPV-based transistors. The higher the density of the silanol group, the poorer the electrical performance. Similarly, they obtained the best performance with PMMA as the dielectric, having no OH groups.

5.2.1 Dielectric Surface Passivation (Treatment) Methods

The most widely used strategies to eliminate traps are to use a suitable modification layer to passivate the hydroxyl groups and desorb/repel the water, thus improving the performance of the organic transistors. For example, by depositing a 0.6-nm-thick calcium (Ca) layer onto the surface of the gate dielectric as an interfacial doping layer, Ahles et al. [716] realized n-type pentacene-based FET characteristics with an electron mobility of ~0.19 $cm^2\ V^{-1}\ s^{-1}$. Benson et al. [717, 718] studied how traces of Ca as an interfacial modification layer function to enable a charge trap-free pentacene/semiconductor interface and therefore the n-type operation of pentacene OFETs. The Ca thin layer could become oxidized as a result of an interface reaction between the atomic Ca and the oxygen substrate components such as -OH groups on the surface of the SiO_2. They found that the oxidized Ca layer, containing -CaOH and -CaO, represented an additional dielectric layer as passivation, which could eliminate the surface traps because of the ionic binding nature of -CaOH, such that Ca atoms could donate electrons and thereby neutralize the hydroxyl groups. The thickness of Ca passivation was associated with the electron mobilities. Only thin Ca layers enhanced the mobility (highest mobility occurred at ~12 Å) because the fraction of metallic Ca from thicker layers remained too high, thus impeding the performance.

SAM modification by using long-chain organosilanes is a general and powerful strategy to reduce the number of electron-trapping hydroxyl groups on the SiO_2 surface. However, the silanol traps cannot be completely passivated because SAMs cannot eliminate all surface SiOH groups (the ideal situation of full surface coverage of SAMs cannot be accomplished) and the ultrathin SAM (<2 nm) can cause charge-carrier tunneling under a high voltage [110, 122, 446, 719–721]. Regarding this, several other masterly methods to achieve the same goal have been proposed, such as polymer encapsulation and gap dielectrics.

5.2.1.1 Polymer Encapsulation of Dielectrics

Yoon et al. [712] used a variety of bilayer gate dielectrics (pretreated SiO_2 with SAMs or polymers before depositing semiconductors) to study the dielectric chemical structure–OFET performance correlations (Figure 5.9a). For different dielectrics, the efficiency of interfacial electron trapping, arranged in an ascending order, was PS, PVA, HMDS, bare SiO_2, and PS–O_x, owing to the increasing density of silanol groups. PS–O_x came from the carbonyl groups, which are much more efficient

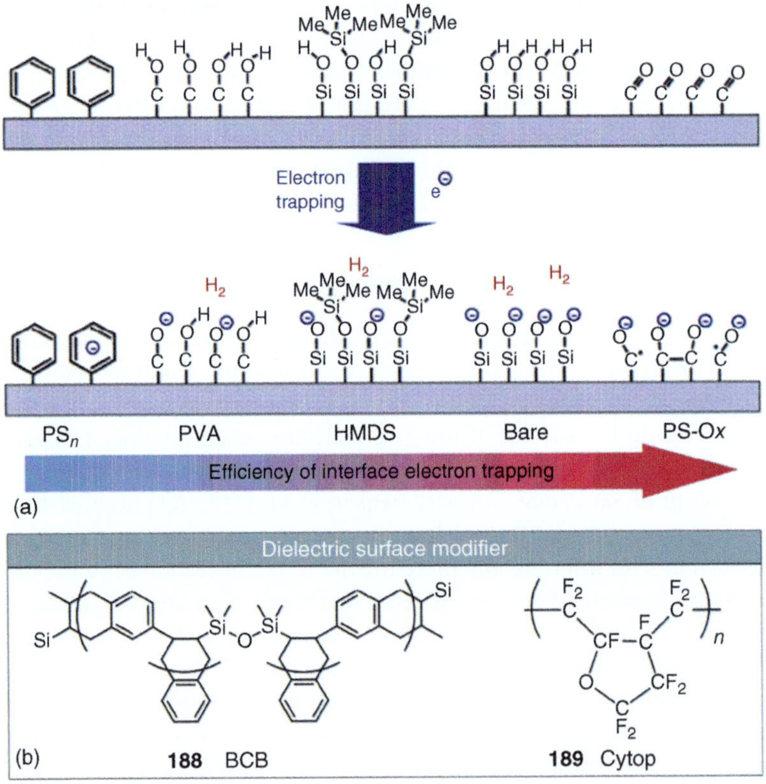

Figure 5.9 Polymer encapsulation of dielectrics dielectrics. (a) Schematic diagram of the electron-trapping efficiency of functional groups on various bilayer dielectrics. (b) Chemical structures of dielectric surface modifier, BCB, and Cytop. Source: (a) Reproduced with permission from Yoon et al. [712], © 2006 American Chemical Society.

electron traps. In the meantime, the OH groups, although they trap charge carriers [722–724], are far less deleterious to charge transport than the silanol groups. PS and PVA encapsulation, with a smooth morphology with minimal nanoscopic pinholes, led to the passivation of surface chemical defects and the stabilization of the charge carriers by aromatic cores. The relatively lower dielectric constant of PS ($k = 2.5$) than PVA ($k = 7.4$) was possibly responsible for the reduced interface trap densities and the enhanced mobility at a dielectric interface with reduced carrier localization/trapping [725, 726].

In another work [727], it was reported that for pentacene OFETs with PVA, PVP, or PS as the dielectric layer, the dissociated protons at the surface of the PVP dielectric (pKa = ~10) formed the electron traps and restricted the formation of an n-type conduction channel, whereas for devices with PVA (having a larger pKa value (~15) or a relatively lower dissociation constant) or with the hydroxyl-free PS as the dielectric, n-type field-effect transport was obtained. When employing Ca as source–drain electrodes and PS as top gate dielectrics, the electron mobility reached up to ~0.077 $cm^2\ V^{-1}\ s^{-1}$.

Chua et al. [722] reported that by using an appropriate hydroxyl-free polymer, BCB (**188**) (Figure 5.9b) [728], as a surface modifier on the thermal SiO_2 dielectric, n-type charge-transport behaviors in many conjugated polymers could be readily achieved, which was rather elusive in the past with SiO_2 or hydroxyl-containing materials acting as gate dielectrics. They clearly observed interfacial electron trapping caused by silanol groups in the spectrum of multiple-reflection attenuated-total-reflection Fourier transform infrared (ATR–FTIR) spectrometry and proposed the electrochemical OH electron-trapping mechanism by a reaction involving the generation of immobile surface Si–O^- ions from Si–OH under n-channel operation, quenching the FET activity. Finally, ambipolar behaviors were obtained in organic transistors with BCB-modified dielectrics by using several different conjugated polymers as active materials. This work clearly demonstrated that the electron trap states, which quench n-type carriers with OSCs, originate from the hydroxyl groups at the SiO_2 interface.

Cytop (**189**) is another widely used dielectric surface modifier, which is chemically inert and highly hydrophobic (Figure 5.9b). By using Cytop as the dielectric layer, the corresponding pentacene transistors did not show long-term gate bias stress effects and had low trap densities [390, 610, 729, 730], which was mainly due to the hydroxyl-free, hydrophobic, and non-reactive properties of Cytop.

Diemer et al. [731] fabricated a dual-gate electric organic transistor by using diF–TES–ADT (**25**) as the channel, with SiO_2 as the bottom dielectric and Cytop as the top gate insulator. They found that when the polymer dielectric was used, the average hole mobility was $1.56 \pm 0.70\ cm^2\ V^{-1}\ s^{-1}$, ten times higher than that when using SiO_2 ($0.17 \pm 0.19\ cm^2\ V^{-1}\ s^{-1}$). This value agreed well with the published values in the literature [693, 732]. The calculated trap densities of state at the diF–TES–ADT/Cytop and the diF–TES–ADT/SiO_2 interfaces showed that the diF–TES ADT/Cytop combination was roughly two orders of magnitude lower than the diF–TES–ADT/SiO_2, whose major source originated from the surface trap states at the SiO_2 interface. In addition, the structural imperfection and molecular misalignment of OSC films grown on different dielectric surfaces can bring additional trapping sites [590]. In fact, the use of a dual gate can also tune the electrical performance of OFETs by adjusting the threshold voltage and functioning as self-contained logic gates, as well as by installing the sensor/memory ability [733–737].

5.2.1.2 Gap Dielectrics

Another strategy for avoiding complicated phenomena and eliminating interfacial effects at the semiconductor/dielectric interface was developed by Menard et al. [738] in 2004 (Figure 5.10a). They introduced an unusual device design by applying a thin free-space gap as a gas (air, nitrogen, and other gases) or vacuum gate dielectric to replace the standard solid dielectric layer. Although the low dielectric strength of the unusual air gate dielectric was approximately $3\,V\,\mu m^{-1}$, the transistors were able to work under higher electric-field operation. It was observed that, under atmospheric pressure, the breakdown voltages were greater than 100 V with a narrow air gap of 1.8 μm. By using p- or n-type organic single crystals as channels, such as rubrene or TCNQ, laminated to a PDMS stamp embossed with source, drain, and gate electrodes in topographic relief, the resulting transistors exhibited good mobilities and normalized sub-threshold slopes thanks to the high quality of the OSC/air-dielectric interface (Figure 5.10b). In fact, the air gap could provide a pristine interface to reduce the interface defects and minimize the Fröhlich polaron effect [739, 740]. Tang et al. [741, 742] developed the use of an OSC ribbon to realize air-stable transistors with air-gap dielectrics (Figure 5.10c). With p-type CuPc and n-type $F_{16}CuPc$ ribbons as channels, a comparable ambipolar transport was obtained, showing electron and hole mobilities of ~0.17 and $\sim 0.1\,cm^2\,V^{-1}\,s^{-1}$, respectively (Figure 5.10d). By applying short-channel air gaps into organic single-crystal transistors, high-speed operation was also enabled [743]. Also, although in the absence of solid material gate dielectric, Chen and Podzorov observed the existence of the bias stress effect in air-gap organic transistors [744]. This is due to a different mechanism from traditional transistors, caused by a small but finite density of localized in-gap tail states in pristine devices.

Rather than using a free-space gap with air or vacuum as gate dielectrics, Xia et al. [745] filled the gap between the OSC and gate electrodes with liquids with varying permittivities (a pure ionic liquid [EMIM][TFSI] (**190**) and a Dow Corning #704 silicone oil) to measure the mobility of OSC crystals or films (Figure 5.10e). Ionic liquids are strongly polarizable and can induce the accumulation of opposite charges under the gate bias, forming an electrical double layer [746–749]. It was found that for single-crystal OFETs based on rubrene and pentacene, considerably higher mobilities were obtained when gated by vacuum (Figure 5.10f, inset). These mobilities decreased when using oil and an ionic liquid, whereas for polymer transistors using PQT-12 and P3HT films, maximum mobilities were extracted when using an ionic-liquid gate dielectric with the largest gate capacitance (Figure 5.10f). The opposite trend originated from the different dependence between the effective mobilities and the carrier concentration of semiconductors due to disorder effects, as well as from the gate liquid dielectric-associated electrochemical doping into the

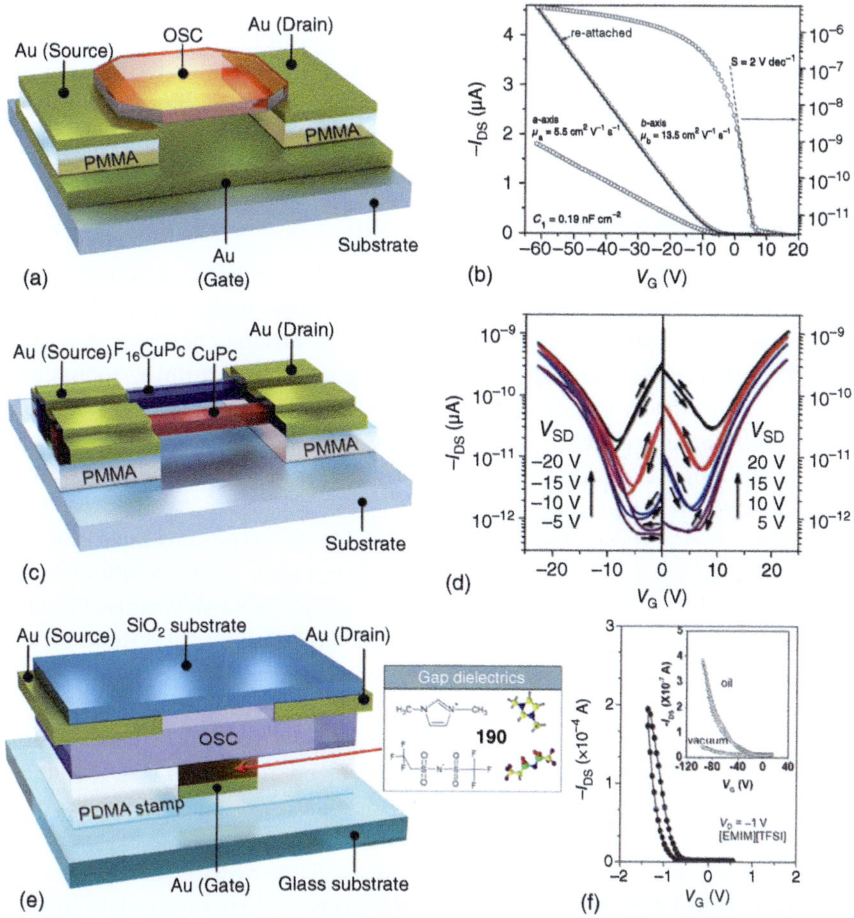

Figure 5.10 OFETS with air-gap dielectrics. (a) Device structure and (b) electrical performance of rubrene single-crystal transistors with free-space gate dielectrics. (c) Device structure and (d) ambipolar transport behavior of a bipolar two-channel single-crystalline device with air-gap dielectrics. (e) Cross section of the device structure of the polymer FETs. The gap could be vacuum or filled with oil/ionic liquid. Inset showing the chemical structure of the ionic liquid [EMIM][TFSI] (**190**). (f) Transfer curve showing the performance of the devices with vacuum, oil, or ionic liquid filled gap dielectrics. Source: (a, b) Reproduced with permission from Menard et al. [738], © 2004 John Wiley & Sons, Inc. (c, d) Adapted with permission from Tang et al. [741], © 2005 John Wiley & Sons, Inc. (e, f) Adapted from Xia et al. [745].

channels. In fact, the liquid-gate dielectric structure provides a more convenient approach to build chemical sensors and biosensors, especially implantable devices [750–752].

As discussed briefly earlier, using a polymer layer as a dielectric encapsulation layer or gap dielectrics can eliminate interfacial traps through full coverage of the inorganic oxide surfaces, and the basic considerations have been explored. However, some shortcomings of these methods hinder their wide application. For polymer encapsulation, the bilayer dielectric will cause complications in capacitance calculation, thus resulting in an overestimation of charge-carrier mobility. For gap dielectrics, the operation voltage is too high to apply in real applications. As a more general method, SAM modification of dielectrics has solved the problems to some extent, which we will discuss in detail in the next Section 5.2.2–5.2.4.

5.2.2 SAM/SiO_2 Dielectrics

Back to the topic of eliminating traps at the semiconductor/dielectric interface, the SAM modification strategy is worth highlighting because it has attracted considerable research interest. SAM molecules, possessing various chemical properties coming from their simple and clear structure, are capable of forming robust and stable linkages to the inorganic oxide surface via covalent bonding. By focusing on the trap-eliminating effect in organic transistors based on using SAM-modified dielectrics, we summarize four aspects to discuss the topic in detail: SAM/SiO_2, SAM/high-*k* dielectric, and SAM–OFETs.

The modification of the surfaces of inorganic oxide dielectric substrates by using SAMs or polymers prior to the deposition of OSCs has been found to influence the film morphology [635, 753]. In the meantime, the function of SAMs was also found to be powerful. In terms of the basic science, they are intriguing model systems for the study of self-organization in 2D matter under the complex interplay of vdW, covalent, and electrostatic interactions [122, 754]. In terms of covalent modification of the oxide surface, different types of molecules have been used, such as silanes, phosphonates, carboxylates, catechols, alkenes/alkynes, and amines [121]. To focus on the SAM modification in organic electronics to eliminate traps on the dielectric surface, two categories of the most commonly used SAM molecules are discussed: (i) silane-based SAM molecules, which prefer to react with the hydroxyl group of –SiOH on the silicon-based dielectric (SiO_2, and Si_3N_4) to form SAMs via covalent linkage of Si—O—Si bonds [122] and (ii) the phosphonic acid ($R–PO_3H_2$) and its phosphonate ester derivative (RPO_3R_2)-based SAM molecules, which can form covalently bound monolayers on a large range of inorganic oxide dielectrics, including high-*k* materials, by multidentate binding of strong covalent P–O–M anchoring (M stands for metal atom) [111, 151, 755, 756]. Under normal ambient conditions, phosphonic acid molecules do not react chemically on the surface of SiO_2 [123, 757].

Surface modification with organosilanes ($RSiX_3$, R_2SiX_2, or R_3Si_X, where R represents an alkyl group and X represents a leaving group, such as alkoxy, hydride, or chloride) (**191–213**) (Figure 2.4), which have been used in diverse areas, have been common methods for a long time to prepare monolayers on a wide range of surfaces bearing –OH groups, such as silica, aluminum oxide, and zinc oxides [124]. Surface modification with silanes can be done by reactions in solution or in the vapor phase. In general, closely packed monolayers are usually formed through the vapor phase approach [758–760]. The challenge is that uniform silane monolayers are still difficult to obtain [121, 761], thus limiting the device performance to some extent.

In addition, the processing condition had a great impact on the quality of the resulting SAM. Mityashin et al. [140] found that the quality of the SAM coverage was very sensitive to the pretreatment conditions owing to the imperfect SAM coverage. In general, vapor-deposited SAMs were likely to result in high but not full coverage, different from the solution-processing procedure requiring more time [171]. Walter et al. [125] showed that an OTS from solution deposition at 55–60% relative humidity would afford OTS monolayers with contact angles in a narrow range, promoting optimum semiconductor growth and enhancing the FET performance. Theoretically speaking, full surface passivation and zero-onset voltage are only possible at near 100% SAM coverage, and if the coverage is incomplete or uncontrolled, high variability in the onset voltage occurs. The authors pointed out two impacts caused by a partially covered gate-dielectric SAM on the transport levels of an OSC. The first significant effect at the surface with the partial coverage was the surface-induced electronic disorder in the adjacent semiconductor, which broadened the distribution of charge-transport states. Consequently, the number of charge carriers above the Fermi level increased and might have led to a reduction or increase in the carrier mobility, depending on the nature of the semiconductor and whether the formation of deep traps or increased charge density prevails [47, 762, 763]. The second effect was the rigid shift of the transport levels of the semiconductor owing to electrostatic interactions between the SAM and the semiconductor, inducing either hole or electron accumulation accordingly.

To get a comprehensive understanding of the relationship between SAM materials, preparation methods, and the FET performances, we have listed the recent works of SAM-modified dielectric surfaces in the following, which is shown in Table 2.1.

5.2.2.1 Provide Efficient Insulating Barrier Height

Since the early report of applying silane-based SAMs into organic transistors to enhance the FET properties by passivating traps, many works have been carried out to achieve high-performance OFETs [722, 764]. Although the SAM modification can change the surface properties, which alters the morphology of OSCs directly grown on substrates and leads to changes in the electrical characteristics, some reports have clearly shown that without a morphology change in the channel films, the SAM acts solely as a passivation layer to eliminate the surface traps and thus realize better performance. Yagi et al. [765] developed a surface treatment method for SiO_2/Si substrates by spin-coating a HMDS (**166**) liquid onto the substrate surface to control the electrical properties of pentacene OFETs. By applying the surface termination of the SiO_2 method, no obvious morphological changes occurred in comparison with the nontreated substrate, and a drastically improved off-current was obtained. The off-current suppression could result from the reduction in the density of charges trapped at the pentacene/SiO_2 interface. The reduction of off-current was also observed by Kumaki et al. [766] who used β–PTS (**167**) to treat the SiO_2 surface. In another work of using β–PTS as SAMs, Miyadera et al. [131] studied the current instability of pentacene OFETs. They found that the increased barrier height in SAM-modified transistors suppressed charge trapping, and the energy distribution depended on the treatment of the dielectric surface, whereas the morphology of pentacene films showed little effect on the trapping barrier.

On the basis of the fact that a densely packed SAM of *n*-alkyltrichlorosilanes with a thickness in the range of 1.9–2.6 nm could show an efficient insulating barrier with a high tunneling barrier of ~4.5 eV [767] Collet et al. [768] fabricated nanometer-scale organic transistors on silicon substrate by using a COOH-terminated organic silane-based SAM as the gate insulator (2 nm thick) and α-6 T as the channel layer, with the shortest length of 30 nm, which could be operated under a very low voltage (<2 V) with a low gate current density of 10^{-6} A cm^{-2}.

In 2000s, similar strategies were reported by some other groups by using SAMs as gate dielectrics. However, it should be mentioned that silicon is easy to be oxide under ambient conditions, forming a very thin SiO_x layer atop the substrate. So it is not proper to claim SAMs work solely as gate dielectrics in these cases. It is actually SiO_2/SAM hybrid dielectric. For example, Halik et al. [769] reported a low-voltage OFET with an amorphous molecular gate dielectric by using a 2.5-nm-thick specially synthesized SAM, PhO–OTS (**203**). The devices were fabricated by first depositing the SAM onto an oxygen plasma-treated silicon wafer. Having an aromatic terminal group, PhO–OTS SAMs were capable of forming top linkages by π–π interaction with the adjacent SAM molecules self-assembled on the SiO_2 surface, resulting in a more closely packed surface than that with linear end groups. With the large-area pinhole-free and molecularly flat SAM as an ultrathin gate dielectric layer and 5-μm-long pentacene as the active channel, a hole mobility of ~0.05 cm^2 V^{-1} s^{-1} was obtained, with a very small gate current density of 10^{-9} A cm^{-2}, and the transistors could be operated with a voltage lower than 2 V.

5.2.2.2 Control Surface Polarity and Carrier Density

The polarity of the dielectric interface has a significant impact on the semiconductor mobility, as extensively analyzed by Veres et al. [725] on an amorphous polymer-based OFET (Figure 5.11a). They investigated a number of gate dielectrics with the varied polarities by using polytriarylamines (PTAA) and poly(9,9-dioctylfluore-cobithiophene) as semiconductors. They found that the device performance showed a significant increase when using insulators with permittivities below 2.5. This was ascribed to an increase in carrier localization polarized by high-*k* dielectrics. As illustrated in Figure 5.11b, the DOS broadening becomes more severe as the semiconductor/dielectric interface becomes more polar, leading to more tail states. The random dipoles are not likely to significantly influence the extended states of a wide band. To qualitatively explain this, the Sirringhaus's group developed a theoretical approach [770]. By using an analytical model that took the distance from the interface into account, they calculated the broadening of the DOS caused by dipolar disorder and found that the interaction of the charge with the dielectric environment was dominated by the nearest dipoles in the dielectric layer. Good quantitative agreement with the experimental data was obtained when the authors modeled the dependence of the field-effect mobility of PTAA on the dipolar disorder broadening by using a Gaussian disorder model [771], even without using a fitting parameter in the model (Figure 5.11c). The results completely supported a previous work from Veres et al. [725] which demonstrated

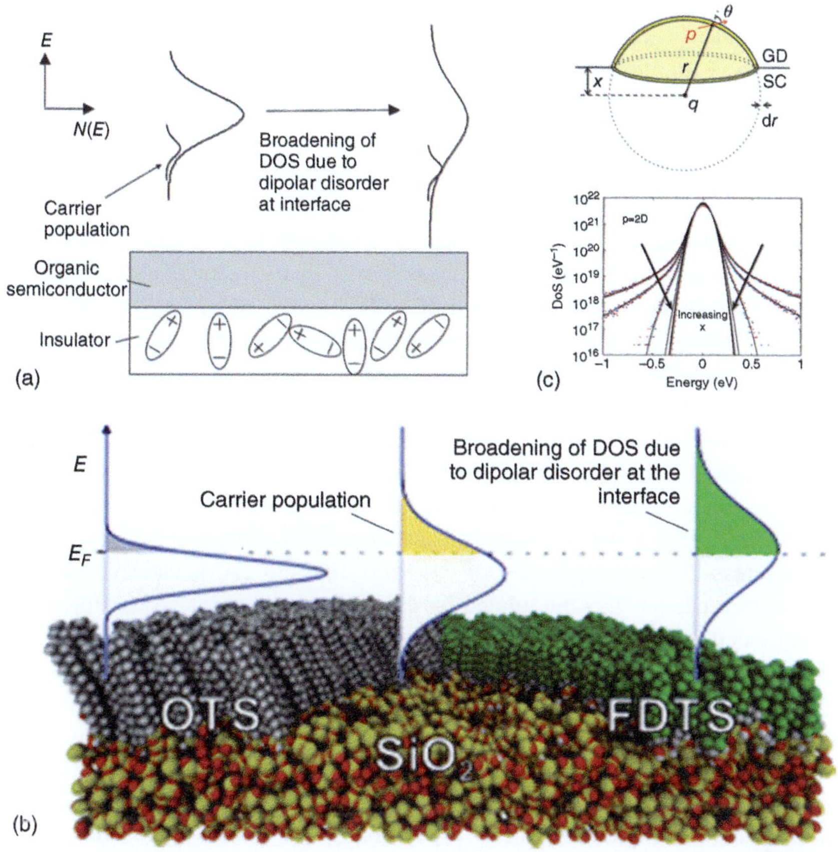

Figure 5.11 Effects of the dielectric interface on the device performance. (a) Suggested mechanism for the enhancement of carrier localization due to polar insulator interfaces. (b) A microscopic model taking into account the disorder and localized states of OSCs caused by SAMs. (c) (top) Schematic illustration of the interaction of charge q in the OSC with dipole moment p in the gate dielectric. The charge in the semiconductor is located at a distance x from the interface. (bottom) Calculated DOS broadening due to the static dipolar disorder in the dielectric with the increasing distance into the semiconductor. Each line corresponds to a step. Source: (a) Reproduced with permission from Veres et al. [725], © 2003 John Wiley & Sons, Inc. (b) Reproduced from Mityashin et al. [140], © 2014 American Chemical Society. (c) Reproduced with permission from Richards et al. [770], © 2008 AIP Publishing.

that the interfacial effect is of critical importance to the device performance as the DOS could be changed by local polarization.

Essentially, the random dipoles present at the interface could modulate the energy of the localized states, leading to an increased energetic disorder. These states would be totally isoenergetic in the absence of disorder. The broadening is attributed to a number of intermolecular interactions, among which fluctuations due to the randomly oriented dipoles are present locally. This may also explain the enhancement of carrier mobility caused by SAM modification, because such a thin "insertion" layer could play a role in reducing the interaction of the charges with the dipoles within the dielectric. As a consequence, if the interface is more polar (fluoroalkylated silane SAM-modification), the DOS broadening becomes more severe [140]. Therefore, carriers in equilibrium will face a higher potential barrier for upward hopping into the denser sites that lie closer, leaving many sites as trap states in the distribution. Actually, as long as localized states are present, localization is likely to be enhanced by dipolar disorder. That is to say, the increase in energetic disorder is greater when the insulator is more polar. This suggests that in most OFETs, the improvement in device performance can be achieved by using low-*k* gate insulators or high-*k* hybrid dielectrics modified by low-*k* alkylated SAMs. Also, an additional benefit of low-*k* dielectrics is reduced hysteresis between forward and reverse gate sweeping and, in some cases, improved threshold voltages [606, 772].

In a further step, Kobayashi et al. [773] showed that the surface polarity and carrier density in OFETs could be tuned by forming different SAMs on SiO_2 gate insulators with three kinds of organosilanes – F–SAM (**211**), CH_3–SAM (**210**), and NH_2–SAM (**206**) – as a result of the molecular dipole effects and weak charge transfer. The difference of drain currents at the zero-gate bias and threshold voltage in both p-channel pentacene and n-channel C_{60}-based OFETs with different SAMs clearly indicated that the surface carrier was modulated by changing the type of SAM. Holes and electrons tended to be accumulated by F– and NH_2–SAMs, respectively, whereas CH_3–SAMs showed no obvious carrier accumulation. Another possible mechanism used for a better explanation was the occurrence of direct charge transfer, that is, electrons were transferred from electron-donating NH_2–SAMs to the channel.

However, the above two studies did not consider the surface energy difference of the insulators. Considering the important role that surface energy plays in the growth of a semiconductor, Jang et al. [774] studied the effects of various permanent dipoles while keeping the surface energy of the insulators constant. They found that the hole mobility increased by a factor of approximately 20 for pentacene transistors based on a dielectric with an electron-withdrawing group and a more positive turn-on voltage ($0.2\,cm^2\,V^{-1}\,s^{-1}$ versus $0.01\,cm^2\,V^{-1}\,s^{-1}$, 1 V versus −6 V), which meant that electrons from the pentacene layer were attracted by SAMs, leaving behind more mobile holes in the channel. As a result, a more positive gate bias was needed to switch off the device; that is, the carrier density within the conducting channel could be well tuned because of the band bending at the semiconductor/dielectric interface.

By using SKPM as a promising tool to characterize the local charge trapping in weakly conducting semiconductor films with strong charge-carrier localization, Zhang et al. [775] investigated the local charge-trapping states at the semiconductor/dielectric interface in an OFET with oligothiophene as the channel and bare or SAM-modified SiO_2 as the gate dielectric. A much lower charge-trapping hysteresis was yielded in the APTES (**168**)-passivated FETs, illustrating that the trap states induced by hydroxyl groups were successfully eliminated by pre-adsorption of a layer of APTES.

In addition, dielectric/semiconductor interface interactions can impact not only the charge-carrier mobility but even the sign of major carriers flowing through the adjacent channel. Nakano et al. [143] recently demonstrated the first example that by using a SAM-modified dielectric, the major carriers in an ambipolar polymer semiconductor can be controlled. With fluoroalkyl–SAM FDTS (**204**) enhancing the hole accumulation in p-channel transistors and amine-based MAPS–SAM (**208**) inducing n-channel unipolarization, the hole mobility was ~0.43 $cm^2\ V^{-1}\ s^{-1}$ (average, ~0.31 $cm^2\ V^{-1}\ s^{-1}$) and the electron mobility was ~0.24 $cm^2\ V^{-1}\ s^{-1}$ in corresponding OFETs, both with a high current on/off ratio of 10^5. Moreover, almost no sign of the influence of the opposite charge carrier was detected with the SAM. The complementary-like inverters fabricated on Si/SiO_2 substrates with two distinct surface regions modified with FDTS– and ODTS–SAMs exhibited clear switching characteristics under ambient conditions and a voltage gain greater than 150.

5.2.3 SAM/High-*k* Dielectrics

High-*k* materials include different kinds of materials, such as the most abundant inorganic oxides [145], and polymers [146, 147, 776, 777], as well as hybrid dielectrics [160, 621, 778, 779]. Herein, we focus on the widely used inorganic oxide high-*k* materials to achieve low-voltage OFETs. Usually, they are prepared by solution processing and vapor deposition, such as the sol–gel and ALD methods, to be integrated into organic transistors. The sol–gel method is generally used to fabricate BG configurations, and the ALD method allows the preparation of both BG and TG configurations [780]. Owing to some distinct characteristics of high-*k* metal oxides, such as the unavoidable charge-carrier trap sites on the surface (usually associated with −OH groups), the induced ionic polarization owing to the interaction between charge carriers and the high-*k* lattice, as well as the surface roughness, they can adversely affect the device performance. In order to mitigate the problems, surface modification is needed for controlling the surface properties of metal oxide dielectrics [781–783], thus tuning the electrical properties of the OSC and subsequent device performance. Besides, because the SAMs are molecular films that spontaneously form by the adsorption of an organic surfactant on a solid substrate just a few nanometers in thickness, their high capacitances will not be destroyed.

By taking into account the fact that silane-based SAMs show instability in air because of their violent reaction with water [784], Hanson et al. [123, 149] demonstrated a class of SAMs by utilizing organophosphonates as the constituent molecules, which were easily prepared in air and able to form well-ordered, strongly covalently bonded films on oxide surfaces. Phosphonic acid (PA) molecules, showing better stability to moisture and less tendency to homocondensation between phosphonic acids, can be chemisorbed readily onto metal oxides to form dense and robust SAMs without being limited by the content of the surface hydroxyl groups. Once the monolayer formed, the SAM showed more resistance to hydrolysis of phosphonate than siloxane [150, 151]. Therefore, PA–SAMs (**214–237**) have been widely used as functional materials for organic electronics as gate dielectric modifiers (Figure 2.5). Also, PA–SAMs can also be used to modify transparent electrodes based on metal oxides, such as ITO, with almost no impact on the transparency [152, 153].

To get a comprehensive understanding of the relationship between SAM materials, preparation methods, and the FET performances, we have listed the recent works of SAM-modified dielectric surfaces in the following, which is shown in Table 2.2.

5.2.3.1 Fundamentals of SAM-Modified High-*k* Dielectrics

The Jen's group studied the applications of PA SAMs, featured with the contribution of anthryl-terminated alkyl PA (π–σ–A) SAMs, in organic electronics (Figure 5.12a) [150, 156, 161]. In 2008, for the first time, they applied sol–gel processing of HfO_2 on silicon wafers as the gate insulator in fabricating OFETs and

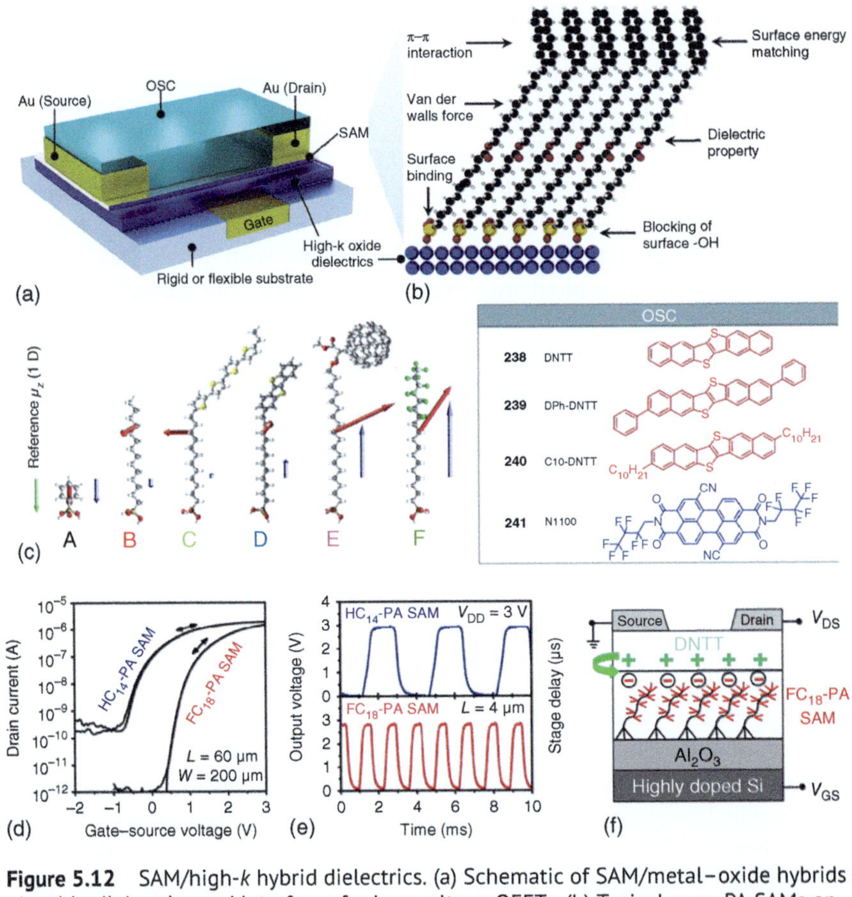

Figure 5.12 SAM/high-*k* hybrid dielectrics. (a) Schematic of SAM/metal–oxide hybrids as ultrathin dielectrics and interfaces for low-voltage OFETs. (b) Typical π–σ–PA SAMs on metal oxides with optimized dielectric and interfacial properties. (c) Chemical structures of the SAMs and OSCs. (d) Transfer curves of n-channel TFTs and (e) output signals of ring oscillators fabricated with AlO_x dielectric modified with alkyl-phosphonic acid (HC_{14}–PA SAM) or fluoroalkyl-phosphonic acid (FC_{18}–PA). (f) Schematic illustration showing the accumulation of positive charges in the OSC layer and immobile negative charges in the SAM at the interface. Source: (a, b) Adapted from Ma et al. [150]. (c) Adapted from Salinas et al. [164]. (f) Reproduced with permission from Aghamohammadi et al. [168], © 2015 American Chemical Society. (d, e) Reproduced with permission from Kraft et al. [169], © 2015 John Wiley & Sons, Inc.

used π–σ–PA-based SAMs to modify the dielectric surface. The π–σ–PA SAM was designed to combine the advantages of PA anchoring groups and anthryl terminals (Figure 5.12b). As mentioned earlier, PA SAMs showed the better stability and less condensation tendency. The reaction between PA and the oxide substrate was not limited by the content of surface hydroxyl groups. In the meantime, the anthryl surface groups, capable of exhibiting π–π interaction (0–50 kJ mol^{-1}) between neighboring molecules, together with the vdW interactions from the aliphatic chains (<5 kJ mol^{-1}), promoted the formation of high-quality chemically and electrically compatible SAMs (Figure 5.12b). Thanks to the high capacitances (~0.69 μF cm^{2}) and low leakage current densities (~6×10^{-9} A cm^{-2}) of the hybrid dielectrics of π–σ–PA–SAM/HfO_2, the corresponding pentacene-based transistors showed a low operation voltage under 1.5 V with very low subthreshold slopes, a high hole mobility of ~0.22 cm^{2} V^{-1} s^{-1}, and high on/off ratio of more than 10^5. The low subthreshold slope and V_{th} were the results of a reduction of charge traps by eliminating surface hydroxyl groups with a SAM modification layer, with a simultaneously improved pentacene/dielectric interface by using the anthryl surface group, in comparison with the methyl surface group of ODPA (**216**).

Some other factors are also crucial to the performance of SAMs, such as alkyl chain length and molecular dipoles. Fukuda et al. [785] investigated the effects of the alkyl chain length in alkyl–PA SAMs on the performance and stability of low-voltage pentacene-based OFETs. The maximum mobility was obtained with a SAM chain length of 14 carbon atoms, with a value of ~0.7 cm^{2} V^{-1} s^{-1} and a high on/off current ratio greater than 10^5. The PA SAM with a length of less than approximately 10 carbon atoms was speculated to form mostly disordered, liquid-like monolayers, possibly due to the lack of cohesive interaction between the molecules [786, 787]. This can explain the large gate currents with C_6–SAM (**214**). C_{16}–SAM (**218**) and C_{18}–SAM (**219**), having larger surface roughness values than C_{14}–SAM (**217**) and larger densities of gauche defects near the top surface of the monolayer [788] Acton et al. [160] drew the conclusion that in pentacene-based transistors with alkyl–PA SAM/HfO_2 hybrid dielectrics, the high field-effect mobility was compromised between disordered SAMs with varying carbon numbers to promote large grains and a thick enough monolayer as a physical buffer to mitigate the adverse effects of the underlying high-*k* surface. The results provided new insights for choosing adequate SAMs as interfacial modifications in OFETs.

Another important issue is molecular dipole because the threshold voltage of an OFET can be tuned by the dipole of PA–SAMs. To investigate the relationship between them, Salinas et al. [164] fabricated transistors with three different OSCs [pentacene, α,α′-dihexylsexithiophene (DH6T), and C_{60}] as channel layers with six different self-assembled monolayer-modified AlO_x as hybrid gate dielectrics to extract the electrical properties of the corresponding devices (Figure 5.12c). They found that the shift of V_{th} correlated linearly with the dipole moment along the molecular axis. The correlation can help optimize the performance of the transistors by tuning the dipole moments of the SAM molecules.

As mentioned in SAM/SiO_2 hybrid dielectrics, surface polarity, carrier density, and even the sign of major carriers can be tuned by SAMs. This principle works for SAM/high-*k* dielectrics as well. For example, alkyl or fluoroalkyl SAMs have different influences on the hole and electron transport in active semiconductor layers. Usually, p-type OSC performs better on alkyl SAMs, while n-type OSC prefers fluoroalkyl SAMs. To obtain the best performance for all OFETs, selective deposition of different SAMs on the same substrate is required, which is difficult in terms of the achievable resolution and patterning accuracy. Therefore, Kraft et al. [169] evaluated how the choice of a single gate-dielectric SAM (alkyl C_{14}–SAM, **206** or fluoroalkyl FC_{18}–SAM, **222**) affected the characteristics of DNTT (**238**), DPh–DNTT (**239**), C10–DNTT (**240**), and N1100 (**241**) OFETs (Figure 5.12c). It was observed that the difference between the threshold voltages of OFETs with the fluoroalkyl SAM and those with the alkyl SAM occurred regardless of the choice of semiconductor (although there was a weak dependence). The mobility decreased significantly with the reduction of the channel length below approximately 20 μm because of the increased relative influence of the contact resistance on the total device resistance. The n-type N1100 transistors showed a width-normalized contact resistance of 23 kΩ cm with an alkyl SAM and 15 kΩ cm when using the fluoroalkyl SAM, of which the latter showed the smallest contact resistance for n-channel transistors on flexible substrates. Hence, DPh–DNTT, which had the smallest contact resistance of the three p-type semiconductors (0.38 kΩ cm on the fluoroalkyl SAM), was chosen for fabricating p-channel transistors. On a ring oscillator based on DPh–DNTT p-channel and N1100 n-channel transistors having a channel length of 1 μm with an FC_{18}–PA SAM/AlO_x gate dielectric, the shortest signal propagation delay per stage, 3.1 μs, was measured under a supply voltage of 5 V (Figure 5.12e). Another effect when using SAMs was pointed out by Aghamohammadi et al. [168] who reported that in transistors with an alkyl SAM, the threshold voltage is determined mainly by the dipolar character of the SAM rather than by the gate dielectric capacitance. In contrast, with a fluoroalkyl SAM, the threshold voltages showed linear dependence on the inversion of the gate-dielectric capacitance (Figure 5.12f).

5.2.3.2 SAM/High-*k* Hybrid Dielectrics for Flexible Substrate

The ultrathin and low-voltage operation properties of PA–SAM/high-*k* metal–oxide hybrid dielectrics make them applicable for flexible organic transistors. Klauk et al. [779] reported an ultralow-power OFET built with *n*-octadecylphosphonic acid SAMs modified oxygen plasma pretreated Al gate electrodes. The total thickness of the hybrid dielectric layers was ~4.7 nm (~2.1 nm for the SAMs and 3.6 nm for AlO_x), providing a large gate capacitance of ~0.7 μF cm^{-2}. Furthermore, thanks to SAM dielectrics, the authors successfully realized the first low-voltage organic complementary circuits (inverters, NAND gates, and ring oscillators) by using pentacene and $F_{16}CuPc$ acting as p-channel and n-channel components, respectively, working under supply voltages between 1.5 and 3 V (Figure 5.13a).

In flexible electronics, the substrates are generally polymers, which cannot withstand high annealing temperatures (>200 °C). It is of great importance to develop low-temperature fabrication method making it compatible with low-cost solution processing on polymer substrates. Regarding this, Park et al. [791] demonstrated a low-voltage OFET by using entirely solution-based fabrication at room

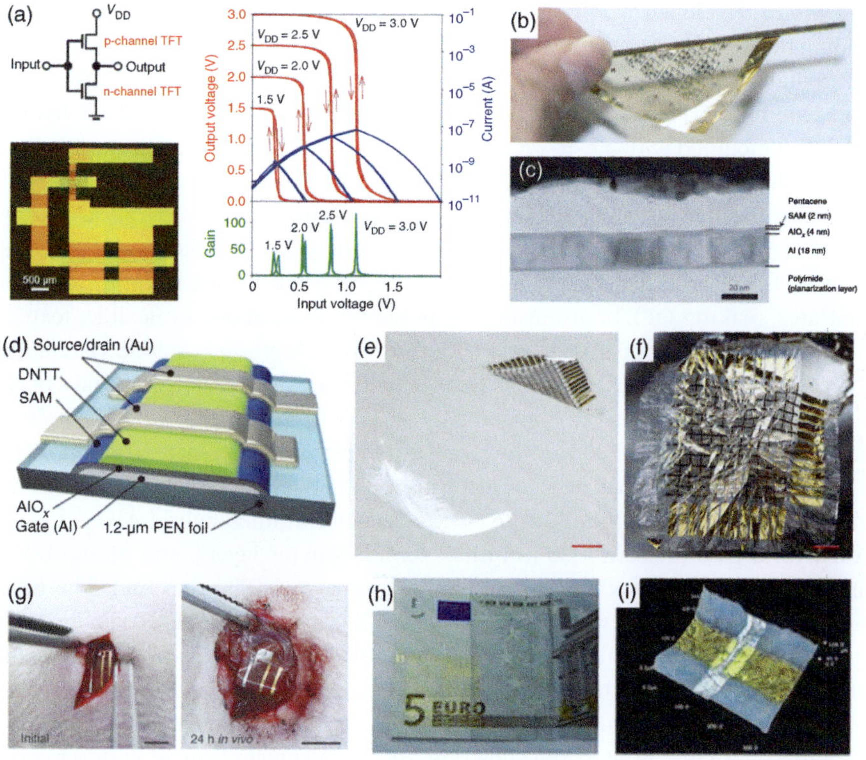

Figure 5.13 SAM/high-*k* hybrid dielectrics for flexible electronics. (a) Complementary inverter with SAM/high-*k* hybrid dielectrics. (b) Photograph of ultra-flexible organic circuits atop ultrathin polyimide substrate. (c) Cross-sectional electron microscopy image of a flexible TFT, which clearly shows the layered structure of the device. (d) Cross-sectional diagram of an ultrathin OFET with Al as gate contact and a mechanically durable and dense hybrid dielectric formed from potentiostatically anodized Al and a phosphonic acid SAM. The complete device including the PEN substrate is 1.3 µm thick. (e) Ultrathin plastic electronic foils, which are extremely lightweight; they float to the ground more slowly than a feather and are therefore virtually unbreakable. The scale bar is 2 cm. (f) 2-µm-thick devices, which are ultra-flexible and can be crumpled like a sheet of paper. The scale bar is 1 cm. Source: Reproduced with permission from Kaltenbrunner et al. [166], © 2013 Springer Nature. (g) An OFET coated with SMP, which conforms to the body tissue of a living rat after being implanted for 24 hours. (h) Photograph of a 5-Euro note with arrays of organic transistors and circuits. (i) Photograph of an individual transistor on a banknote. Source: (a) Reproduced with permission from Klauk et al. [779], © 2007 Springer Nature. (b, c) Reproduced with permission from Sekitani et al. [789], © 2010 Springer Nature. (d–f) Reproduced with permission from Kaltenbrunner et al. [166], © 2013 Springer Nature. (g) Reproduced with permission from Reeder et al. [790], © 2014 John Wiley & Sons. (h, i) Reproduced with permission from Zschieschang et al. [163], © 2011 John Wiley & Sons.

temperature. They deposited an ultrathin layer of ODPA–SAMs (**219**) (3.1 nm) onto a sol–gel ZrO_x layer (~10 nm) to obtain a hybrid high-*k* dielectric. By using poly(2,5-bis(3-tetradecylthiophen-2yl(thieno[3,2-*b*]thiophene) (PBTTT–C_{14}) as an active channel layer, the as-prepared transistor could operate in a 3-V range and presented a mobility of ~0.18 $cm^2\ V^{-1}\ s^{-1}$ with the high-*k* hybrid material. The

Jen's group developed a low-temperature ultraviolet-assisted sol–gel process for yielding high-quality amorphous HfO_x to expand the universality of PA–SAM/HfO_x hybrid dielectrics that were compatible with flexible substrates [158, 159]. By using pentacene and TIPS–pentacene as active channels, the OFETs based on hybrid dielectrics modified with a π–σ–PA SAM could be operated under 2 V and presented average mobilities of ~0.32 and ~0.38 $cm^2 V^{-1} s^{-1}$, respectively.

Because the development of flexible electronic circuits is emerging as a new research area [792–794], one urgent need is the reduction of gate dielectric thickness down to 10 nm and even below, as well as incorporation with ultrathin flexible substrates, making OFETs promising for potential applications in flexible, textile, and stretchable electronics, an area where the Someya's group has done systematic work. In 2010, they fabricated OFETs and organic complementary circuits on a polyimide substrate that had a thickness of only 12.5 μm (Figure 5.13b) [789]. To reduce the surface roughness, a 500-nm-thick polyimide planarization layer was deposited by spin coating and cured at a temperature of 180 °C. On the almost atomically smooth surface, 20-nm-thick Al gate electrodes, a 6-nm-thick hybrid AlO_x/organic SAM hybrid gate dielectric, 30-nm-thick semiconductor layers, and 50-nm-thick Au source/drain contacts were sequentially constructed (Figure 5.13c). The total thickness of the OFETs was only ~106 nm. The devices operated reliably with excellent performance while being folded into a bending radius as small as 100 μm. Recently, the authors further fabricated an active-matrix array with resistive tactile sensors directly on ultrathin polymer foils [166]. A polyethylene naphthalate (PEN) foil (1 μm thick) was used as the substrate; the circuits were formed with a hybrid anodic aluminum oxide [795, 796] a phosphonic acid SAM [779] gate dielectric, and DNTT was employed as an air-stable OSC (Figure 5.13d). Organic transistors with an ultrathin (19 nm) and ultradense hybrid gate dielectric formed at room temperature produced sophisticated large-area electronic foils with unprecedented mechanical and environmental stability. They could withstand repeated bending to radii of 5 μm and less and could be crumpled like paper (Figure 5.13e,f). This strategy to manufacture ultra-flexible devices could be essential to realize low-cost, wearable electronic applications. By exploiting the features of shape-memory polymer substrates, the Someya's group further demonstrated shape changing and organic-electronic softening, which softly conformed to 3D shapes after exposure to a stimulus [790, 797]. Low-voltage operation of ~3 V was achieved by utilizing an ultrathin hybrid dielectric SAM/AlO_x layer. When implanted in physiological tissue, these initially planar devices could be softened to conform to the biological environment, and they withstood 24 hours of fluid exposure (Figure 5.13g). The deformations occurred in association with implantation in living tissue while maintaining the electrical properties without obvious degradation.

Besides polyimide and PEN, some other substrates, including foils [798–800], fibers [801, 802], and papers [803, 804], have been readily used to fabricate workable OFETs. The ability to build organic circuits on banknotes may create the possibility of outfitting banknotes with active electronic security features [163, 805, 806]. This requires the development of techniques that could allow device fabrication without damaging the banknotes. Hence, the operating voltage and power consumption of the circuits should be sufficiently low to allow power to be supplied by radio-frequency coupling with low magnetic fields [807]. Zschieschang et al. [163] utilized SAM/AlO_x gate dielectrics to realize low-voltage functional OFETs and inverters on banknotes (Figure 5.13h), with good electrical performance and a switching frequency of up to 2 kHz. Because of the small thickness (5.7 nm) and large capacitance (700 nF cm^{-2}) of the SAM/AlO_x gate dielectric, the OFETs could be operated at low voltages of approximately 3 V. Despite the small dielectric thickness, the gate leakage current at the maximum gate bias was below 10^{-9} A, confirming that the plasma-grown AlO_x and the solution-processed alkylphosphonic acid SAM formed a continuous, conformal coating dielectric layer with useful insulating properties on Al gate electrodes (Figure 5.13i).

Despite huge progress in SAM/high-*k* hybrid dielectric-based flexible electronics, great challenges still exist, such as the biocompatibility of inorganic dielectric materials within human tissues, the degradability of dielectrics and polymer substrates *in vivo*, and most importantly, the long-term and stable operation in the body. All of these challenges call for a holistic design of the device.

5.2.4 Self-assembled Monolayer Field-Effect Transistors (SAMFETs)

The rigid, covalent attachment of SAMs containing semiconducting cores onto an inorganic, high-*k* dielectric surface provides a direct method for the development of miniaturized devices of SAM organic field-effect transistors (SAMFETs) with low operation voltages and high electrical performance by reducing the dielectric thickness and improving the dielectric quality.

5.2.4.1 Molecule Design for SAMFETs

For an organic semiconducting monolayer growing on a rigid or flexible substrate to construct a SAMFET (Figure 5.14a), three golden rules should be taken into consideration: (i) anchor group was chosen based on the dielectric used; (ii) chained tail

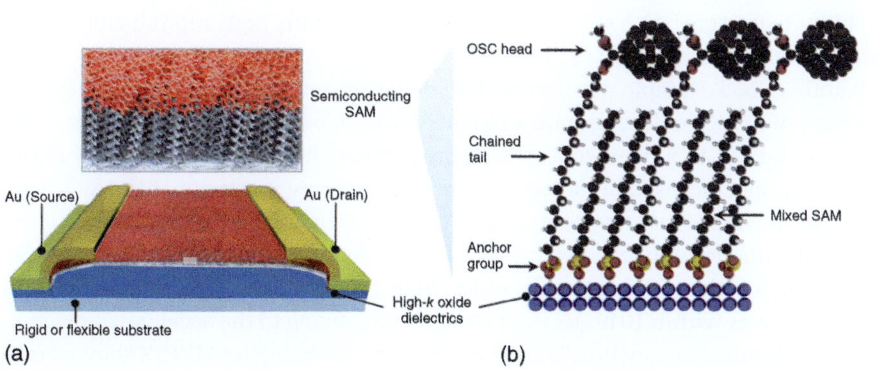

Figure 5.14 Organic semiconducting molecular design for SAMFEFs. (a) Schematic illustration showing a SAMFET device configuration. (b) Molecule design strategy of a typical SAMFET semiconductor. Source: (a) Reproduced from Schmaltz et al. [42], © 2019 American Chemical Society.

was chosen considering interaction; and (iii) head group was chosen to integrate the functionality (Figure 5.14b). We have summarized in Figure 5.14 and Table 3.1 some widely used OSCs and their performances in SAMFETs.

By directly attaching OSCs to gate oxides through covalent bonding, the Nuckolls' group reported the *in situ* assembly of monolayer FETs [43, 170]. A tetracene derivative (**242**) that had one of its terminal rings functionalized with ortho-hydroxyl as a catechol was used to spontaneously form upright monolayers on AlO_x as active channels in transistors. Tightly packed behavior in the monolayer was confirmed by synchrotron X-ray reflectivity (XRR) measurements [43]. The monolayer of self-assembled semiconductor between Au source and drain electrodes (with a 30–100 nm spacing) showed p-type operation. However, the current–voltage (I–V) curves did not allow quantitative extraction of the transistor parameters. Mottaghi et al. [44] described the fabrication of SAMFETs by using a bifunctional molecule, 4T–MEOA (**244**), comprising a short alkyl chain linked to an oligothiophene moiety that acts as the active semiconductor and a -COOH terminal group as an anchor on alumina. By using patterned Pt–Ti source and drain electrodes, an estimated mobility of ~0.0035 $cm^2\ V^{-1}\ s^{-1}$ and on/off ratios reaching 1800 at a gate voltage of −2 V were achieved in a submicrometer channel.

Early attempts to fabricate SAMFETs resulted in relatively poor performance at channel lengths of more than 1 μm because of the high contact resistance and the lack of long-range order of the semiconductor films. A remarkable breakthrough was achieved by Smits et al. [171] in 2008. To obtain efficient charge transport in long channels, they fabricated bottom-up organic SAMFETs and integrated circuits by using a liquid-crystalline molecule (**245**) consisting of a π-conjugated oligo-thiophene (5T) core separated by a long aliphatic chain from a monofunctionalized anchor group. This design was crucial because with only one Si—Cl bond for covalent binding on the SiO_2 surface, the defects caused by uncontrolled self-condensation were prevented, facilitating the formation of long-range intermolecular π–π coupling in the monolayer. The dense, smooth monolayer fabricated on atomically smooth SiO_2 allowed the SAM–OFET to exhibit bulk-like carrier mobility (~0.04 $cm^2\ V^{-1}\ s^{-1}$ for a 40-μm channel) with high reproducibility and large current modulation.

Another good example was demonstrated by Parry et al. [179] recently. They synthesized two OSC molecules with a monochlorosilane (**246**) or trichlorosilane (**247**) anchor attached to a phenylene–thiophene semiconducting mesogen. They found that the monochlorosilane molecule showed a very low reactivity at the SiO_2 substrate surface and could not form complete monolayers, in contrast to the previously reported quinquethiophene derivative, whereas the other trichlorosilane molecules resulted in a SAM with long-range order, high coverage, and good connectivity on SiO_2 substrates within 10 hours from the solution, owing to the accelerated reaction of the trichlorosilane anchor. Therefore, the corresponding SAMFETs showed unified performance over large areas for channel lengths up to 100 μm and exhibited high mobilities of $\sim 1.7 \times 10^{-2}\ cm^2\ V^{-1}\ s^{-1}$ with a high on/off ratio over 10^6.

Besides silane-based/carboxylic acid-based SAMFETs, PA-based semiconducting SAMs can also be used to fabricate organic transistors, as demonstrated by the

Halik's group [173, 175, 490, 808]. Based on 4T (**248**) and C_{60} (**249–250**) units, they synthesized functional SAMs featuring alkyl PA anchors as the active layer, with channel lengths up to 10 μm, and obtained hole and electron transport operating at low supply voltages (−4 V to +4 V) with mobilities of 10^{-5} and 10^{-4} $cm^2\ V^{-1}\ s^{-1}$, respectively, in ambient atmosphere [173]. Besides, it is the first time that n-type SAMFETs have been reported.

To overcome the challenge of achieving efficient contact between the metal electrodes (such as Au) and the SAM semiconductors, Cernetic et al. [180] designed new SAM molecules, MTB4TC11 (**251**) and MTB4TC12 (**252**). The phosphonic acid binding group ensured covalent bonding with the dielectric surface of the metal oxides and allowed for the formation of robust and highly ordered SAMs, and the SAM molecules contained methylthiobutyl as a terminal group, designed for the realization of efficient charge injection owing to the Au–S interaction. After annealing, an increase of two orders of magnitude in charge-carrier mobility was obtained. Based on a HfO_2 sol–gel dielectric under optimized processing conditions, the resulting transistors showed hole mobilities as high as ~0.015 $cm^2\ V^{-1}\ s^{-1}$ at a low voltage.

N-type SAMFETs that are self-assembled from solution have been demonstrated by Ringk et al. [176]. They described n-type SAMFETs based on a heterosubstituted perylene bisimide derivate, PBI–PA (**253**), that could be covalently fixed to an aluminum oxide dielectric via a phosphonic acid linker. Because of the formation of a nano-crystalline PBI–PA monolayer, highly reproducible transistors were demonstrated with channel lengths up to 100 μm, showing electron mobilities of $\sim 1.5 \times 10^{-3}$ $cm^2\ V^{-1}\ s^{-1}$ and on/off current ratios up to 10^5. By integrating a PBI–PA SAM-based n-type transistor and a p-type transistor into a single device, a complementary inverter with a large noise margin of 7 V and a high gain of up to 15 solely based on SAMFETs was successfully demonstrated [171]. The authors also reported a SAM molecule designed for reducing steric repulsion, PBIF–PA (**254**), as the active layer of SAMFET [177]. In this design, the linear fluorinated alkyl tail increased the electron transport stability. With a homogeneous and dense monolayer, transistors built on rigid or flexible substrates showed electron mobilities of $\sim 10^{-3}$ $cm^2\ V^{-1}\ s^{-1}$.

It is worth noting that by utilizing the strong interaction between SAM molecules and the substrate surface, SAMFET-based sensors can be developed by utilizing non-covalent interactions [809–813]. For instance, Chen et al. [184] designed core-clad liquid-crystalline pentathiophene (2R5T2R) as a semiconductor to form a densely packed, long-range-ordered monolayer that worked stably even under aqueous conditions and served as a melamine sensor. He et al. [571, 573] fabricated SAMFETs by using C_8–BTBT grown through the vdW epitaxy method mentioned previously.

5.2.4.2 Morphology Control of SAMFET

Besides molecule design, morphology control also plays an important role in the charge transport of SAMFETs. Jäger et al. [808] combined MD simulations and device fabrication/characterization to describe and optimize complex systems of SAMs on the surfaces (Figure 5.15a). They found that in pure SAM of **249**, the long

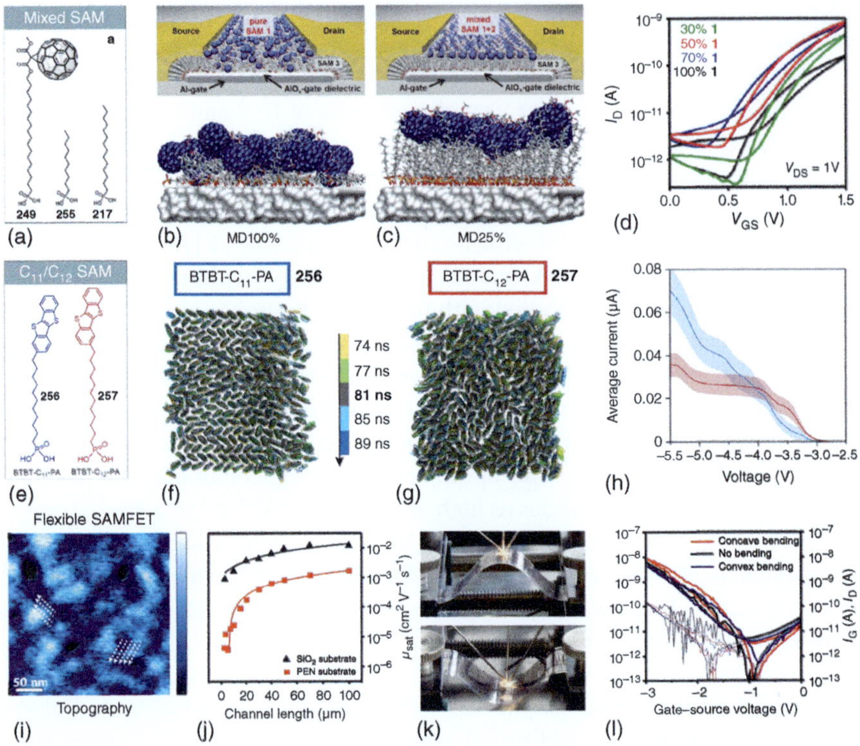

Figure 5.15 Morphology control of SAMFETs. (a–d) Effect of mixed SAMs on morphology and charge transport. (a) Molecular structure of SAMs. (b, c) Schematic SAMFET setups (top) and SAM morphology of pure **1** (left) and 25% mixed SAM derived from molecular-dynamics simulations. (d) Electrical performance of SAMFETs with different content of **1**. (e–h) Effect of alkylchains on structure disorder and charge transport. (e) Chemical structures of BTBT−C_{11}−PA and BTBT−C_{12}−PA. (f, g) Simulated results showing the dynamic changes of morphology. (h) Averaged IV-characteristics for BTBT−C_{11}−PA (blue) and BTBT−C_{12}−PA (red), respectively. (i–l) Flexible SAMFET. (i) AFM images showing a dense packing of monolayer. (j) FET mobility as a function of channel length of the SAMFETs on rigid and flexible substrate. (k) Photographs showing SAMFETs on PEN substrate bended convexly (top) and concavely (bottom). (l) Electrical performance of flexible SAMFET measured during convex (blue), concave (red), and without bending (black). Source: (a–d) Reproduced with permission from Jager et al. [808], © 2013 American Chemical Society. (e–h) Reproduced with permission from Schmaltz et al. [42], © 2017 American Chemical Society. (i–l) Reproduced with permission from Schmaltz et al. [175], © 2013 John Wiley & Sons.

flexible alkyl chain (~2.5 nm) allowed the molecules to twist, collapse, and sink onto the substrate. The resulting SAM has reduced thickness with a disordered occupancy of C_{60} (Figure 5.15b). However, in mixed SAMs, molecules **255** or **217** with a shorter alkyl chain (~1.4 nm) were incorporated below the bulky C_{60} head groups and occupied free sites on the dielectric, preventing them from collapsing onto the substrate (Figure 5.15c). This could result in more well-aligned fullerenes, creating more effective pathways induced by an improved morphology for electron transport for achieving the best device performance (Figure 5.15d). It was found that in a mixed SAM at a ratio of 50 : 50, an almost one order of magnitude increase in mobility was achieved in comparison with the devices with the pure semiconducting SAM as the active channel.

Schmaltz et al. [42] demonstrated that the morphology is governed by relatively weak van der Waals interactions between molecules. They have built SAMFETs based on SAM molecules that consist of a BTBT core linked with a C_{11}–SAM (**256**) or C_{12}–SAM (**257**) alkylphosphonic acid (Figure 5.15e). They found that the crystallization behaviors and the charge-transport properties vary considerably in these two systems. While the C_{11}–SAM forms extensive crystalline domains (Figure 5.15f), the C_{12}–SAM shows only a few, small crystallites (Figure 5.15g). This could be explained by the different tilt angles of the BTBT groups. Less tilted functional end groups are expected to promote the crystallization of π-domains, which can only attain full π–π overlap between the BTBT moieties if these are perpendicular to the surface plane. Besides, MD results revealed that fluctuations were more pronounced in amorphous regions than in crystalline domains, which meant that a high degree of order and crystallinity was important not only for the sake of a high degree of static order but also to limit the amount of dynamic disorder. In this regard, excellent charge-transport performance was found in C_{11}–SAM, exhibiting a threefold higher mobility of $\sim 3 \times 10^{-2}$ cm^2 V^{-1} s^{-1} (Figure 5.15h).

The Halik's group also fabricated SAMFETs on flexible substrates based on SAM **246**, which was considered difficult to achieve [175]. This could be attributed to the densely packed monolayer with strong alignment of the π-domian, which was revealed by AFM imaging (Figure 5.15i). The maximum saturation mobilities of SAMFETs fabricated on different substrates with channel lengths larger than 70 μm were measured to be $\sim 1.2 \times 10^{-2}$ cm^2 V^{-1} s^{-1} on SiO_2 and $\sim 1.7 \times 10^{-3}$ cm^2 V^{-1} s^{-1} on polyethylene naphthalate (PEN) (Figure 5.15j). The latter was almost comparable to the best values obtained in SAMFETs with very flat surfaces. Further bending experiments (Figure 5.15k) demonstrated mechanical flexibility in SAMFETs (Figure 5.15l).

5.3 Integrating New Functionalities

As discussed earlier, charge carriers are generated and transported at the semiconductor/dielectric interface. Modifying the interface not only significantly improves the device performance but also installs new functions. Despite these considerable achievements, there are still no commercially available functional OFET-based electronic devices. To satisfy the requirements for real applications, the development of novel molecular devices with specific functions is a prerequisite. In this section, we focus on reviewing the ever-developing trends of integrating molecular functionalities into OFETs at OSC/dielectric interface (Figure 5.16), including (i) superconductors; (ii) pressure sensors; (iii) thermal sensors; (iv) photoswitches; and (v) photo memories and other nonvolatile memories. For different functions resulted from individual components, both "dielectrics" and molecular "modifiers" are equally important due to the fact that the proposed functions could be affected not only by

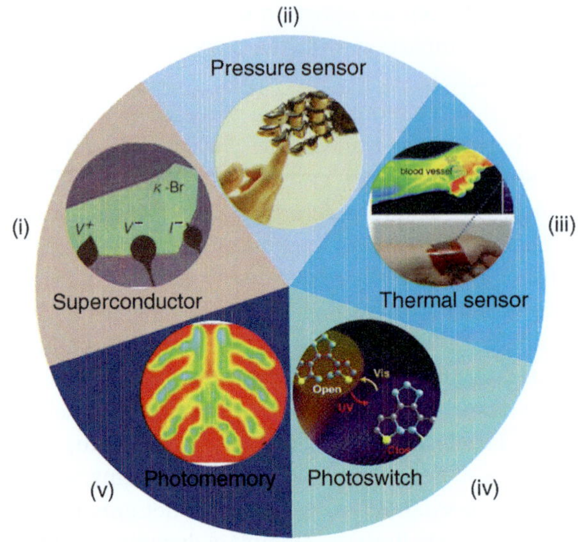

Figure 5.16 Schematic description of current trends in developing different types of functional OFETs.

molecular electrical characteristics but also by the synergetic effect at the interface. Therefore, we will pay specific attention to the design of the molecular structures as well as the device constructions and their influence on the device functions.

5.3.1 Photoresponsive Dielectrics

So far, photocontrol of the electrical properties of OFETs has been achieved through the incorporation of photochromic organic molecules, which are capable of undergoing a photoisomerization [405, 406, 814–816]. Changes in the molecular dipole moment or energy-level alignments associated with photoisomerization upon light irradiation generate an additional electric field that modulates the drain current and threshold voltage. The first example involves a novel way to modify the dielectric/semiconductor interface through a photochemical process, as demonstrated by Guo's group [817, 818]. Specifically, the SiO_2 surface was treated with SAMs of photochromic spiropyran (SP) molecules, as shown in Figure 5.17a. Such chemical systems were able to undergo a reversible transition between a neutral, colorless form and a zwitterionic, colored form upon exposure to different wavelengths of light (Figure 5.17a, inset). This transition caused a considerable difference in the electric dipole moment of the SP molecules. Because of this, the electrical characteristics of functional OFETs have been reversibly and optically controlled. Figure 5.17b demonstrates the switching cycles of the drain current as a function of time, showing the reversible photo-switching behaviors under the irradiation of UV and visible light.

In another work, Suda et al. [819] demonstrated carrier-doping–induced superconductivity in an organic Mott insulator with a photoinduced electric double layer (EDL) based on a photochromic SP–SAM (Figure 5.17c). EDL transistors, with ionic liquids or electrolytes as gate dielectrics, have been used in superconducting OFETs as nanogap capacitors [821–825]. As mentioned earlier, SPs can

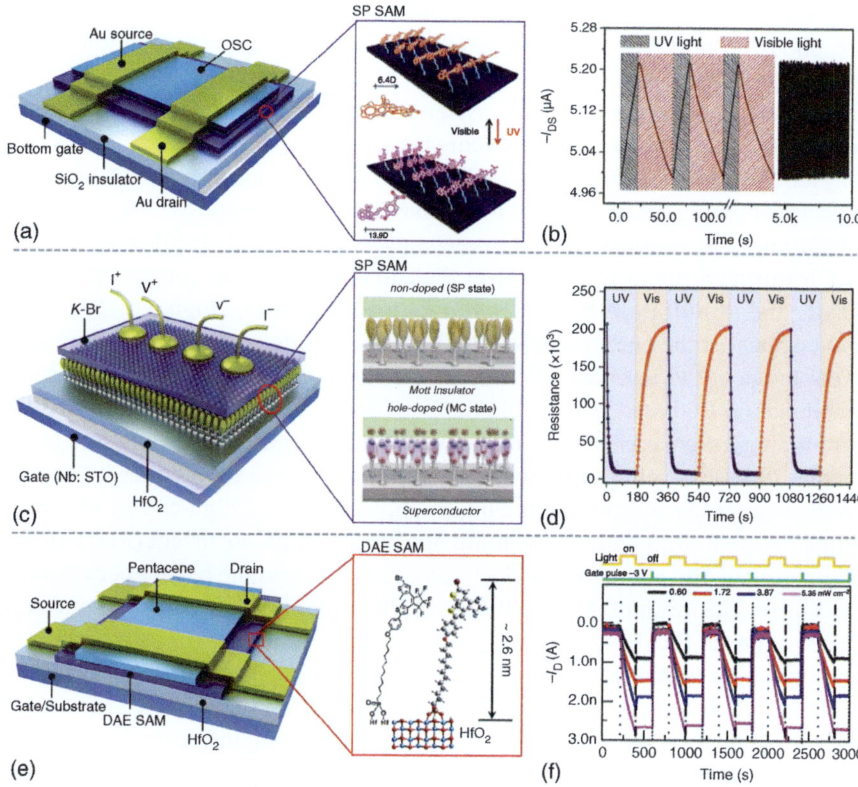

Figure 5.17 Photoresponsive semiconductor/dielectric interfaces. (a) Schematic structure of an OFET with photochromic SP–SAMs on SiO_2 gate substrates. The SP monolayer can isomerize reversibly by UV and visible light irradiation at the interface. (b) Photomodulation of the drain current. $V_D = -30$ V; $V_G = -15$ V. (c) Schematic cross section and optical images of an EDL transistor. (d) Time trace of the resistance for the device with alternate photoirradiation of UV and visible light. (e) Schematic cross section of a diarylethene-based photoactive hybrid bilayer dielectric (PHBD) organic memory transistor. (f) Five representative programming and erasing cycles of a device as a function of light intensity. $V_D = -100$ mV; $V_G = 0$. Source: (a, b) Reproduced from Zhang et al. [817], © 2011 American Chemical Society. (c, d) Reproduced with permission from Suda et al. [819], © 2015 American Association for the Advancement of Science. (e, f) Reproduced from Chen et al. [820], © 2016 American Chemical Society.

isomerize reversibly between nonionic and zwitterionic isomers upon illumination with UV or visible light irradiation (Figure 5.17c, inset). Therefore, an EDL-like large electric field across the film could be induced or eliminated by UV light or visible light irradiation, respectively. Hence, a highly aligned SP monolayer is a promising candidate for a photoinduced EDL system that could directly switch superconductivity only by photoirradiation. During UV irradiation, the resistance value drastically decreased, and the final state persisted even after the irradiation stopped. Moreover, the reduced resistance value was recovered to nearly the initial value by visible light irradiation and lasted even after the irradiation stopped (Figure 5.17d). The advantage of the photochromic EDL system over a conventional one is that it retained modulation capability even at low temperatures because these photochromic reactions proceeded with photon-energy dissipation under nonequilibrium conditions.

By contrast, in other photochromic systems such as diarylethene (DAE) families, a negligible dipole moment change occurs between their photoisomers. Actually, a totally different working mechanism exists when these photochromic systems affect the electrical properties of OFETs with similar device architectures [588, 826, 827]. To determine the possible working mechanism, Chen et al. [820] described an OFET featuring a photo hybrid bilayer dielectric (PHBD) that comprised a SAM of DAEs and an ultrathin layer of solution-processed HfO_2 (Figure 5.17e). They photo-engineered the energy levels of DAE SAMs, aiming at facilitating the charge and discharge at the interface of the hybrid dielectrics, which could further form a functional OFET-based nonvolatile memory device. Light signals were used for programming, whereas electrical signals were used for erasing ($\lesssim$3 V); thus, a large, reversible threshold-voltage shift with long retention times were produced. This can be explained by the selective tunneling of photo-generated high-energy electrons through the DAE SAM and the retention of a persistent photocurrent (PPC). In the DAE-open state, the photoexcited electrons could not cross the barrier between the LUMO of pentacene and DAE into the DAE/HfO_2 interface. UV light (365 nm) was required to preset the device so that it could function as a memory cell. UV irradiation caused the isomerization from DAE-open to DAE-closed, thus lowering the DAE LUMO with respect to pentacene. As a result, the high-energy electrons could tunnel through the DAE SAM and charge the DAE/HfO_2 interface. In addition, because the LUMO energy of DAE-closed was lower than that of pentacene, the trapped electrons at the DAE/HfO_2 interface were efficiently prevented from leaking away through the pentacene/DAE interface. Along with the thermal stability of DAE-closed, these results demonstrated the significant nonvolatile memory capability of these PHBD OFETs. Figure 5.17f shows five representative programming and erasing cycles of I_D as a function of light intensity (P_{light}), demonstrating the reproducibility and programmability of the nonvolatile memory capacity.

It is a straightforward method to construct such a photo-engineered semiconductor/dielectric interface. However, the limited switching speed of photochromic molecules (from seconds to minutes) makes these switched or memory devices uncompetitive to Si-based devices. Besides, unsatisfied long-term stability and reproducibility also hinder their actual application. Potentially, the ultrathin nature of the hybrid dielectrics makes them suitable for application in disposable and wearable electronics, where a compromised performance is acceptable. In the meantime, by combining photo operation with traditional electrical operation, a hybrid functional device could be constructed, reaching a balance between performance and functionality.

5.3.2 Other External Stimuli-Responsive Dielectrics

If organic transistors show responses to external stimulation such as light, pressure, thermal, or magnetic changes [828, 829], the output current of the channel would be affected directly, providing a design strategy to fabricate external stimuli-responsive OFETs. In fact, the possibility of fabricating flexible and stretchable pressure and thermal sensors has attracted many researchers because of their potential applications in electronic artificial/robotic skin and human healthcare, and various strategies have been developed [30, 829–834]. It is worth mentioning that there are different device structures for fabricating OFET-based mechanical sensors. Because of the promising applications of the related sensors, here we list a few representative works, although some have rather complex compositions, where organic transistors were used for signal recording (output current) rather than sensing.

5.3.2.1 Pressure Sensor

The pioneering work presented by Someya et al. [835, 836] demonstrated the first flexible pressure sensor arrays by integrating organic transistors with pressure-sensitive rubber, in which OFET active matrixes were used to read pressure data from piezoresistive elements. The resistance of the pressure-sensitive rubber decreased with increasing pressure, and thus the transconductance as well as the measured current increased.

A new type of OFET-based pressure-sensing device structure was presented by Mannsfeld et al. [837] in which the dielectric layer, which consisted of thin, structured PDMS rubber with regular pillars or pyramidal shapes, was compressible and enabled the device to show high pressure sensitivity (Figure 5.18a). The variations in current were attributed to the presence of deformable air voids inside the film, which changed capacitance with applied pressure. Moreover, the sensors could operate stably at a low bias voltage of −20 V and showed a sensitivity of 0.55 kPa^{-1} (Figure 5.18b). In a subsequent work, the authors fabricated the flexible pressure-sensitive OFET on a BCB-coated polyimide film by using a high-mobility conjugated polymer, polyisoindigobithiophene-siloxane (PiI2T–Si), as the conductive channel [838]. By combining the microstructured PDMS dielectric with the high-mobility PiI2T–Si in a monolithic transistor design, improved sensitivity at low pressure was observed, with the devices operating in the subthreshold regime. Finally, a maximum sensitivity of 8.4 kPa^{-1} with a fast response time of less than 10 ms was obtained for the flexible pressure sensor, demonstrating its potential usage in mobile health monitoring and remote diagnostics.

Zang et al. [839] reported a novel pressure sensor to overcome the elastic limitation of the rubber dielectric layer. They designed a flexible suspended gate OFET (SGOFET) for ultrasensitive pressure detection, which was formed from an air dielectric layer and suspended gate electrodes (Figure 5.18c). Au electrodes were first deposited on a glass or PI substrate, followed by the semiconductor layer (PDPP3T) protected with Cytop. Then, with laminated 40-μm-thick Kapton tape as the support layer, the flexible PI/Al foil was incorporated. Because of the ultralow capacitance of the air gap, no obvious modulation was observed without applying pressure, whereas only under pressure would the FET curve appear, presenting a mobility of up to 0.34 $cm^2\ V^{-1}\ s^{-1}$ with an on/off ratio of 10^4. By fine-tuning the properties of the suspended gate to optimize SGOFETs, an unprecedented sensitivity of 192 kPa^{-1} [limit-of-detection (LOD) pressure

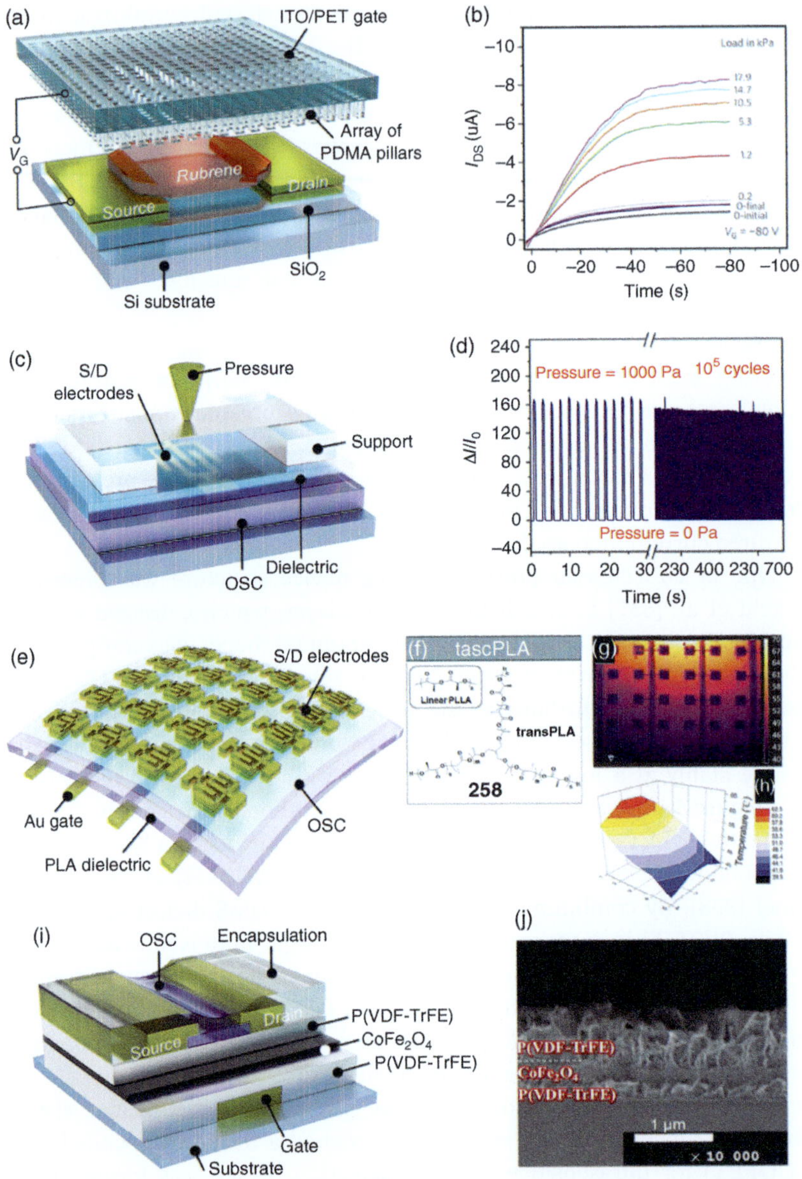

Figure 5.18 Other functional dielectrics. (a) Layout of pressure-sensing organic single-crystal transistors, consisting of thin rubrene single crystals and structured PDMS dielectric films. (b) Output curve of an OFET-based pressure sensor with different external pressures applied. (c) Schematic representation of a pressure sensor based on suspended gate organic thin-film transistor (SGOTFT). (d) Durability test of a SGOTFT under a pressure of 1000 Pa. (e) Schematic representation of a flexible thermal sensor array. (f) Chemical structure of thermal responsive polymer dielectric transPLA (**258**). (g) Photograph of the flexible temperature sensor with 5 × 4 temperature pixels and showing 2D temperature detection in (h). (i) Schematic structure of the flexible ME–FET with P(VDF–TrFE)/$CoFe_2O_4$ NPs/P(VDF–TrFE) hybrid gate dielectrics. (j) Cross-sectional field emission-scanning electron microscopy (FE-SEM) image of ME gate dielectrics. Source: (a, b) Reproduced with permission from Mannsfeld et al. [837], © 2010 Nature Publishing Group. (c, d) Reproduced from Zang et al. [839], © 2015 Springer Nature / CC BY 4.0. (e–h) Adapted from Yokota et al. [840]. (i, j) Reproduced with permission from Triet et al. [841], © 2015 Springer Nature.

of <0.5 Pa] with an ultralow power consumption of <100 nW was achieved (Figure 5.18d).

5.3.2.2 Thermal Sensor

Organic transistors can serve as a key element in high-end sensor applications for thermal imaging. For example, Zirkl et al. [842] integrated a transducer configuration with low-voltage operating OFETs with an optothermal light sensor element composed of a ferroelectric polymer P(VDF–TrFE) layer sandwiched between two thin Al electrodes. With or without illumination of an infrared laser on the top Al electrodes, the current was modulated in the circuit, exhibiting stable on/off switching. Generally speaking, the OSCs showed the intrinsic dependence of the FET mobility on temperatures ranging from 4 to 300 K but only very limited thermal sensitivity beyond room temperature [843–846]. To overcome this limitation, Wu et al. [847] introduced polylactide (PLA) as a new dielectric layer into OFETs (Figure 5.18e). PLA contains polar groups to provide an interfacial interaction with charge carriers of the semiconductor layer (Figure 5.18f), which could be affected by environmental changes such as temperature. Hence, the authors were able to make use of the polar group-induced interfacial charge-trapping effect to enhance the thermal sensitivity of biocompatible BGTC transistors for temperature sensing, where three-arm stereocomplex PLA (tascPLA, **258**) was used for better thermal stability (up to 200 °C). The tascPLA–OFETs, with high thermal sensitivity and biocompatibility, were further utilized to construct skin-like temperature sensor arrays (5×4 OFETs) (Figure 5.18g), and they presented reliable 2D temperature detection (Figure 5.18h).

5.3.2.3 Magnetic Sensor

OFET-based magnetic sensor devices can also be achieved by taking advantage of the research results above. For example, Zang et al. [848] demonstrated for the first time a sensitive, flexible magnetic sensor with magnet-functionalized suspended gate electrodes, where iron oxide (Fe_3O_4) NPs and Ag NWs were selected as the magnetic and conductive components, respectively, embedded in a PDMS rubber. The magnetic sensing property came from the introduction of magnetic NPs in the composite gate. When no magnetic field was applied, the transistor exhibited a very low source–drain current, with no obvious modulation owing to the ultralow capacitance. As soon as an external magnetic field of 150 mT was loaded, the current increased by two orders of magnitude and displayed good FET responses. Triet et al. [841] developed a new type of flexible magnetoelectric OFET (ME-OFET) in which a magnetic-responsive stacked 2–2 horizontal heterostructure of alternating layers, comprising a combination of piezoelectric and magnetostrictive components, was used as a multifunctional gate dielectric layer (Figure 5.18i). As a matter of fact, the heterostructure consisted of a sandwich layer of Co ferrite ($CoFe_2O_4$) NPs with sizes in the range of 30–50 nm and the highly crystalline copolymer P(VDF–TrFE) [P(VDF–TrFE)/$CoFe_2O_4$ NPs/P(VDF–TrFE)] as the key components (Figure 5.18j). The sensing mechanism of the ME-OFETs was based on the ME coupling effect, that is, the coupling between the magnetostrictive ($CoFe_2O_4$) and piezoelectric

(P(VDF–TrFE)) phases. When varying the magnetic field applied to the gate dielectric, the ME effect caused a change in the effective gate field, thus enabling the modulation of the channel conductance of ME-OFETs.

5.3.2.4 Multifunctional Sensor

Multifunctional sensors, such as bifunctional sensors for pressure and temperature, have been fabricated based on integrated FET circuits with various sensor elements. Graz et al. [849] designed a nanocomposite foil made of ceramic ferroelectric nanoparticles, lead titanate (PbTiO), embedded in a polymeric ferroelectric matrix, P(VDF–TrFE), which was aligned and laminated with OFETs on a flexible backplane to achieve bifunctional pressure- or temperature-sensitive sensory cell arrays. By a sequential, area-selective, two-step poling process, the polarization directions in the two components were adjusted independently, enabling sensor elements that were only piezoelectric or only pyroelectric to act as pressure/temperature subcells, where almost no cross-sensitivity between adjacent subcells was observed. Moreover, the authors pointed out that by adding magnetic or photosensitive nanoparticles, the sensing capability of single-material frontplanes to magnetic fields and light might be enhanced.

A transparent and flexible OFET (tf-OFET) composed of a ferroelectric copolymer [P(VDF–TrFE)] gate dielectric as a sensing layer was presented by Tien et al. [850, 851] to function as a multimodal sensor. The sensing ability depended on the modulation of the remnant polarization inside the ferroelectric polymer gate insulator, which showed responsivity to multiple external stimuli such as pressure and strain as well as infrared (IR) light, owing to the transparent nature of the device. Tien et al. [852] incorporated a nanocomposite material as the gate dielectric in OFETs to act as physically responsive FETs. The functional piezo- and pyroelectric nanocomposite gate material enhanced the electrophysical coupling effects and allowed simultaneous multimodal sensing of pressure/strain and temperature changes. Also, the integration of FET arrays with multimodal sensing elements could reduce the complexity of structural integration, minimize the signal interferences coupled with strain, and improve the stability of their stimuli-responsive properties.

5.3.3 Integrating Memory Effect at the Dielectrics

Nonvolatile memory device is one of the most widely studied applications utilizing functionalized semiconductor/dielectric interface. It is regarded as the closest application toward commercialization. In general, nonvolatile memories are classified into three basic types depending on their adopted architectures: capacitor-, resistor-, and transistor-based memories, as well as various combinations [853]. The capacitor-based memory is vulnerable to losing the stored data because the medium between the electrodes is merely dielectric, with nothing to stabilize the programmed state [854]. The organic resistive memory devices consist of organic-resistive layers between two electrodes, which typically act as either electrically insulating components or electrically conducting components under appropriate voltage conditions. However, they require the transistors for addressing signals in 2D memory arrays because they do not possess third electrodes for signal addressing [855]. The third memory architecture that can be adopted is OFET-based memory devices. Such transistor-based memories are appealing because they allow direct integration of the memory element within the standard device structures used in complementary integrated circuits.

To date, nonvolatile organic memory transistors reported previously adopted structures such as a floating gate or multilayer dielectrics, both of which can store injected charges in the floating gate or at the dielectric interface upon the application of the gate voltage [416, 856, 857]. Among them, the floating gate has achieved notable success in manipulating the V_{th} of OFETs, with various layers inserted between the dielectric and semiconductor layers [734, 858]. In the floating gate architecture, as shown in Figure 5.19a, additional metal and dielectric layers are placed in series between the gate dielectric and semiconductor layers of OFETs. Upon the application of a writing voltage, carriers may be written onto the floating gate layer via thermionic emission or tunneling, where they remain trapped between the floating gate metal and thin encapsulating dielectric. These trapped carriers on the floating gate will screen the gate voltage, enabling a shift in the gate voltage required to turn on OFETs (Figure 5.19b). Although tremendous progress has been made in the floating gate memory devices, it is still a significant challenge to scale down MOS flash memory to sizes below 10 nm per data cell [859–864]. To this end, Busche et al. [865] showed that core–shell polyoxometalate (POM) molecules could act as candidate storage nodes for MOS flash memory (Figure 5.19c). Embedding $[(Se(IV)O_3)_2]^{4-}$ as an oxidizable dopant in the cluster core allowed oxidation of the molecule to a $[Se(V)_2O_6]^{2-}$ moiety containing a Se(V)—Se(V) bond and revealing a new oxidation state for selenium (Figure 5.19d). This new oxidation state could be observed at the device level, affording a new type of memory called "write-once-erase" by the authors. Taken together, these results demonstrate that POMs have the potential to function as realistic nanoscale flash memory (Figure 5.19e). This work also suggests a route to the practical integration of configurable molecules in MOS technologies as lithographic scales approach the molecular limit.

By combining the floating gate layers and tunneling layers in traditional floating gate OFETs into one single charge-trapping layer at the semiconductor/dielectric interface, new structures of memory cells are produced with even better performance (Figure 5.20a). The most frequently used charge-trapping layers [866, 867] can be divided into two types: nanofloating gate (Figure 5.20b) and polymer electret (Figure 5.20c).

Nanofloating gate OFET-based memory can be obtained as the traditional floating gate layer is replaced with discrete nanostructured metallic or semiconductor materials, whose size and density of charge trapping can be finely determined [868]. Metal nanoparticles (NPs) capped with insulators are the most frequently used materials, which can be easily and directly separated in the device during the device fabrication process to serve as charge traps. As the active elements of the device, the performance of the memory is determined by every aspect associated with NPs. Different

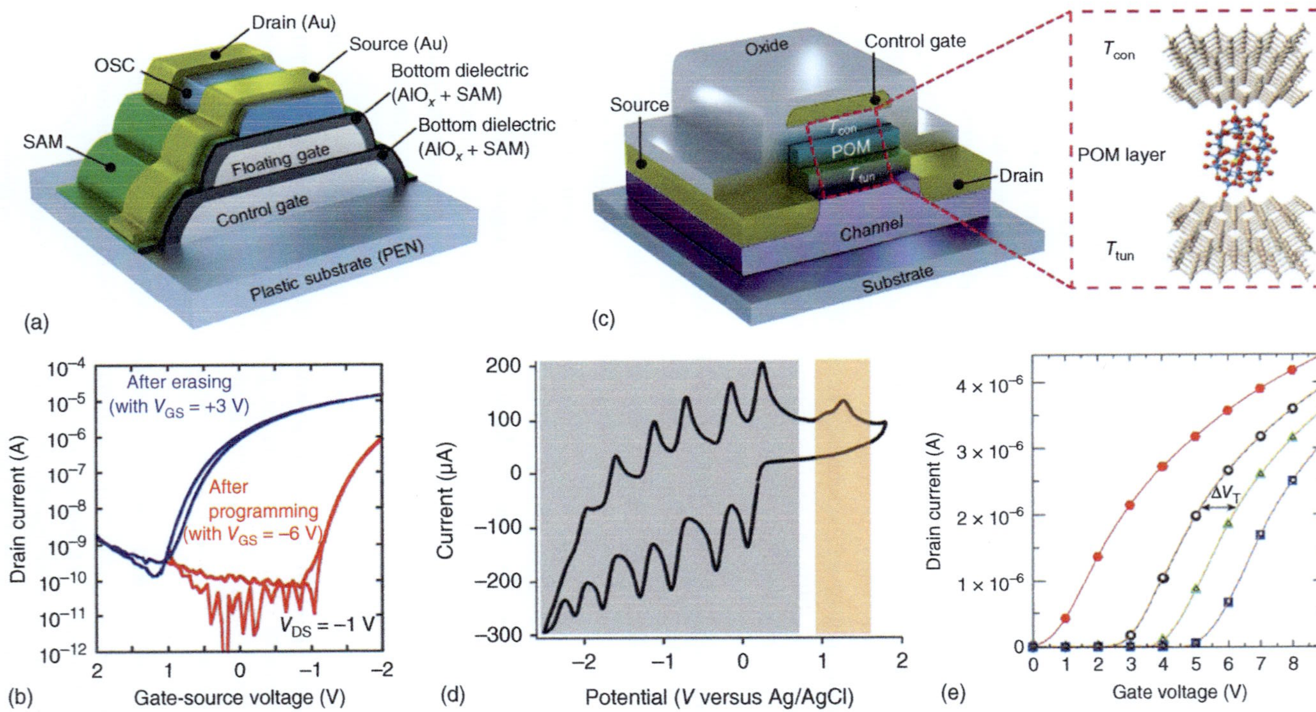

Figure 5.19 Conventional floating gates. (a) Schematic of a floating gate structure using Al–AlO_x for both devices and floating gates. (b) Transfer characteristics for a p-channel OFET after programming and erasing steps. Erasing: $V_G = +3$ V; programming: $V_G = -6$ V. (c) Schematic diagram of a single-transistor nonvolatile memory cell, indicating the aimed substitution of the poly-Si floating gate with an array of POM clusters. (d) Electrochemical properties of the POM clusters. (e) Drain current versus gate voltage with a drain bias of 50 mV on a linear scale for a bulk molecular flash cell. Source: (a, b) Reproduced with permission from Sekitani et al. [858], © 2009 American Association for the Advancement of Science. (c) Adapted and (d, e) Reproduced with permission from Busche et al. [865], © 2014 Nature Publishing Group.

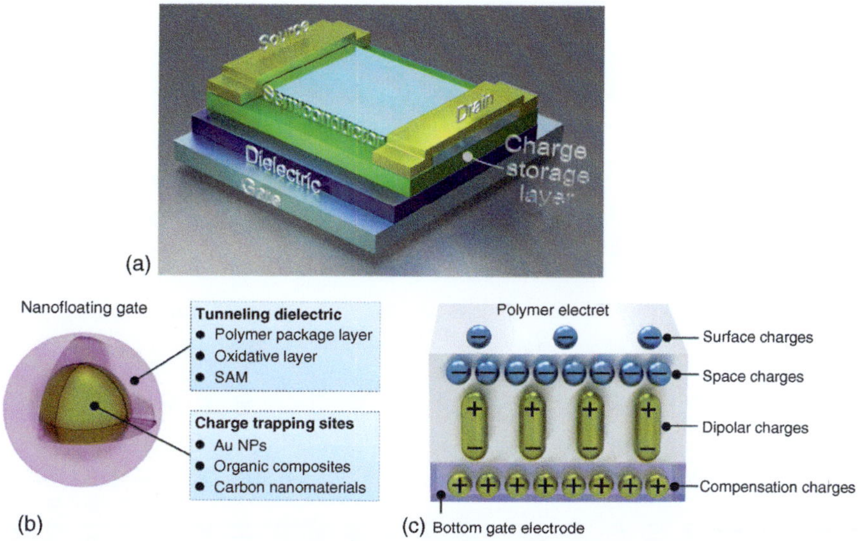

Figure 5.20 Charge storage layers. (a) Schematic architecture of an OFET structure with a charge storage layer. (b) Nanofloating gate charge storage layer. (c) Polymer electret charge storage layer.

elements of NPs influence the behavior [869, 870]. Generally, Au is the most frequently used NP, and sometimes organic composites are used [871–873]. With the same NPs, the different concentrations and diameters show different impacts on the charge trap characteristics [874, 875]. Also, nanostructured carbon materials can serve as charge storage sites and memory devices based on C_{60} and graphene NPs or islands, as the nanofloating gate has been fabricated [594, 876, 877].

The polymer package layer, such as PS [874, 878], PMMA [869], PVP [879], and oxidative layer [880] are the universal insulator layers functioning as the tunneling dielectric, which also can help stabilize the NPs. Metal–oxide NPs can also be used as the nanofloating gate [875, 881]. Several universal methods to create the charge trap layer have been carried out by using NPs with physical processes such as thermal evaporation and chemical processes such as self-assembly methods and synthesizing them together with a block polymer. The thickness of the NPs encapsulated with polymers and the kind of polymer used are two crucial factors for enhancing the device storage performance because a dielectric layer helps retain charge and avoid leakage [869, 873, 878]. The NPs could also be capped with a SAM on the surface before being packaged with a polymer so that their memory characteristics would be tuned using the influence of the SAM dipole [882–884]. In addition, if a double floating gate architecture in the memory device is utilized, a higher charge retention capability can be achieved [877].

As for other types of charge-trapping OFET memory, a dielectric polymer with a long-term charge-storing ability is used, denoted as a polymer electret. This polymer electret possesses both charge-trapping and tunneling characteristics [752, 885]. The memory mechanism is very similar to the floating gate memory;

Table 5.1 Summaries of OFET-based memory cells by using various charge-trapping layers.

Charge-trapping layer	OSC	Memory ratio	Memory window (V)	Sweep voltage (V)	Endurance test (cycle)	Retention time (s)	References
Au PMMA NPs	P(NDI2OD-T2)	10^5	51.7	−80 to 80	100	10^7	[869]
Ag OA-Pt PVP NPs	Pentacene	$>10^5$	18.7	−25 to 15	1000	10^5	[870]
PAH/Ferritin NPs	Pentacene	10^4	20	−30 to 30	200	10^8	[871]
CuPc-PS_4	Pentacene	10^7	21	−50 to 10	100	10^5	[872]
PS_{56k}-*b*-$P4VP_{8k}$	Pentacene	10^5	21.1	−30 to 50	400	10^7	[873]
Au PS NPs	F8T2	$>10^4$	30	−90 to 30	—	hours	[874]
8 nm $CoFe_2O_4$ NPs	Pentacene	10^3	73.84	−60 to 60	1000	10^4	[875]
N-C_{60}-7/CuPc	Pentacene	10^3	4.4	−5 to 5	500	10^4	[877]
GQDs	Pentacene	10^6	37.6	−60 to 60	100	10^4	[594]
Au PS NPs	Pentacene	10^5	70	−80 to 80	—	10^6	[878]
Au@SiO_2 NPs	Pentacene	—	23.6	−40 to 40	2200	10^6	[880]
ZnOPS30 NPs	Pentacene	10^5	31.26	−50 to 60	200	10^5	[881]
ZnOPVPK30 NPs	Pentacene	10^6	60.25	−80 to 60	200	10^5	[881]
C_{18}-Au(55 Å)-NPs	Pentacene	10^4	107.66	−80 to 80	—	2×10^3	[882]
Au-NH_2 NPs	P3HT	10^4	10.6	−10 to 15	100	10^4	[883]
Au-L6 NPs	F8T2	$>10^3$	41	−30 to 30	100	10^5	[884]
P3	Pentacene	10^5	41.8	−100 to 100	100	10^4	[887]
PVTT	BPE-PTCDI	10^3	81	−100 to 100	—	10^4	[890]
PI(BPDA-DAP)	Pentacene	10^6	81	−100 to 100	100	10^4	[891]
PI(APS-BPA)	BPE-PTCDI	104	81.49	−100 to 100	100	10^4	[892]
PI(APSP-6FDA)	BPE-PTCDI thin film	10^3	63	−60 to 80	200	10^4	[893]

Table 5.1 (Continued)

Charge-trapping layer	OSC	Memory ratio	Memory window (V)	Sweep voltage (V)	Endurance test (cycle)	Retention time (s)	References
PI(APSP-6FDA)	BPE-PTCDI nanowires	10^3	82	−60 to 80	200	10^4	[893]
MH(4Py-BT-4Py)$_{1.5}$-*b*-PPyMA$_{10}$	Pentacene	$>10^6$	34	−50 to 30	100	10^4	[896]
PCBM:N(PTPMA)$_3$	BPE-PTCDI	102	33.55	−30 to 50	200	10^4	[898]
MHPy4	Pentacene	$>10^5$	52.7	−40 to 40	200	10^4	[899]
PCBM:P(StFl)$_4$	Pentacene	106	35.59	−50 to 30	100	10^4	[900]
PF-*b*-PSA	Pentacene	—	35.8	−50 to 30	—	10^4	[901]
PF(SWCNTs)-*b*-PSA	Pentacene	$>10^4$	49.2	−50 to 30	240	10^4	[901]
Au-OH PS-P2VP NPs	P3HT	10^4	41	−80 to 20	720	4×10^4	[902]
PS-*b*-(Py-Pt$_{1.0}$@P4VP)	Pentacene	10^5	26.4	−30 to 50	300	10^5	[903]
PAD/TOA-Au NPs	Pentacene	10^6	145	−100 to 100	200	10^5	[904]
Au PMMA NPs	Pentacene	10^4	2.1	−5 to 0	1000	10^5	[905]
Au NPs	Pentacene	10^5	63	−40 to 80	100	10^4	[906]
AzoC6H-AuNPs	Pentacene	10^2	102.5	−80 to 80	—	10^3	[907]
Amorphous C NPs	Pentacene	10^6	35	−40 to 40	1000	10^4	[908]
Amorphous C NPs	PTCDI-$C_{13}H_{27}$	10^5	60	−40 to 40	1000	10^3	[908]
MH-*b*-PStFl$_1$	Pentacene	$>10^6$	40	−50 to 30	180	10^4	[909]
Star-PTMPMA	BPE-PTCDI	10^5	54	−100 to 60	—	2×10^4	[74]
PI-B	PTCDI	10^5	22.1	−10 to 40	—	5×10^3	[910]
PI(BAPF-6FDA)	Pentacene	10^6	77	−100 to 75	100	10^4	[911]
P(St-Fl)$_3$	Pentacene	10^5	133	−100 to 80	100	10^4	[912]
L2-FM50	Pentacene	102	78.4	−100 to 100	100	10^4	[913]

the only difference is where the charges are stored. Both single-polymer compounds and polymer composite mixtures can be used as polymeric electrets. The performance of a memory device based on a polymer electret is surely influenced by the chemical structure of the polymer, such as hydrophobic/hydrophilic side chains, polymer conjugation length, molecular ratio of different moieties, and donor–acceptor strength in a D–A-type co-polymer [752, 866, 886, 887]. It was suggested by Chou et al. [752] that hydrophobic polymers showed better memory performance than hydrophilic polar polymers such as PVA. This could be attributed to hampered electron accumulation in the channel caused by hydroxyl or polar groups acting as electron traps [886, 888]. With a broader conjugated plane, the electron-storing efficiency is higher [889]. For example, the memory performance differed with different thiophene chain lengths because thiophene is a stronger electron-donor than fluorine or phenyl, and thus it has a large memory window [890]. As for the D–A-type polymers, such as commonly used D–A polyimide, the change in aromatic cores between imide groups and donor moieties, which results from the change in the interaction strength between the donor and the acceptor, impacts the device performance [891–893]. Bio-based materials are also good candidates as electrets because of their charge storage ability, along with the advantages of abundance, renewability, and environmental friendliness. For example, oligosaccharide-based memories have been fabricated, which could also be used as D–A co-polymer materials [894–896]. When using polymer mixtures as the charge storage layer, there are many choices, such as blends of polymers with varieties of small molecules and carbon-based materials [897–901]. The occurrence of a charge-transfer interaction when one of the compounds in the mixture acts as the donor and the other acts as the acceptor or hydrogen bonding when strong polar groups with O or N atoms are included in the blends would further influence the electrical behaviors of the memory devices [866, 867].

Last but not least, we have summarized in Table 5.1 the different trapping layers at OSC/dielectric interface and compared their nonvolatile memory performances, which include memory ratio, memory window, sweep voltage, endurance test, and retention time.

6

Semiconductor/Environment Interface

Three different interfaces – the interface within the semiconductor layers, the semiconductor/electrode interface, and the semiconductor/dielectric interface – have a significant impact on charge-transport behavior, which more or less affects the stability of an organic field-effect transistor (OFET) device. In this chapter, we focus on the semiconductor/environment interface because the charge-transport behaviors within the channel layer are sensitive to external stimuli. When the channel is directly exposed to the external environment, it allows the corresponding OFET to work as a platform for sensing applications. Also, one of the most essential advantages of OFET-based sensors is that the organic semiconductor (OSC) molecules can be chemically functionalized via covalently bonded side groups to guarantee more favorable adsorption of the analytes of interest, thereby improving the selectivity. In fact, among all the interfaces existing in OFETs, the semiconductor/environment interface is very attractive because OFETs have emerged as state-of-the-art sensing platforms for flexible, versatile, and disposable biosensors. Various OFET-based sensors, such as physical, optical, chemical, and biological OFET detectors, have been proposed for practical applications, including photodetection, humidity control, pH measurement, chemical and biological species detection, cell growth monitoring, and even drug delivery [12, 15, 18, 828, 914].

First, we start with the most classical OFET-based chemical sensor, where the charge transport can be modulated by noncovalent attractions, such as H-bonds, donor–acceptor interactions, and π–π interactions between the analytes and the OSCs in the active channel layer. These interactions alter the device characteristics, such as mobility, on/off ratio (I_{on}/I_{off}), threshold voltage (V_{th}), because of carrier density changes and/or carrier transport properties. As a result, binding events of analytes can be converted into easily recorded/analyzed electrical signals by using OFETs as a transducer. Many active organic materials, such as pentacene, substituted thiophene-based oligomers/polymers, metalloporphyrins (MPs), and metallophthalocyanines (MPCs), have come into use in OFETs for detecting nitro-based explosives [915], ammonia [916], nitric oxides [917], and peroxides [917]. The analytes induce a V_{th} shift depending on the charges of the analytes injected into or withdrawn from the OSC. It is worth noting that the modulation of carrier transport usually requires the diffusion of analytes to interact with the

Interface Engineering in Organic Field-Effect Transistors, First Edition. Xuefeng Guo and Hongliang Chen.
© 2023 WILEY-VCH GmbH. Published 2023 by WILEY-VCH GmbH.

first few layers of the semiconducting molecules, thus ensuring high sensitivity. Generally speaking, a semiconductor sensor can be considered to operate with two parts: one is the receptor layer (sensor area), where current variation occurs to produce signals owing to adsorbing and electrically interacting with the analytes; the other is the charge-transport layer (transducer area), which is responsible for generating the background current and presenting the corresponding current variations. Both significantly influence the sensor performance [918, 919].

6.1 Device Optimization to Improve Sensing Performance

Considering the fact that a couple of well-written reviews introducing OFET-based chemical/biological sensors have been published elsewhere earlier [12, 15], we herein choose several recent representative works and highlight the most important developments of OFET-based sensors. To better understand the sensing ability of OFET-based chemical sensors, we would like to discuss the general approaches used to optimize the device structure for improving the sensing performance.

6.1.1 Monolayer Functionalization

In principle, the most direct method for achieving higher sensitivity is to use a "buried" conductive channel as the direct interaction interface, that is, to make analyte molecules able to be directly adsorbed onto the conductive channel without diffusion through a dense semiconductor film. Shaymurat et al. [920] utilized an OFET with an air dielectric layer to address this requirement (Figure 6.1a). By exposing the conductive channel to the analyte, direct interaction between the two occurred at the analyte/semiconductor interface. The enhanced interaction can be verified by observing the improved sensitivity for sulfur dioxide (SO_2) detection. The air dielectric device showed a high sensitivity of 119% at a low detection limit of 500 ppb (Figure 6.1b).

A promising approach to manipulate the sensing capacity of OFETs has been developed, that is, downscaling the OSC active layer to a minimum number of molecular layers or even a monolayer (Figure 6.1c) [17, 184, 703, 921]. This is because the exposure of the first few layers in the active channel at the dielectric/OSC interfaces offers direct interaction between the analyte and the conductive channel, which is preferred for improving the sensitivity and response speed. For example, Zhang et al. [922] employed n-type naphthalene diimide derivatives (NDIs) to construct ultrathin film devices. Unlike conventional OFETs, the ultrathin devices possessed faster responses and higher sensitivity to the analytes. The response signal to 100 ppm ammonia (NH_3) was one order of magnitude stronger than that of a thick-film (70 nm) device (Figure 6.1d). Li et al. [923, 924] presented a versatile dip-coating process for the growth of dendritic microstrips with four to six molecular layers. As expected, the ultrathin sensors exhibited sensitivities as high as 1000 ($I_{DS\text{-}NH3}/I_{DS\text{-}air}$) for 100 ppm NH_3 and higher than 30 ($I_{DS\text{-}NH3}/I_{DS\text{-}air}$)

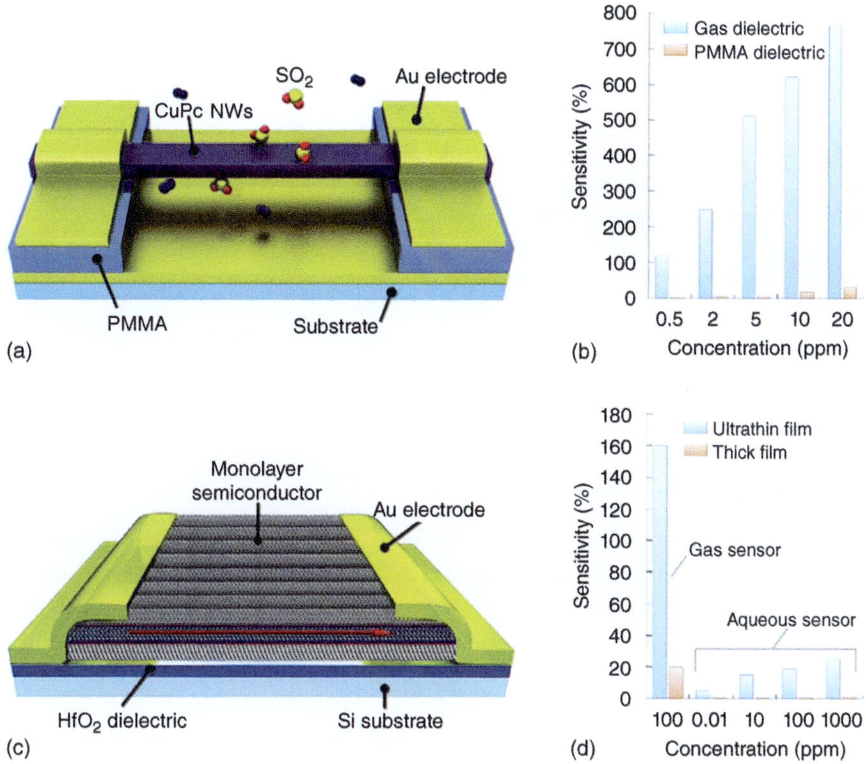

Figure 6.1 Device engineering. (a) Gas dielectric FETs based on CuPc nanowires and (b) high sensitivity for SO_2 detection. (c) Monolayer FETs based on core-cladding liquid-crystalline pentathiophene and (d) high sensitivity for melamine detection in water. Source: (a, b) Adapted from Shaymurat et al. [920]. (c, d) Reproduced with permission from Chen et al. [184], © 2015 John Wiley & Sons, Inc.

for 10 ppm NH_3 and short response/recovery times of 15–31 s/35–50 s. As a control, the thick-film devices showed low sensitivities of only 10 ($I_{DS\text{-}NH3}/I_{DS\text{-}air}$) and long response/recovery times of 70–130 s [924].

In addition to gas sensors, ultrathin-film OFETs possess potential applications as sensors in aqueous media. However, these devices are hard to construct and suffer from instability problems in aqueous media. To solve this problem, Chen et al. [184] recently presented a monolayer field-effect transistor (MFET) with core-cladding liquid-crystalline pentathiophene as the semiconductor, where the conductive pentathiophene core was sandwiched by two insulating alkyl chains. As a result, the MFET device displayed impressive *in situ* detection capability of trace amounts of chemical and biological analytes with long-term operational stability in aqueous solution. It also demonstrated a sensitivity that was three orders of magnitude higher, as low as 0.01 ppm, and a faster response time of 2 s than thick transistors when detecting melamine in water (Figure 6.1d). The excellent outcome of ultrathin OSC film devices such as MFET suggests that it is a highly beneficial and promising method to enhance the sensor performance of electrical chemo-/biosensors.

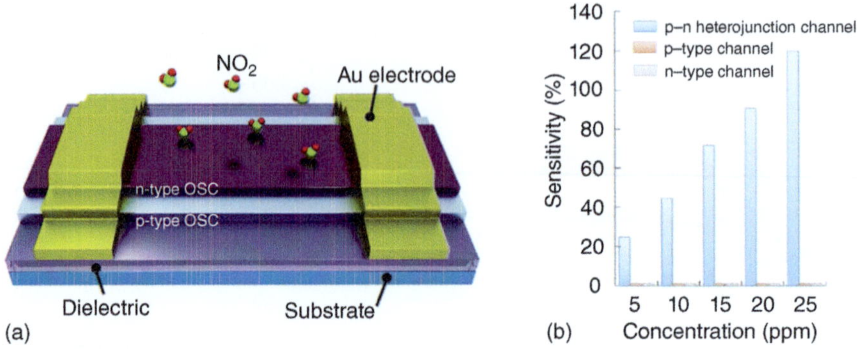

Figure 6.2 Bilayer heterojunction sensor. (a) p–n junctions sensor based on PTCDI–Ph/*p*–6P and (b) high sensitivity for NO_2 detection. Source: Adapted from Ji et al. [925].

6.1.2 Bilayer Heterojunction Approach

OFETs with heterojunction OSCs as the active layer have also been proven to be an effective solution to enhancing the sensitivity owing to the accumulated charge density at the heterojunction interface (Figure 6.2a) [153, 925–927]. Ji et al. [925] reported an ultrathin organic heterojunction OFET sensor that enabled room-temperature detection of nitrogen dioxide (NO_2). They introduced n-type *N*,*N*′-diphenyl perylene tetracarboxylic diimide (PTCDI–Ph) films (1 nm) on top of p-type para-hexaphenyl (*p*-6P) films (5 nm). The heterojunction effect of the *p*-6P/PTCDI–Ph caused a high density of electrons to accumulate in the n-type thin layer. In addition, the asymmetric thickness made the electron density much larger at the n-type OSC/environment interface. NO_2, as a strong oxidizing gas, was able to easily capture electrons from the electron-rich n-type OSC layer. Furthermore, the ultrathin characteristic (1 nm) shortened the diffusion process of NO_2 in the n-type thin layer, thus realizing fast room-temperature detection. A long-term stable high-performance NO_2 sensor was successfully fabricated with a limit-of-detection (LOD) of 5 ppm (Figure 6.2b). Moreover, based on the idea of controlling the sensitive-layer thickness and making good use of the charge-transfer effect, the authors added a third layer (continuous or discontinuous) of vanadyl phthalocyanine (VOPc) onto the heterojunction, thus producing a double heterojunction. The response sensitivity was enhanced by five times without affecting the fast response.

Recently, Wang et al. [919] realized ultrahigh-performance NO_2 OFET-based sensors with crystalline 6,13-bis(triisopropylsilylethynyl) (TIPS)–pentacene/*p*-6P films as active heterojunction channels. The 5-nm-thick crystalline *p*-6P served as the molecular growth template for the deposition of high-quality crystalline TIPS–pentacene films. It showed short response/recovery times (within 200 s/400 s) with an extremely low LOD of 20 ppb. By comparing these results with those of pentacene, they suggested that a vital reason why the highly conductive OSCs only showed lower sensitivity might be the high original carrier concentration, hindering the gas-on conductivity. The authors finally pointed out the important impact of charge transport of the semiconducting molecules on the sensor performance, that is, the combination of high mobility and low initial carrier concentration is the key to the fabrication of ultrasensitive sensors for oxidizing gases such as NO_2.

6.1.3 Remote Floating Gate

So far, most of the OFET sensors discussed have been based on the idea of modifying the electrical characteristics of the transistors by physisorbing analyte molecules on the OSC or responsive layer. However, OSCs are fragile to moisture, oxygen, and other reactive species. This limitation is common to all detection mechanisms, as the main drawback is the possible degradation of the OSCs during the measurements in liquids. To solve this problem, Lai et al. [928] designed a unique ultralow-voltage OFET sensor for label-free detection of DNA with a floating gate that applied the bias through a control capacitor. The sensing area was remote from the semiconductor layer, allowing the detection of DNA hybridization without exposing the OSC to the aqueous environment where DNAs were hosted (Figure 6.3). DNA negative charges determine a threshold voltage shift of the OFET, which in turn results in a current change of the transistor, thus, leading to direct, label-free hybridization detection of DNAs. Moreover, thanks to the separation between the sensor area and the OFET area, the same group was able to tailor the sensing performance of an OFET-based biosensor by optimizing the transistor structure and the sensing performance of the device independently [929]. In addition, monitoring pH values in aqueous solutions can also be realized by applying this device configuration [930, 931]. In a study of label-free DNA detection published by Park et al. [932], they demonstrated an Au floating electrode over an active layer of pentacene to achieve high sensitivity. The thiolated ssDNA probe deposition and hybridization could be detected on the transistors, achieving a minimum of 100 nM of the immobilized poly-A oligomer. The drain current change was due to the negative charge nature of the DNA, affecting the p-channel operation of the pentacene OFETs, positively shifting the threshold voltage and decreasing the mobility.

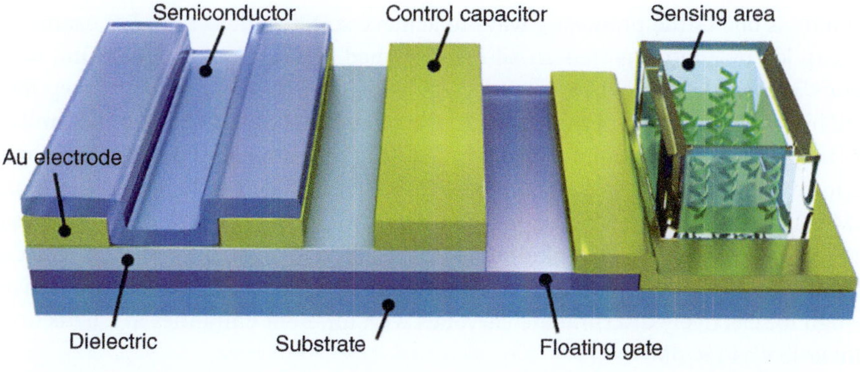

Figure 6.3 Remote floating gate device for a DNA sensor with two organic charge-modulated FETs (OCMFETs) (one sensor and one reference). Source: Adapted from Lai et al. [928].

Minami et al. [933] presented an OFET-based sensor with an artificial receptor modified gate for the detection of Hg^{2+} ions in water. They chose pBTTT–C16 as the active channel and encapsulated it with a passivation layer, providing stable and reproducible operation. The gate electrode was modified with a thiolated dipicolylamine (DPA) derivative that could coordinate with Hg^{2+}. When DPA captured Hg^{2+}, a surface potential shift was observed, further affecting the channel conductance owing to the potential difference between the Ag/AgCl reference and the extended-gate electrode. Hence, the output current of the channel was changed. The LOD of the device to sense Hg^{2+} was estimated to be 9.9 ppb with high selectivity. The same group also demonstrated phosphate detection in water by using an extended-gate-type OFET [934]. By making good use of the extended solution gate, Song et al. [935] reported label-free detection of glial fibrillary acidic protein (GFAP) with OFET-based biosensors. By applying the novel extended solution gate, the sensing area and the OSC (pentacene) were separated without the need for a reference electrode for the first time, improving the stability and simplicity of the device. On the sensor area, i.e. the extended solution gate, anti-GFAP was immobilized on the 1-ethyl-3-(3-dimethylaminopropyl)carbodiimide (EDC)/*N*-hydroxysuccinimide sodium (NHS)-treated polystyrene-*co*-methacrylic acid (PS–MA) layer, providing a smooth surface. Different molecular weight polyethylene glycols (PEGs) were mixed into the bioreceptor layer to alter the Debye screening length. The increased effective Debye screening length led to an increased drain current, owing to the reduced dielectric constant and low ion density environment for the binding proteins after mixing with higher-molecular-weight PEGs. A higher sensitivity was obtained with an increase in gate voltage from −5 to −2 V. Then, the authors fixed a gate voltage of −2 V for all other tests when recording real-time current vibration. Moreover, good stability and selectivity were achieved when sensing GFAP, even in the presence of different interferents in the solutions on the devices.

6.2 OECT-Based and EGOFET-Based Sensors

Chirality is one of the promising ways to achieve selective detection. For example, a group led by Torsi reported an ad-hoc-designed OSC endowed with chiral side groups as the selective chiral sensing layer, which provided a significant step forward in the development of a chirality-sensing platform [936]. The detection limitation they obtained was in the tens of ppm concentration range. Recently, the same group reported a water-gated organic FET (WGOFET) sensor [937, 938] with an SAM of porcine odorant-binding proteins (pOBPs) on its gate [939, 940]. Because (S)-(+)- and (R)-(−)-carvone enantiomers were differentially bound to immobilized odorant-binding protein (OBP) SAMs, the transduction was remarkably sensitive enough to selectively discriminate carvones with different chiralities at a detection limit as low as 150 pM.

The selectivity and sensitivity of organic sensors can be enhanced by interface modification with specific host–guest molecules such as calixarene derivatives. According to the inner cavity size and molecular interactions, perallyloxyCB[6] ((allyloxy)12CB[6]) [941] and calix[8]arene (C[8]A) [942] showed highly selective discrimination of acetylcholine (ACh^+) and common organic solvents such as methanol, ethanol, DI water, and hexane. This strategy, based on host–guest molecular recognition, can be naturally extended to achieve ion detection as well. For example, an ion-sensitive FET (ISFET) on flexible substrates was designed to measure the concentration of K^+ and H^+ ions with high specificity [943]. The K^+ sensor used a valinomycin ionophore composed of an 18-crown-6 ionophore, whose hydrophilic interior was the right size to selectively accommodate potassium ions. Furthermore, when tantalum pentoxide (Ta_2O_5) was used as the high gate dielectric and H^+ ion selective layer, both K^+ and H^+ reached detection limits as low as 6.68–9.5 and 33 mM, respectively.

High selectivity can be easily achieved by choosing selective membranes, especially in organic electrochemical transistors (OECTs). For example, Malliaras's group used bilayer lipid membranes (BLMs) incorporated with gramicidin ion channels to control the gating effect of OECTs [944]. The valence-dependent permeability of gramicidin enabled the sensors to discriminate between monovalent and divalent ions. This work demonstrated that ion channels could effectively improve the selectivity of organic transistor-based sensors. Moreover, based on an electrolyte-gated P3HT thin-film transistor and a state-of-the-art sodium ion (Na^+) selective membrane, different detections of Na^+ and other cations, including K^+, Ca^{2+}, and Ag^+, were realized at 10^{-6} and 10^{-4} M levels [945, 946]. Prakash and coworkers reported an electrochemical transistor that consisted of two platinum electrodes separated by 100 μm spacing and bridged with an anodically grown film of polycarbazole [947]. They found that the conductivity of the polycarbazole film changed upon the addition of Cu(II) ions, which was attributed to the conformational change in the polymer phase owing to the occupation of Cu(II) ions. The conformational change offered a detection limit of 2.5×10^{-6} M for Cu(II) ions. This work proved that the conformational change could be a promising strategy to realizing specific detection, and thus it could be used in biological systems. For example, Jen's group demonstrated that genetically engineered peptides for inorganics (GEPIs) formed a functional layer at the semiconductor/dielectric interface and adjusted the threshold voltages of OFETs [948]. The device performance varied according to the variation of the produced dipoles of amide bonds and side chains when the peptide conformation or structure changed by adjusting the pH of the solutions. Further integration of biosystems into an OFET structure was demonstrated by fabricating phospholipid or protein layers between the gate dielectric and the OSC [949]. The platform bench tests involved OFETs that included phospholipids and bacteriorhodopsin. When exposed to 1–5% anesthetic doses that revealed drug-induced changes in the lipid membrane, the devices showed an extremely high LOD (as low as 10 ppt).

Sensors based on OFETs have attracted much attention, and the development of such sensors is booming. However, the fabrication of functional sensors with specific recognition is still in the early stage, and more effort is needed. Here, we introduce some important OFET-based sensors with high selectivity using different working mechanisms, such as chirality interaction, coordination interaction, selective membrane and channel, conformation change, host–guest molecular recognition, charge interaction, chemical reaction, and others, as shown in Table 6.1.

Table 6.1 Summary of OFET-based sensors for selective sensing.

Type	Mechanism	Analyte	Receptor	Active layer	Detection limit	References
OFET	Coordination	SO_2, gas	CuPc	Cupc nanowire	0.5 ppm	[920]
OFET	Coordination	NH_3, gas	Tris-(pentafluorophenyl)borane	CuPc/CoPc	350 ppb	[950]
OFET	Chemical reaction	NH_3, gas	ROMP-dielectric materials	rr-P3HT	100 ppm	[951]
OFET	Chemical reaction	NH_3 (HCl), gas	NDI(2OD)(4tBuPh)-DTYM2/ pentacene	NDI(2OD)(4tBuPh)-DTYM2/ pentacene	10 ppm	[952]
	Chemical reaction					
OFET		NO_2, gas	DNA	Pentacene	10 ppm	[953]
OFET	Chirality	Carvone (R/S), in solution	D-glucose	PTA	5 ppm	[936]
EGOFET	Chirality	Carvone (R/S), in solution	pOBP-mutant-F88W	PBTTT-C14	150 pm	[937]
OFET	Coordination	DMMP/MeOH, gas	CuPc/F_6CuPc	CuPc/F_6CuPc	68 ppm	[954]
Ext-OFET	Coordination	Phosphate anions, in solution	Zinc(II)dipicolylamine derivative	pBTTT	NA	[934]
EGOFET	Selective membrane	Na^+, in solution	Na^+ selective membrane	P3HT	10^{-6} M	[945]
OECT	Selective membrane	$K^+/Ca^{2+}/Ag^+$, in solution	Ionophore-based membrane	PEDOT:PSS	10^{-4} M	[946]
OECT	Selective membrane	Monovalent and divalent cation, in solution	Bilayer lipid membrane	PEDOT:PSS	NA	[944]
OECT	Conformation change	Cu^{2+}, in solution	Polycarbazole	Polycarbazole	2.5×10^{-6} M	[947]
OFET	Conformation change	Peptides basic/acidic solutions	QBP peptide	Pentacene	—	[948]
OFET	Conformation change	Biotin, in solution	Bacteriorhodopsin or streptavidin	P3HT	10 ppt	[949]
		halothane/diethylether, gas			2%	
OFET	Host–guest recognition	Methanol, ethanol, deionized water, hexane	Calix[8]arene	Azide-functionalized P3HT	1 vol%	[942]
OFET	Host–guest recognition	Acetylcholine	PerallyloxyCB[6]	DDFTTF	10^{-12} M	[941]
ISFET	Host–guest recognition	H^+/K^+, in solution	Ta_2O_5/valinomycin ionophore	P3HT	6.68–9.5 mM/33 mM	[943]

OFET	Host–guest recognition	DMMP, in solution	Two-layer heterostructure	6PTTP6 (bottom)/6PTTP6–HO6OPT(top)	10 ppm	[955]
OFET	Host–guest recognition	Hg^{2+}, in solution	DNA	DDFTTF	NA	[956]
OFET	Host–guest recognition	Hg^{2+}, in solution	DNA	Isoindigo-based polymer	10 μM	[957]
OFET	Charge interaction	SO_4^{2-} , in solution	Sulfate binding protein	PDTT	1 mM	[958]
Ext-OFET	Charge interaction	pH, in solution	Thioaminic group	Pentacene	NA	[930, 931]
OFET	Charge interaction	NH_3 , gas	P3HT/PS Blends	P3HT	5 ppm	[959]
OFET	Charge interaction	Urea, in solution	Urease	DNTT	NA	[960]
OFET	Charge interaction	Alkylators	Pyridine group	Polythiophene analogue	NA	[961]
OFET	Charge interaction	H_2O	Poly(METAC-co-TSPM	Pentacene	NA	[962]
OFET	Redox reaction	$POCl_3$ /DECP, gas	TTF derivative	TTF derivative	10 ppb	[963]
Ext-OFET	Redox reaction	Nitrate, in solution	Nitrate reductase	pBTTT	45 ppb	[964]
OFET	Redox reaction	Lactate, in solution	Lactate oxidase	pBTTT-C16	66 nM	[965]
OFET	Redox reaction	H_2O_2 /DTBP	Metal-phthalocyanine/ naphthalocyanine	Metal-phthalocyanine/ naphthalocyanine	0.64 ppm for DTBP	[966]
OFET		Acetic acid/acetone/IPA/water/H_2O_2/ DMMP/hexane/toluene	8-NTCDI/DMP-NTCDI/CuPc	8-NTCDI/DMP-NTCDI; CuPc	NA	[967]

Note: Ext-OFET: Extended gate OFET /floating gate OFET; ISFET: Ion-sensitive FET.

7

Interfacing Organic Electronics with Biology

One key application of organic field-effect transistor (OFET)-based functional devices is to explore their use of electrical stimulation to reflect biological events and build an efficient bridge to connect the worlds of biology and electronics. The unique characteristics of organics, such as facile chemical modification, low-temperature processing, allowance for oxide-free interfaces with aqueous electrolytes, and efficient ion transport at room temperature, together with other features such as flexibility, biocompatibility, functionality, and solution processability, make organic electronics a promising platform in biosensors and neural implants [968, 969]. To date, there are a variety of bioelectronic devices available that can improve healthcare, provide environmental protection, and accelerate the pace of scientific progress. Recently, bioelectronics has been in the news, especially since the launch of the EU Flagship Initiative called the "Human Brain Project," the US equivalent "Brain Activity Map" [970], and the "China Brain Project" [971]. In this regard, we aim to provide an invaluable overview of the organic device/biology interface and its newest developments in the field so far, providing readers with a broad view from fundamental understandings to technological and experimental techniques.

Specifically, OFETs/organic electrochemical transistors (OECTs) [972], which adopt organic semiconductors (OSCs) as their active layer or employ anchored receptors along the organic-semiconductor surface facing the aqueous medium, have been explored as biosensors in various applications, such as to detect and monitor DNA [956], protein [973], glucose [974], virus [975], dopamine [976], and enzyme [977]. There are many research publications as well as reviews focusing on this area [750, 751, 978]. We do not want to collect data or repeat the clichés here. A comprehensive summary of the application of organic electronics in biology can be found in several previous review articles [751, 979, 980]. However, establishing platforms of OFET/OECT sensors that target sensing and monitoring of specific biological signals by interfacing with cells or tissues for diagnostics and therapy within future healthcare systems remains a great challenge.

In this chapter, works and advances related to the coupling of OFET/OECTs with live mammalian cells will be summarized, addressing issues such as tissue compatibility, long-term stability, water stability, and ionic–electronic transduction [981].

Interface Engineering in Organic Field-Effect Transistors, First Edition. Xuefeng Guo and Hongliang Chen.
© 2023 WILEY-VCH GmbH. Published 2023 by WILEY-VCH GmbH.

This chapter is divided into two sections: integration with nonelectrogenic cells for monitoring toxicology/diagnostics and integration with electrogenic cells such as cardiomyocytes and neurons for electrophysiological measurements.

7.1 Integration of OFETs/OECTs with Nonelectrogenic Cells

Biocompatibility is the main advantage of organic devices [750]. Beyond that, the redox states of active channels can be reversibly altered by electrochemical methods, which will in turn tailor the surface chemistry of the sensing areas. On the biotic side, surface properties are an important determinant for biological events such as cell adhesion, differentiation, and proliferation. As a result, electrochemical modulation of the active layers has a huge impact on the activities of the anchored cells [982]. Studies have proven that patterned poly(3,4-ethylenedioxythiophene) polystyrene sulfonate (PEDOT/PSS) electrodes under an electric field can effectively control the conformation of proteins and regulate the proangiogenic capability of tumor-associated adipogenic stromal cells [983]. In other words, by applying an electric field, a cell culture dish with two adjacent PEDOT/tosylate electrode surfaces would become reduced and oxidized, respectively. The reduced polymer surface could efficiently promote cell proliferation, whereas the oxidized polymer would alter the conformation of functional fibronectin and severely compromise the viability of cells.

Bearing this in mind, when employed as the active layer, semiconducting/conducting polymer films could control cell adhesion and proliferation by electrochemical modulation. Actually, two approaches can be applied, namely, the counterion approach and the electric field approach. For the counterion approach, counterion-modulated conducting polymer surfaces for cell cultures have been a mainstay of organic bioelectronics. Herland et al. [984] used the principle of heparin "doping" to guide stem cell differentiation (Figure 7.1a). In their work, fibroblast growth factor-2 (FGF2) was bound to the heparin polymers, and this composite was used as the counterion in electropolymerized PEDOT. Stem cells remained in their predifferentiated state controlled by FGF2, resulting in undifferentiated cells when reducing the PEDOT to the (near) neutral state (Figure 7.1b). Oxidization of PEDOTs allowed the cells to begin differentiation, which was caused by a "retraction" of the heparin–FGF2.

Bolin et al. [985] reported the electric-field modulation of OECT channels to control the cell-density gradients (Figure 7.1c). In their work, Madin–Darby canine kidney cells were seeded and cultivated along the channel of an OECT. Depending on the redox potential (controlled by V_D and V_G) of discrete areas of the active channel, they observed differential cell adhesion, illustrating the potential for CPs to electrically control the adhesion of cells. On the other hand, the location of cells in the channel can also be predicted by changing the operating voltages (Figure 7.1d). This work may open up a way to precisely control the cell gradient characteristics of tissues.

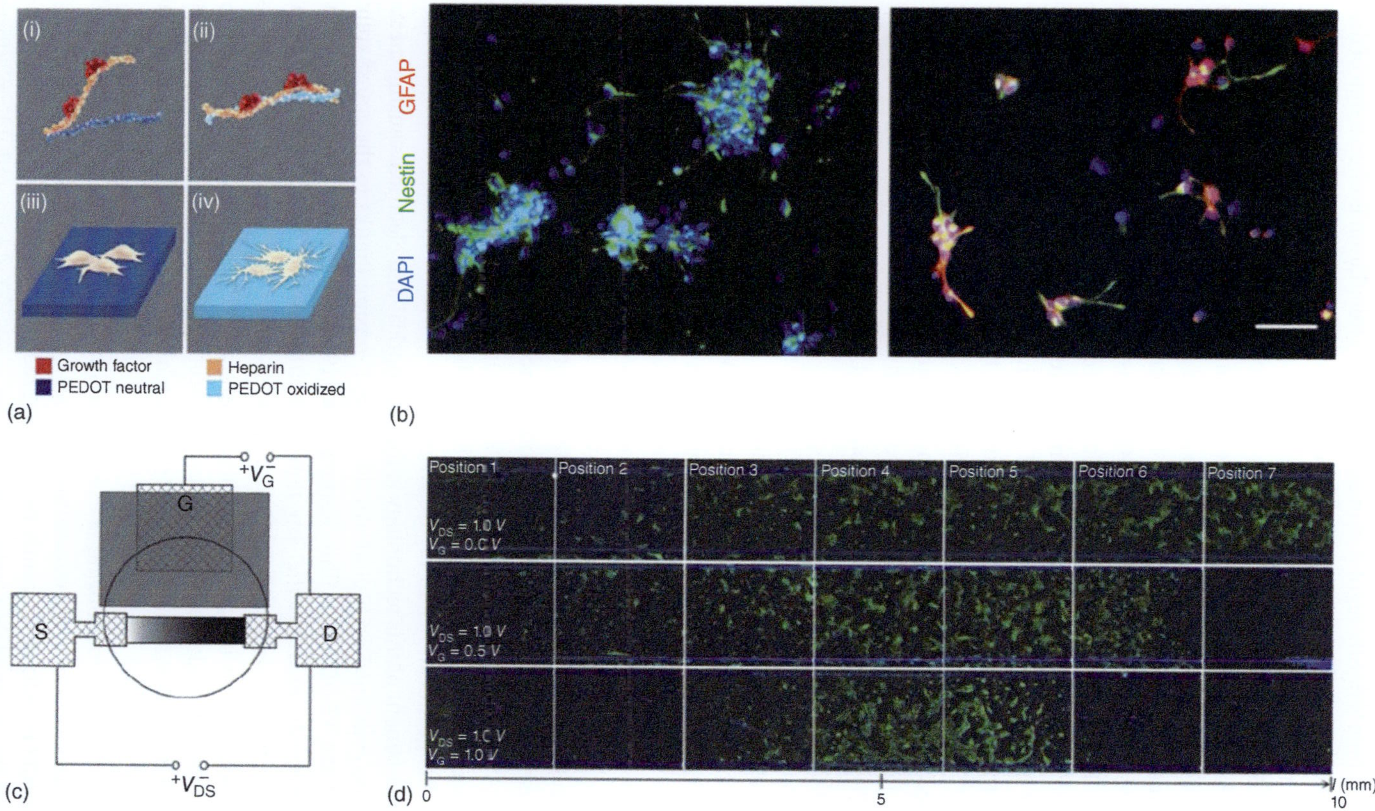

Figure 7.1 Electrochemical control of cell culture on substrates. (a) Neutral (reduced) PEDOT causes heparin-growth-factor compounds to be exposed/available to stem cells. (b) Neural stem cells cultured for four days on PEDOT: heparin surfaces were kept at open circuit (left) or oxidized with live cells (right). Glial fibrillary acidic protein (GFAP) staining (red) indicates differentiated cells. 4′,6-diamidino-2-phenylindol (DAPI) (blue) indicates cell nuclei, and nestin (green) indicates neural stem cells. The scale bar is 75 μm. (c) Schematic illustration of the electrochemical transistor showing the voltage bias configuration. The ring area represents a container wall that confines the cell culture medium and the active channels of the transistor. (d) Cell adhesion along the electrochemical gradient as a function of different gate potentials at a constant source–drain potential. Source: (a, b) Reproduced with permission from Herland et al. [984], © 2011 John Wiley & Sons. (c, d) Reproduced with permission from Bolin et al. [985], © 2009 John Wiley & Sons.

Subsequent integration of OECTs with live cells and even mammalian cells has focused on increasing the sensitivity of the devices to changes in biological ion flux. The flow of ions is tightly regulated in tissues, and dysregulation is often a sign of disease or dysfunction. As a result, this parameter can be used to monitor the integrity of cells. Lin et al. [986] reported a cell-based biosensor by using PEDOT/PSS OECT (Figure 7.2a–c). They used this method to detect cell attachment/detachment by shifting the V_G and off-current via a mechanism similar to that used by them to detect antibody/DNA binding [750]. The sensors that were integrated with cells on the channel regions were characterized in culture medium and showed excellent biocompatibility. After trypsin treatment, because the cancer cells detached from the active layer, the transfer characteristic of the device shifted to a lower gate voltage of approximately 150 mV, as shown in Figure 7.2b. The sensing mechanism of an OECT cell sensor could be attributed to the interfacial electrostatic interaction between the cells and the active layer. Normally, the cells attached to the surface of PEDOT/PSS shift the transfer curve (I_D versus V_G) of the device to a more positive value because of the negative charge on the cell surface. This experiment has demonstrated a new technique for monitoring cell activities. And the operation stability of the OECT in cell culture medium was confirmed, as well as their ability to support cell growth.

However, direct coupling of cells' dynamic physiological ion transport with OECTs electrical readout remains challenging. In a similar configuration, Yao et al. [987] showed the integration of human airway epithelial cells with the OECT array, where the dynamics of transepithelial ion transport could be monitored in real time through an electrical readout. Cells were directly seeded onto an OECT array; however, the cells above the channel are postulated to be suspended over the channel with a gap formed below (Figure 7.2d). The authors investigated the dose response of transepithelial ion transport to forskolin, an agonist that causes opening of the cystic fibrosis transmembrane conductance regulator (CFTR) channel. The transport of Na^+ ions from the basolateral compartment to the apical compartment leads to changes in channel current, which has been converted to an effective gate voltage change and shown in Figure 7.2e.

Another benefit of this technology is that the optical and electrical sensing methods can be combined on a platform during cell sensing thanks to the transparent nature of the PEDOT/PSS film. Ramuz et al. [988] demonstrated the possibility to continuously measure the ion flow in epithelial cells coupled with optical imaging of the cell layer on the device (Figure 7.2f). A corollary of this work is that high-resolution imaging of cells is possible on PEDOT/PSS films, not only in a bright-field mode but also for fluorescence imaging, which would be highly valuable for multi-parametric monitoring of molecular mechanisms in biological systems.

Multi-parametric monitoring is essential to improving the predictive quality of the data used to inform the clinical studies that follow. In addition, future drug discovery could benefit significantly from more predictive and multi-parametric readouts from *in vitro* models. Curto et al. [989] have presented a microfluidic platform integrated with OECT cell sensors (Figure 7.2g). First, they generated a platform

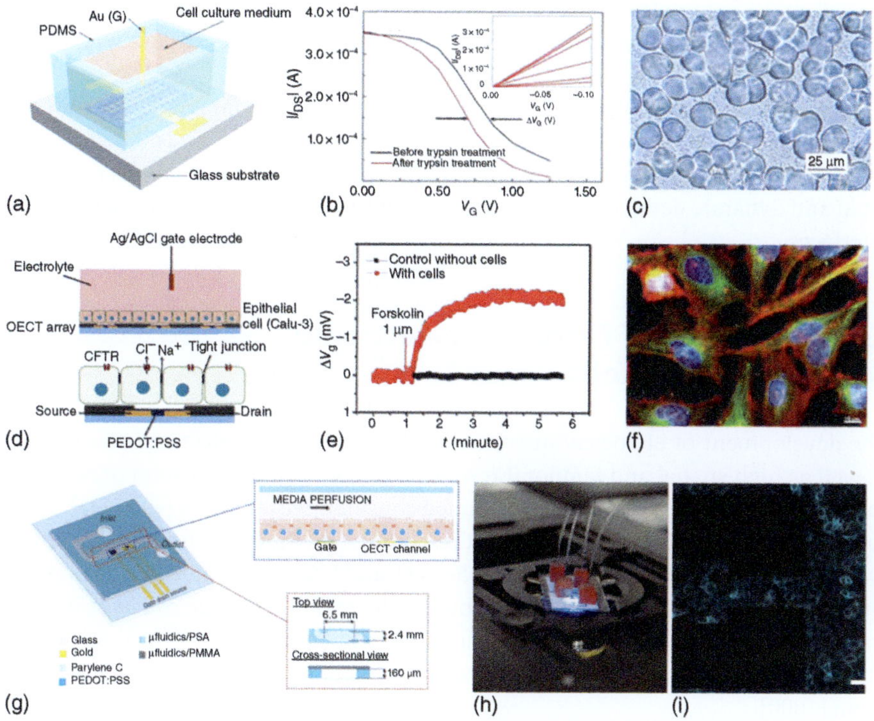

Figure 7.2 Cell sensors. (a) Schematic illustration of a cell-based OECT biosensor. (b) Transfer curves of the sensor with cancer cells before (black) and after (red) trypsin treatment. Inset: output curves of the device before (black) and after (red) trypsin treatment. (c) Optical image of cancer cells cultured on PEDOT/PSS surface. (d) Side view of polarized Calu-3 cells with tight junction sitting on an OECT. (e) *In situ* response with (red) and without (black) Calu-3 cells upon the addition of the 1 mM cystic fibrosis transmembrane conductance regulator (CFTR) agonist forskolin. The transistor channel current change was converted to the effective gate voltage change. (f) Illustrative example (HeLa cells) of high-resolution fluorescence imaging possible on OECT cell sensor. (g) A platform integrating OECTs with microfluidics. Top right: illustration of the OECT and the cell layer lining the bottom surface of the microfluidic channel; bottom right: top and cross-sectional views of the microfluidic device. (h) Image of a fully integrated microfluidic platform located on a microscope stage, featuring inlet and outlet ports and tubing. (i) Fluorescence image of a fully confluent layer of Madin-Darby canine kidney cells from the distal tube of the nephron (MDCK II) cells transfected with red fluorescent protein-labeled F-actin (pLifeAct) grown inside the microfluidic channel integrated with a planar OECT. The scale bar is 100 μm. Source: (a–c) Reproduced with permission from Lin et al. [986], © 2010 John Wiley & Sons. (d, e) Reproduced with permission from Yao et al. [987], © 2013 John Wiley & Sons, Inc. (f) Reproduced with permission from Ramuz et al. [988], © 2014 John Wiley & Sons. (g–i) Reproduced from Curto et al. [989], © 2017 Springer Nature/CC BY 4.0.

to host cells in a more physiologically relevant environment (using physiologically relevant fluid shear stress [FSS]). Second, they showed efficient integration of multiple methods for assessing cell morphology, differentiation, and integrity. These included fluorescence imaging, impedance monitoring, metabolite sensing, and a wound-healing assay (Figure 7.2h–i). They illustrated the possibility and versatility of such multi-parametric monitoring in terms of validating differentiation of cells toward a physiological profile, which will result in more accurate data when used for drug discovery.

In conclusion, the examples cited earlier have involved the integration of the cells on an OECT cell sensor, detecting the cell fate or monitoring the cell response to additives. OECTs cell sensors show great potential for broad applications requiring rapid and dynamic detection of variations in ion flux at additive/cell/device ternary interfaces.

7.2 Integration of Flexible Bioelectronics with Electrogenic Cells

The development of biological instruments that can achieve electrical stimulation as well as high spatial and temporal resolution recordings of activation of neurons, cardiomyocytes, and muscle cells has gained considerable attention [990–993]. Typically, these recordings are performed with electronic methods, such as punch clamps [994], multielectrode arrays (MEAs) [995–997], and Utah micro-needle arrays [998]. Quietly but significantly, OECTs have entered the realm of bioelectronics as biosensors and neural implants, as demonstrated by their utilization in detecting electrophysiological activities such as electrocardiograms [999] and seizure activity in the brain [1000].

Recent progress in flexible electronics has shown that high-performance electronic properties have begun to be decoupled from rigid mechanical behaviors, which has enabled the fabrication of novel electronic devices with flexible, stretchable, and textile properties (Figure 7.3a) [793, 992, 1001, 1002]. Soft electronic devices have been demonstrated with the ability to address healthcare problems that require chronic device implantation. They have already been demonstrated for stimulating as well as recording signals of cardiac electrophysiology [1003] and mapping of the central [996, 1000, 1004] and peripheral [1005, 1006] nervous system. Among all the biological functions, ion flux serves as the main media for transferring information and regulating biological systems in balanced conditions [1007]. Therefore, the ability to interact with ion fluxes in biological systems is critically important for organic bioelectronics. Interestingly, semiconducting polymers afford the conduction of both ions and electronic carriers (i.e. electrons and holes) [1008]. This allows the exchange of ions between the biological milieu and the bulk of the electronic material. Figure 7.3b [1007] shows the schematic structure of an OECT, which is usually used as the platform in neural signal recording, where the channel is in direct contact with an electrolyte and the neurons (Figure 7.3c) [968]. Commonly, an OECT uses an OSC material as the active channel and takes advantage of an oxide-free interface with the electrolyte and the ability of organic materials to transport ions in order to yield a 3D interaction with the biological milieu. In this device configuration, ions from the electrolyte penetrate the bulk polymer film, change its conductance, and result in an electrical readout (Figure 7.3d) [968].

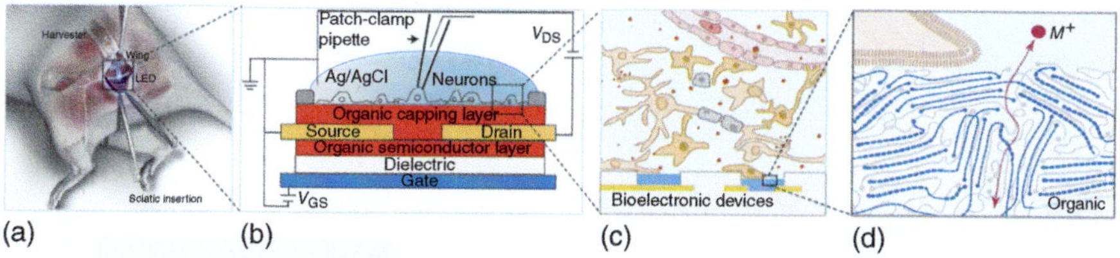

Figure 7.3 Organic bioelectronics. (a) Schematic illustration of bioelectronic devices being implanted in the tissue of a living rat. (b) Schematic of the experiment for extracellular/intracellular stimulation and intracellular/extracellular recording. (c) Schematic of an organic transistor interfaces with neuron cells. (d) Schematic of an OSC, PEDOT, at the interface with an electrolyte solution. Source: (a) Reproduced with permission from Park et al. [992], © 2015 Springer Nature. (b) Reproduced with permission from Benfenati et al. [1007], © 2013 Springer Nature. (c, d) Reproduced with permission from Rivnay et al. [968], © 2014 American Chemical Society.

Figure 7.4a,b demonstrate that OECTs were able to measure a wide range of typical clinical physiological signals, including eye movement (electrooculography, EOG) (Figure 7.4c), cardiac cycle (electrocardiography, ECG) (Figure 7.4d), and neurological rhythms (electroencephalography, EEG) (Figure 7.4e). These electrical signals ranged from a millivolt to tens of microvolts, corresponding to the human's behavior of eye movement, cardiac rhythm, and brain activity, respectively.

EOG can be used for eye tracking [1010] and has applications in research on the visual system [1011], in psychiatry [1012], and in cognitive linguistics [1013]. Two OECTs were wired as shown in Figure 7.4c to simultaneously track both the horizontal (blue traces) and vertical (red traces) movements of the eyes of a human volunteer. In ECG applications (Figure 7.4d) [999, 1009], the gate of the OECT was placed directly on the skin close to the heart at a constant positive potential ($V_g = 0.5\,\text{V}$) relative to the ground potential of the body, while the transistor channel was placed on the forearm at a negative potential ($V_d = -0.3\,\text{V}$) relative to the ground. Every heartbeat creates an additional potential, which modulates the $V_{G,}$ and results in a clear dedoping of the channel (Figure 7.4d, bottom). In addition, Gu et al. [1014] have demonstrated the monitoring capability of a long-term stable 16-channel OECT array to map the action potential conduction of a primary rat cardiomyocyte monolayer. With the ability to map the action potential in a 2D form, it is able to reproduce the irregular action potential propagation on a rat cardiomyocyte model, paving the way for studying cardiac electrophysiological diseases as well as drug screening. The last clinical diagnostic technique is EEG (Figure 7.4e). The synchronized activity of large numbers of cortical neurons in EEG is called a rhythm. This rhythm on an awake human volunteer can be measured by using the recording configuration. A sample extracted from a recording is shown in Figure 7.4e, bottom, which demonstrates that spontaneous brain activity induces drain current changes of the order of tens of nanoamperes.

Specifically, for deeper brain activity, electrophysiology recording is more complicated than EEG techniques. Two other techniques, namely, electrocorticography (ECoG), which utilizes electrodes in direct contact with the brain surface, and stereoelectroencephalography (SEEG), which utilizes electrodes that penetrate deep in

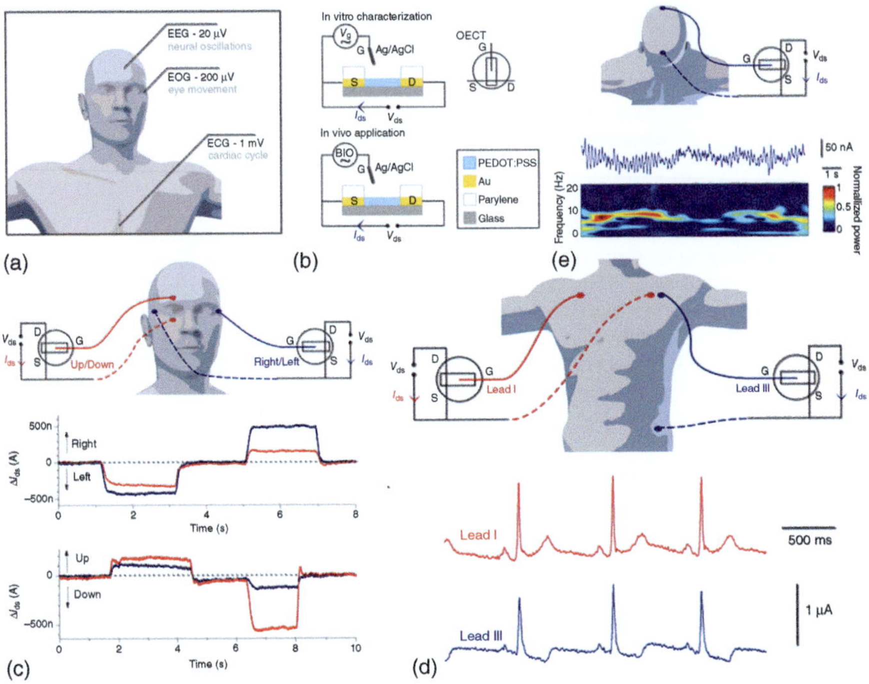

Figure 7.4 OECT-based electrophysiological measurements. (a) Cartoon showing different electrophysiological measurements and typical signal amplitude. (b) Schematic of an OECT and the wiring diagrams used *in vitro* (top) and *in vivo* (bottom). (c) Wiring configuration for EOG measurement (top) and recording of electrical activity during eyeball movements (bottom). (d) Wiring configuration for ECG measurement (top), and recordings of spontaneous heart activity (bottom). (e) Wiring configuration for EEG measurement (top) and recording of spontaneous brain activity (middle) showing the alpha rhythm and associated time-frequency spectrogram (bottom). Source: Reproduced with permission from Leleux et al. [1009], © 2015 John Wiley & Sons.

the brain, are also essential [1015]. Selection of the brain monitoring techniques, EEG, ECoG, or SEEG, depends on the nature of the signal of interest, the size of the neural population to be interrogated, or the invasiveness/goal of the measurement. A primary challenge is to form a well-established mechanical, flexible, and biocompatible interface with the brain. Currently, most of the electrodes used are relatively inorganic and inflexible, anchored in the skull, and cannot afford the movements of the brain. What's worse, the recording quality of the signal usually deteriorates over time because of tissue injury and reaction of the immune system to the implanted electrodes [1016]. The challenge in this area lies in obtaining high-quality and stable overtime recordings, that is, through the use of more biocompatible materials, and of course amplification of neuronal signals necessary to detect low-magnitude signals of interest.

As a proof of concept, Khodagholy et al. [1000] demonstrated an implantable OECT sensor for *in vivo* ECoG recordings on the brain surface. The device, consisting of OECT arrays and integrated electrodes, was fabricated on top of a

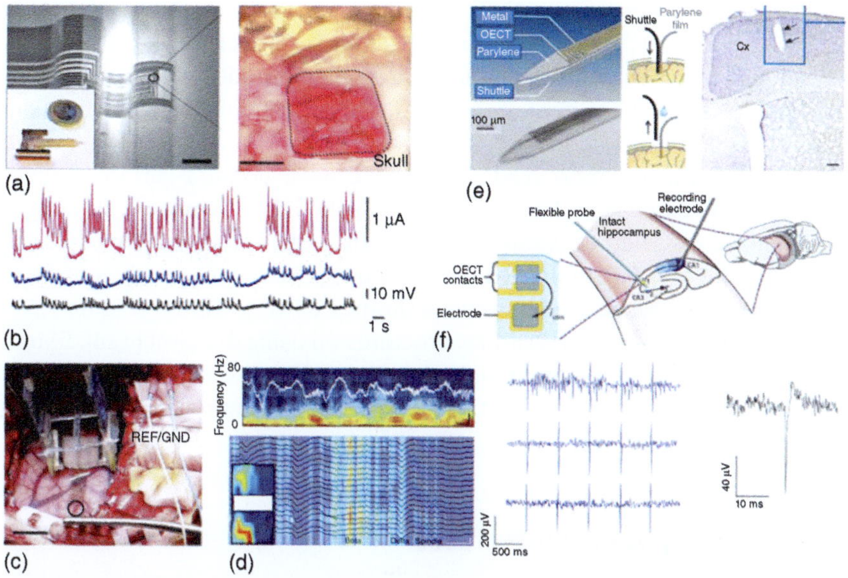

Figure 7.5 Device/brain interfaces. (a) Left: picture of the probe conforming to a curvilinear surface; right: picture of the ECoG probe placed over the somatosensory cortex, with the craniotomy surrounded by dashed lines. The scale bar is 1 mm. (b) Electrical recordings from an OECT (pink), a PEDOT/PSS surface electrode control (blue), and an Ir-penetrating electrode control (black). (c) Surgical placement of the NeuroGrid on the human brain surface (open circle). Inset shows an optical image of a 256-electrode NeuroGrid. The scale bar is 100 μm. (d) Top: sample time-frequency spectrogram of intraoperative recordings under anesthesia from an epilepsy patient. The scale bars are 500 ms × 500 μV. Bottom: sample multichannel LFP recording during intraoperative anesthesia (black traces). (e) Schematic illustration of the OECT probe and corresponding tissue histology. Left: layout and optical micrograph of the entire probe showing the parylene and the shuttle parts; middle: principle of delamination; right: histology showing the absence of a noticeable glial scar after a one-month-long implantation in the rat neocortex. (f) Top: stimulating pyramidal neurons in intact hippocampal preparations; bottom: recorded multiunit response during stimulation at sites 1, 2, 3 and selected action potential from stimulation at site 3. Source: (a, b) Reproduced with permission from Khodagholy et al. [1000], © 2013 Springer Nature. (c, d) Reproduced with permission from Khodagholy et al. [991], © 2014 Springer Nature. (e, f) Reproduced with permission from Williamson et al. [1017], © 2015 John Wiley & Sons.

2-mm-thick polymer substrate and placed on the surface of the brain of an epileptic rat (Figure 7.5a). It showed that the signal-to-noise ratio of the OECT was far superior in comparison with those of the electrodes (Figure 7.5b). Furthermore, the OECT could record low-amplitude and fast-response signals from the interior of the brain that the electrodes were unable to detect, which up to then had only been recordable by depth probes (SEEG). In a further work, the same group demonstrated 4-μm-thick PaC-based probes by using Au interconnects and PEDOT/PSS sensing nodes for both active [1000] and passive [991] ECoG arrays (Figure 7.5c). When the OECT arrays were placed on the surface of the brain, the obtained local field potentials (LFPs) and action potentials reflected the brain activity by measuring the record-breaking signal-to-noise ratio (Figure 7.5d). In addition, the OECTs could detect low-level activity, which is rather difficult for surface electrodes. In its most recent iteration, the NeuroGrid probe integrates 256 electrodes, which can record action potentials from the surface of the brain, and it has been validated in human patients intraoperatively (Figure 7.5c, inset) [991].

For neural interfacing, most readily the probe must lie on a surface without penetrating bulk tissue, just like the case for subdural or epidural 2D arrays, although guided insertion of ultrathin penetrating probes has been demonstrated [1017–1019]. Kim et al. [1020] showed that electrode grids could be made conformal for ECoG recordings when the substrate is thinned down to 2.5 μm (Figure 7.5e). By patterning the array into a mesh-like structure and using dissolvable silk fibroin films as a transient carrier support, adhesion forces due to water capillarity were enhanced, enabling conformal contact on a cat's cortex and improving the recording of sleep spindles (Figure 7.5f).

7.3 Light/Cell/Device Interfaces

Last but not least, we want to briefly talk about recent progresses on light/cell/device ternary interface, at which a new technology named optogenetics could play a critical role. Optogenetics, where light is used as a tool to modulate the cell properties [1021–1023], has attracted tremendous attention because of its unique properties over the patch-clamp technique, such as small tissue damage and high spatial (<1 μm) and temporal (<1 ms) resolutions [1024]. Optogenetics is a new field that combines genetics and optics [1025]. By bioengineering cells with the ability to respond to specific frequencies of light, researchers can use light pulses to switch light-sensitive cells or some related processes inside the light-sensitive cells on and off [1002, 1018, 1026]. This should offer new methods to manipulate cell properties by using light and thereafter convert the light excitation into an electrical, chemical, or thermal stimulus via a photoactive material placed in close proximity to the cell. Meanwhile, make the light/cell/device ternary interface more complicated.

Organic photovoltaic blends have been used to stimulate neurons via a photoexcitation process [1027, 1028]. Ghezzi et al. [1029] documented the use of a semiconductor polymer P3HT film to trigger the neuronal response upon illumination (Figure 7.6a,b). Moreover, they demonstrated that this bio-organic interface restored light sensitivity in explant retinas from rats with light-induced photoreceptor degeneration (Figure 7.6c). These findings suggest the potentially important role of all-organic devices in future subretinal prosthetic implants [924].

Similarly, Lee et al. [1030] realized an ultra-flexible, high-temporal-resolution electrophysiology array by integrating OECTs and OFETs (Figure 7.7a). The integrated devices, with a total thickness of 2.0 μm, were laminated on the surface of a rat's gracilis muscle (Figure 7.7b). To distinguish the evoked biological signal of the muscle from an artifact resulting from electrical stimulation, they used a transgenic rat that expressed a light-sensitive gene, channelrodopsin-2 (ChR2), in the motor nerves and their terminals [1031] and induced muscle movement by

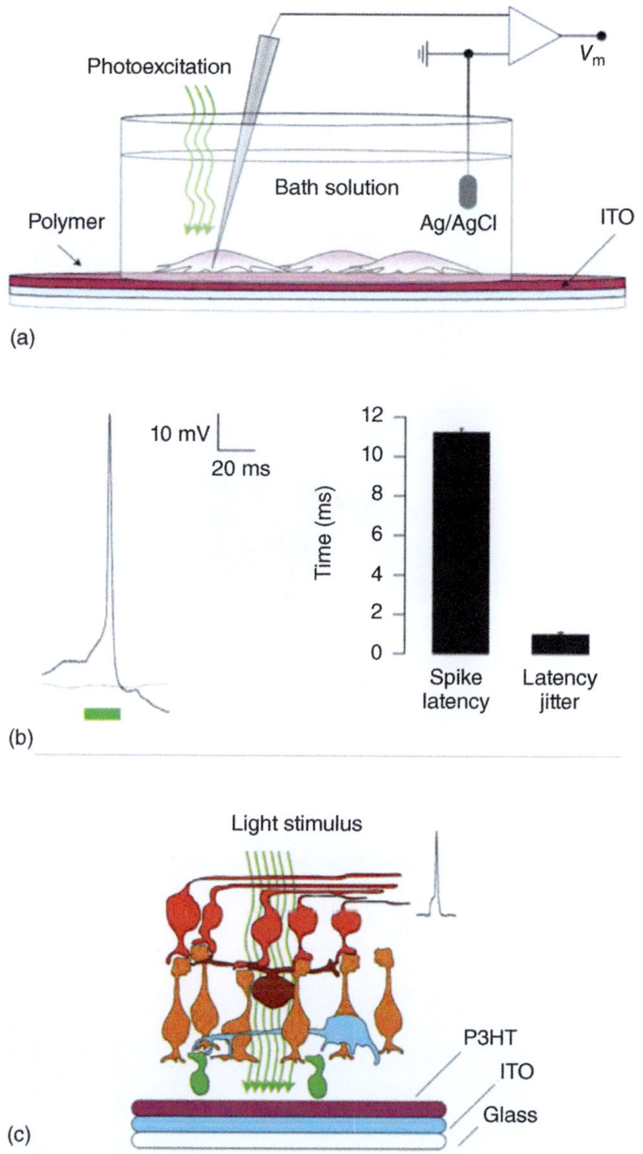

Figure 7.6 Photoresponsive cells/device interfaces. (a) Schematic showing the stimulating and recording paradigm for the photo-stimulus generated at the cell/P3HT interface. (b) Left: light pulse (green bar) used to fire action potentials for a neuron cultured on a P3HT-coated (black) and a control (gray) glass:ITO substrate. Right: mean spike latency peak with respect to light onset and a negligible latency jitter. (c) Schematic illustration of the photoreceptor layer in the retinal structure that was replaced in the degenerate retina by the organic polymer P3HT. Source: Reproduced with permission Ghezzi et al. [1029], © 2014 Springer Nature.

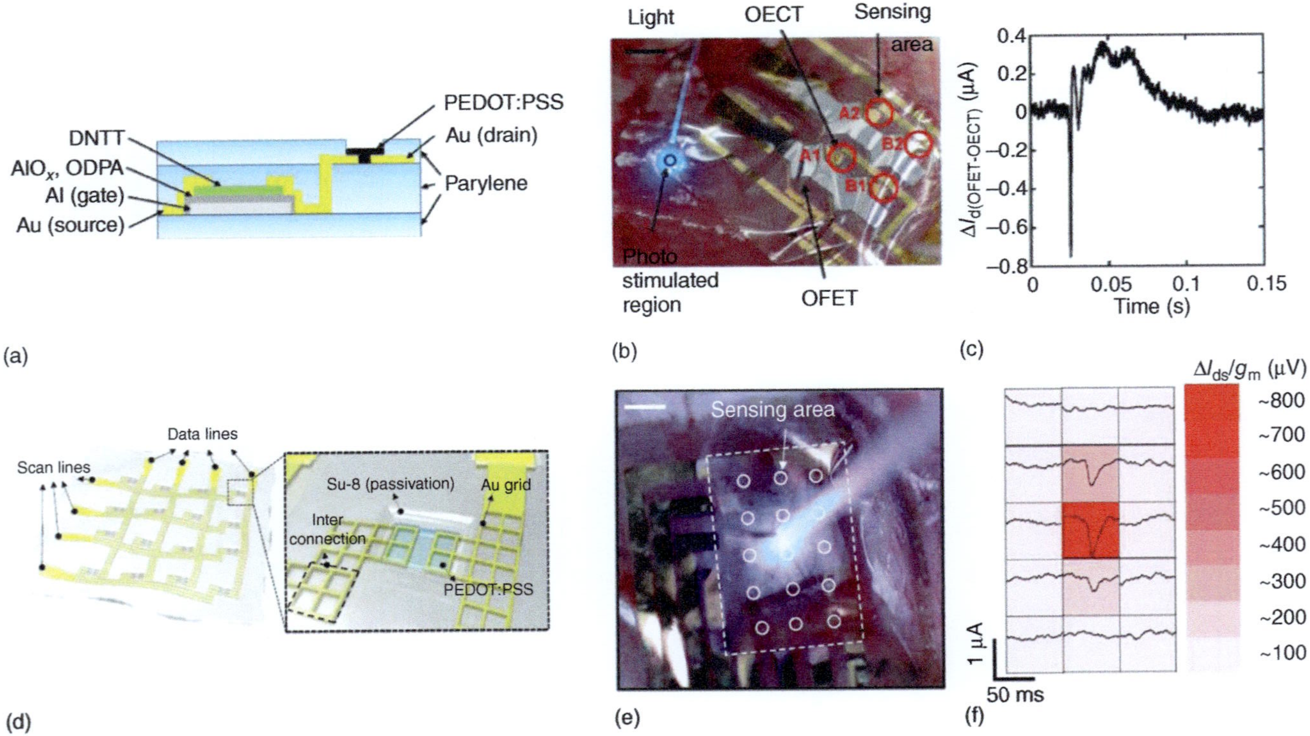

Figure 7.7 Integrated OFET-OECTs. (a) Structure of the OFET-OECT integrated device. (b) Photograph of a 2 × 2 electrophysiology array on muscle tissue. The scale bar is 2 mm. (c) Compound muscle action potential evoked by a laser flash. (d) Transparent electrophysiology OECT array on a parylene substrate. (e) Photograph of the transparent electrophysiology array in a neuron-concentrated area on the cortical surface of optogenetic mice with blue laser continuous stimulation through optical fiber (500 μm diameter). The scale bar is 1 mm. (f) The spatial distribution of the brain signal intensity measured by a 3 × 5 transparent electrophysiology array. The electric potential was calculated by using each OECT's transconductance. Source: (a–c) Reproduced with permission from Lee et al. [1030], © 2016 John Wiley & Sons. (d–f) Reproduced with permission from Lee et al. [1033], © 2017 Proceedings of the National Academy of Sciences PNAS.

laser-coupled optical stimulation at a wavelength of 473 nm [1021, 1032]. Owing to its ultralow thickness, the device exhibited good conformability and negligible mechanical interference during the dynamic motion of the muscles (Figure 7.7). In a further work, they developed mechanically flexible, transparent, and active MEAs integrated with OECTs and transparent Au grid wirings for local signal amplification and high spatial resolution (Figure 7.7d) [1033]. The transparent active MEA (Figure 7.7e) realized the spatial mapping of electrocorticogram electrical signals from an optogenetic rat with 1-mm spacing and showed lower light artifacts than noise level (Figure 7.7f). This technology would open up the possibility of precise investigation of neural network systems with direct light stimulation.

Another remarkable work that interfaced organic electronics with biology came from the Bao's group, who demonstrated that its artificial skin could mimic the sense of touch and also transmit the signal to a component of the nervous system [29]. They presented a skin-inspired mechanoreceptor with a flexible organic transistor circuit that transduced pressure into digital frequency signals directly. Furthermore, they proved that the electronic signal could be recognized by a biological neuron by adapting the optogenetics technique [1002, 1018, 1026]. They fabricated an organic oscillator based on printed organic ring oscillator circuits. The organic oscillator was subsequently integrated with a sensitive flexible pressure sensor, which was able to convert pressure signals into quantifiable direct current signals (Figure 7.8a).

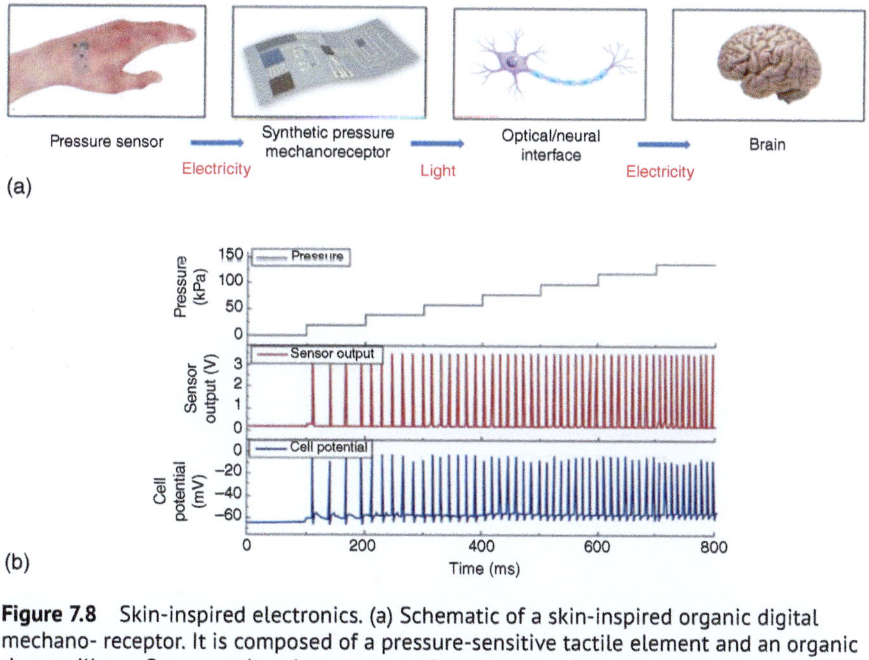

Figure 7.8 Skin-inspired electronics. (a) Schematic of a skin-inspired organic digital mechano- receptor. It is composed of a pressure-sensitive tactile element and an organic ring oscillator. Optogenetic pulses were used to stimulate live neurons. (b) Signal relationship among pressure, sensor output, and cell potential (color online). Inset: the enlarged section showing the neurons' action potentials that closely follow the light pulses. Source: Reproduced with permission from Tee et al. [29], © 2015 American Association for the Advancement of Science – AAAS.

Voltage pulses were subsequently generated by the organic oscillator in order to be transmitted into light pulses (Figure 7.8b). The optogenetic-engineered neurons were then activated by light pulses translated from the electronic pressure signals from the artificial skin. This exciting work has integrated large-area organic electronic devices and optoelectronic devices to mimic the function at the environment/device interface as well as the optical/neural interface [30]. It can be hailed as a milestone in the progress of organic transistor-based brain–computer interfaces.

8

Concluding Remarks and Outlook

In summary, remarkable advances in developing high-quality sophisticated interfaces to control charge transport in organic field-effect transistors (OFETs) provide novel and useful ways to create new high-performance functional optoelectronic devices toward potential applications. On the basis of fundamental working mechanisms of different heterogeneous interfaces, including interface within the semiconductor layers, semiconductor/electrode interface, semiconductor/dielectric interface, and semiconductor/environment interface, from the concept of interface engineering, discrete efficient strategies that have been utilized for improving device performance, realizing new functions, and reducing fabrication costs are thoroughly summarized in this book, with a particular focus on highlighting the importance of interfacial effects in the performance of OFETs. These demonstrations obviously strengthen our fundamental understanding of the correlation between the molecular structure, assembly, and emergent functions at the molecular level and consequently offer novel insights into designing new generations of practical multifunctional optoelectronics and sensors. In spite of these significant achievements, there are still numerous opportunities as well as formidable challenges to overcome for future research. For example, (1) new approaches should be developed to facilitate the growth of wafer-scale organic single crystal films for practical uses, (2) more effort should be devoted to improve better understanding of the interfacial charge transport mechanisms in order to enhance the performance, stability, and reproducibility, and (3) integration strategies need to be extended to other substances required for specific applications as well as new types of device architectures and junctions. In Sections 8.1–8.6, we envision several directions to which we should pay particular attention in order to further advance the science and technology of functional organic electronics, as follows.

8.1 New Challenges in Molecular Design

There are many different classes of organic materials, some already existing and some being developed, that can be applied as the active semiconductor layer or functional layer. Very few of them meet all the requirements throughout the fabrication

Interface Engineering in Organic Field-Effect Transistors, First Edition. Xuefeng Guo and Hongliang Chen.
© 2023 WILEY-VCH GmbH. Published 2023 by WILEY-VCH GmbH.

process, including the performance requirement, compatibility between each layer, robustness, and long-term stability. Another important aspect is the biocompatibility of organic materials with long-term toxicity, which is of great importance for future studies, particularly for the more invasive applications such as implanted electronics in human tissues. All these challenges should be addressed through a holistic consideration of material design, assembly, device architecture, and interface engineering. In addition, we have already shown how the fabrication of a device on a flexible substrate improves the contact with the skin, which is desirable for recordings of freely moving subjects. As a result, the design of biodegradable semiconductors as well as dielectric materials should be put on the agenda for the use of implantable devices. In a further step, waterborne organic semiconductors (OSCs) should also be a promising direction to reduce the use of organic solvents for practical industrial applications.

8.2 High-Quality OSC Films: Self-Assembly Control

High-quality OSC films and even single-crystal films are an everlasting pursuit to satisfy the demands of high-performance electronic and optoelectronic devices. However, defects and traps are inevitable in current solution-processed OSC thin films. Although thermal or solvent annealing has been widely used to decrease defects to obtain high-quality films, it is still far from good enough to match the performance with inorganic 2D materials. Considering various weak interactions within organic materials, such as π–π interaction of the conjugated backbones and hydrophobic interaction of the side chains, unique assembly strategies should be developed to directly grow large-scale and high-quality OSC thin films. Some polar interactions, such as siloxane or phosphate substitution, and halogenation on the side chains, are promising approaches to explore more complicated self-assembly processes. It is highly possible that, by conducting further investigations into the growth thermodynamics and kinetics of the self-assembly process, we could achieve high-crystallinity 2D-layered forms or 1D nano-/micro-wires from these interesting OSC materials by the solution process.

8.3 High-Performance Scalable Flexible Optoelectronics

As discussed in Section 3.1, the achievement of fast growth of OSC films with a high crystallinity is a very important aspect toward their use in flexible devices, yet tough challenges remain. Great efforts have been paid to the development of film-forming technologies, such as off-center spin coating, solution shearing, bar coating, and roll-to-roll coating. However, few technologies can meet the practical fabrication standards, which include high film quality, time efficiency, high throughput requirement, and large-scale demands. In order to produce solution-processed high-quality films directly on flexible and transparent substrates, the following three points need to be addressed: (1) the development of flexible substrates that have high solvent and temperature resistance, biocompatibility and biodegradability, (2) a holistic design of the functional device architecture that has balanced device performance and thickness, and (3) the development of quick, effective, and reproducible fabrication technology. Other special requirements of the substrates, such as air permeability or impermeability and water permeability or watertightness, should also be taken into consideration. In most cases, a hybrid design of flexible substrates should be proposed, considering the practical needs.

8.4 Exploration of Novel Structures: Organic/2D Heterostructures and Vertical Structures

As mentioned in Section 3.2, although we can easily fabricate lateral and vertical organic heterostructures through the solution process or a multistep physical deposition method, their integration with 2D materials still remains a big challenge. 2D materials could serve as excellent epitaxial substrates for the growth of high-crystallinity OSC films. This method can grow atomically sharp junctions at the interfaces of the heterostructures. A viable alternative is to take advantage of OSC/2D interfaces to create well-ordered junctions with improved transport performance. One can further envision the following two points of OSCs that can be incorporated to create functions that cannot be found in all-2D heterostructures: (1) in many OSCs, a singlet exciton has a significantly higher energy than a triplet exciton because of the strong exchange interaction. This results in singlet fission that might improve the optoelectronic performance beyond the Shockey–Queisser limit by using 2D materials to extract the multiple carriers generated by the singlet fission at the interface. (2) Because of the weak spin-orbit coupling of OSCs, they usually have a long spin relaxation time. This means that an excited electron in OSCs would maintain the same spin state for a long period of time that would combine with the spin and pseudospin properties of transition metal dichalcogenide (TMD) crystals to produce next-generation spintronic devices. Although there are many works devoted to understand the charge transfer mechanisms at all 2D-heterostructures, relatively few studies have been done to understand the dynamical processes at the OSC/2D material heterointerfaces.

8.5 Instability: Stability in Aqueous Media and Thermal Stability in Hygienic Applications

The excellent mechanical flexibility of organic electronics is expected to open up a range of new opportunities in medical applications. However, one of the major remaining issues for organic devices is their instability, especially their thermal instability, because low melting temperatures and large thermal expansion of organic materials cause thermal degradation. If OFET-based sensors are integrated into medical apparatuses, they must be able to withstand the hash sterilization

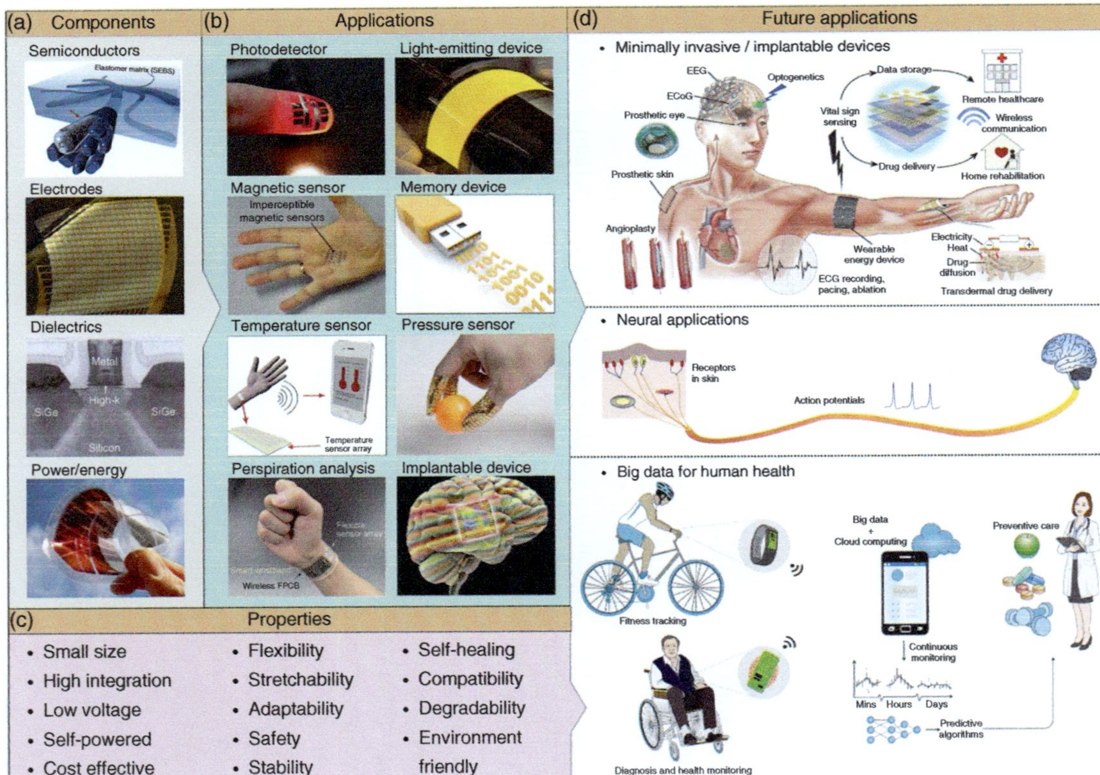

Figure 8.1 Multifunctional sensor systems and future applications. (a) Development of each component in OFETs. Source: The Sihong Wang Research Group; Travis Ross and Yun Soung Kim, University of Illinois at Urbana-Champaign; ©Fraunhofer ISE. (b) Current applications of functional OFETs. Source: Holst Centre; Michael Melzer et al. (2015)/Springer Nature/CC BY 4.0; Alexander/Adobe Stock; IEEE; Takao Someya Group, The University of Tokyo; Wei Gao; Science RF/Adobe Stock. (c) Properties needed for applications. (d) Future applications of OFETs. Source: (d, top) Reproduced with permission from Choi et al. [1036], © 2016 John Wiley & Sons, Inc. (d, middle) Reproduced with permission from Chortos et al. [30], © 2018 Springer Nature. (d, bottom) Reproduced with permission from Bariya et al. [1035], © 2018 Springer Nature.

conditions, which usually involve the exposure to temperatures over 100 °C. At such high temperatures, organic sensors generally degrade, either as a result of the thermally induced oxidation of the OSC, or due to changes in the crystal structure or morphology of thin films. Furthermore, for practical sensing applications in an aqueous detection environment, the device reliability and long-term operational stability are serious problems that have to be solved. One straightforward approach for fabricating water-stable or thermally stable organic sensors involves the use of water- or heat-resistant functional materials. Another approach is to encapsulate the active channel; however, limitations still exist. Water- or heat-resistant OSCs still suffer from accelerated oxidation at operating voltages or high temperatures. In addition, the encapsulation was at the cost of mechanical flexibility. To solve these challenges, the device fabrication should be considered as a holistic process.

8.6 Multifunctional Sensor Systems

Multifunctional sensors (Figure 5.11) have been fabricated based on integrated OFET circuits with various sensor elements [13]. Integration of OFET arrays with multimodal sensing elements could reduce the complexity of structural integration, minimize the signal interferences coupled with strain, and improve the stability of their stimuli-responsive properties [1034].

The practical application of interface-engineered OFETs is still at an early stage and has a long way to go. The exploration of novel strategies for comprehensive control of interfacial charge transport still deserves further attention. That is to say, the development of new materials (Figure 8.1a), device geometries (Figure 8.1b), fabrication processes, and unique properties (Figure 8.1c) is of crucial importance to realize high-performance, multifunctional practical optoelectronics, particularly in the area of emerging soft, stretchable bioelectronics. In combination with other burgeoning biotechnologies, such as optogenetics [1029] and gene-editing techniques [30], big data, and artificial intelligence [1035] (Figure 8.1d), this highly interdisciplinary field could revolutionize organic optoelectronics by providing access to current diagnostic and therapeutic techniques in healthcare and creating new fields and applications yet to be imagined [1036]. There is still a long way to go.

References

1 Kroemer, H. (2001). Nobel Lecture: quasielectric fields and band offsets: teaching electrons new tricks. *Rev. Mod. Phys.* 73 (3): 783.

2 Murphy, A.R. and Frechet, J.M.J. (2007). Organic semiconducting oligomers for use in thin film transistors. *Chem. Rev.* 107 (4): 1066.

3 Wen, Y.G., Liu, Y.Q., Guo, Y.L. et al. (2011). Experimental techniques for the fabrication and characterization of organic thin films for field-effect transistors. *Chem. Rev.* 111 (5): 3358.

4 Zaumseil, J. and Sirringhaus, H. (2007). Electron and ambipolar transport in organic field-effect transistors. *Chem. Rev.* 107 (4): 1296.

5 Wang, C.L., Dong, H.L., Hu, W.P. et al. (2012). Semiconducting π-conjugated systems in field-effect transistors: a material odyssey of organic electronics. *Chem. Rev.* 112 (4): 2208.

6 Di, C.A., Liu, Y.Q., Yu, G., and Zhu, D.B. (2009). Interface engineering: an effective approach toward high-performance organic field-effect transistors. *Acc. Chem. Res.* 42 (10): 1573.

7 Mishra, A., Ma, C.Q., and Bauerle, P. (2009). Functional oligothiophenes: molecular design for multidimensional nanoarchitectures and their applications. *Chem. Rev.* 109 (3): 1141.

8 Liu, D.Q. and Miao, Q. (2018). Recent progress in interface engineering of organic thin film transistors with self-assembled monolayers. *Mater. Chem. Front.* 2 (1): 11.

9 Liu, C., Xu, Y., and Noh, Y.Y. (2015). Contact engineering in organic field-effect transistors. *Mater. Today* 18 (2): 79.

10 Hepp, A., Heil, H., Weise, W. et al. (2003). Light-emitting field-effect transistor based on a tetracene thin film. *Phys. Rev. Lett.* 91 (15): 157406.

11 Liu, C.F., Liu, X., Lai, W.Y., and Huang, W. (2018). Organic light-emitting field-effect transistors: device geometries and fabrication techniques. *Adv. Mater.* 30 (52): 1802466.

12 Torsi, L., Magliulo, M., Manoli, K., and Palazzo, G. (2013). Organic field-effect transistor sensors: a tutorial review. *Chem. Soc. Rev.* 42 (22): 8612.

13 Li, H., Shi, W., Song, J. et al. (2019). Chemical and biomolecule sensing with organic field-effect transistors. *Chem. Rev.* 119 (1): 3.

Interface Engineering in Organic Field-Effect Transistors, First Edition. Xuefeng Guo and Hongliang Chen.
© 2023 WILEY-VCH GmbH. Published 2023 by WILEY-VCH GmbH.

14 Simon, D.T., Gabrielsson, E.O., Tybrandt, K., and Berggren, M. (2016). Organic bioelectronics: bridging the signaling gap between biology and technology. *Chem. Rev.* 116 (21): 13009.

15 Zhang, C.C., Chen, P.L., and Hu, W.P. (2015). Organic field-effect transistor-based gas sensors. *Chem. Soc. Rev.* 44 (8): 2087.

16 Hines, T., Díez-Pérez, I., Nakamura, H. et al. (2013). Controlling formation of single-molecule junctions by electrochemical reduction of diazonium terminal groups. *J. Am. Chem. Soc.* 135 (9): 3319.

17 Zang, Y.P., Huang, D.Z., Di, C.A., and Zhu, D.B. (2016). Device engineered organic transistors for flexible sensing applications. *Adv. Mater.* 28 (22): 4549.

18 Guo, Y.L., Yu, G., and Liu, Y.Q. (2010). Functional organic field-effect transistors. *Adv. Mater.* 22 (40): 4427.

19 Kolesov, V.A., Fuentes-Hernandez, C., Chou, W.F. et al. (2017). Solution-based electrical doping of semiconducting polymer films over a limited depth. *Nat. Mater.* 16 (4): 474.

20 Kang, K., Watanabe, S., Broch, K. et al. (2016). 2D coherent charge transport in highly ordered conducting polymers doped by solid state diffusion. *Nat. Mater.* 15 (8): 896.

21 Xu, J., Wang, S.H., Wang, G.J.N. et al. (2017). Highly stretchable polymer semiconductor films through the nanoconfinement effect. *Science* 355 (6320): 59.

22 Lussem, B., Keum, C.M., Kasemann, D. et al. (2016). Doped organic transistors. *Chem. Rev.* 116 (22): 13714.

23 Tang, C.C., Ang, M.C.Y., Choo, K.K. et al. (2016). Doped polymer semiconductors with ultrahigh and ultralow work functions for Ohmic contacts. *Nature* 539 (7630): 536.

24 Allain, A., Kang, J.H., Banerjee, K., and Kis, A. (2015). Electrical contacts to two-dimensional semiconductors. *Nat. Mater.* 14 (12): 1195.

25 Sirringhaus, H., Kawase, T., Friend, R.H. et al. (2000). High-resolution inkjet printing of all-polymer transistor circuits. *Science* 290 (5499): 2123.

26 Wang, B.H., Huang, W., Chi, L.F. et al. (2018). High-k gate dielectrics for emerging flexible and stretchable electronics. *Chem. Rev.* 118 (11): 5690–5754.

27 Ortiz, R.P., Facchetti, A., and Marks, T.J. (2010). High-k organic, inorganic, and hybrid dielectrics for low-voltage organic field-effect transistors. *Chem. Rev.* 110 (1): 205.

28 Torsi, L., Dodabalapur, A., Sabbatini, L., and Zambonin, P.G. (2000). Multi-parameter gas sensors based on organic thin-film-transistors. *Sens. Actuators, B* 67 (3): 312.

29 Tee, B.C.K., Chortos, A., Berndt, A. et al. (2015). A skin-inspired organic digital mechanoreceptor. *Science* 350 (6258): 313.

30 Chortos, A., Liu, J., and Bao, Z.N. (2016). Pursuing prosthetic electronic skin. *Nat. Mater.* 15 (9): 937.

31 Ebisawa, F., Kurokawa, T., and Nara, S. (1983). Electrical-properties of polyacetylene polysiloxane interface. *J. Appl. Phys.* 54 (6): 3255.

32 Tsumura, A., Koezuka, H., and Ando, T. (1986). Macromolecular electronic device – field-effect transistor with a polythiophene thin-film. *Appl. Phys. Lett.* 49 (18): 1210.

33 Hoshino, S., Nagamatsu, S., Chikamatsu, M. et al. (2002). Device performance of an n-channel organic thin-film transistor with LiF/Al bilayer source and drain electrodes. *Jpn. J. Appl. Phys., Part 2* 41 (7a): L808.

34 Pak, K., Seong, H., Choi, J. et al. (2016). Synthesis of ultrathin, homogeneous copolymer dielectrics to control the threshold voltage of organic thin-film transistors. *Adv. Funct. Mater.* 26 (36): 6574.

35 Moon, H., Seong, H., Shin, W.C. et al. (2015). Synthesis of ultrathin polymer insulating layers by initiated chemical vapour deposition for low-power soft electronics. *Nat. Mater.* 14 (6): 628.

36 Vericat, C., Vela, M.E., Benitez, G. et al. (2010). Self-assembled monolayers of thiols and dithiols on gold: new challenges for a well-known system. *Chem. Soc. Rev.* 39 (5): 1805.

37 Casalini, S., Bortolotti, C.A., Leonardi, F., and Biscarini, F. (2017). Self-assembled monolayers in organic electronics. *Chem. Soc. Rev.* 46 (1): 40.

38 Xue, Y.R., Li, X., Li, H.B., and Zhang, W.K. (2014). Quantifying thiol-gold interactions towards the efficient strength control. *Nat. Commun.* 5: 4348.

39 Zhao, K., Khan, H.U., Li, R. et al. (2016). Carrier transport enhancement in conjugated polymers through interfacial self-assembly of solution-state aggregates. *ACS Appl. Mater. Interfaces* 8 (30): 19649.

40 Liu, D., Xu, X., Su, Y. et al. (2013). Self-assembled monolayers of phosphonic acids with enhanced surface energy for high-performance solution-processed N-channel organic thin-film transistors. *Angew. Chem. Int. Ed.* 52 (24): 6222.

41 Xu, X., Yao, Y., Shan, B. et al. (2016). Electron mobility exceeding 10 $cm^2 V^{-1} s^{-1}$ and band-like charge transport in solution-processed n-channel organic thin-film transistors. *Adv. Mater.* 28 (26): 5276.

42 Schmaltz, T., Gothe, B., Krause, A. et al. (2017). Effect of structure and disorder on the charge transport in defined self-assembled monolayers of organic semiconductors. *ACS Nano* 11 (9): 8747.

43 Tulevski, G.S., Miao, Q., Fukuto, M. et al. (2004). Attaching organic semiconductors to gate oxides: in situ assembly of monolayer field effect transistors. *J. Am. Chem. Soc.* 126 (46): 15048.

44 Mottaghi, M., Lang, P., Rodriguez, F. et al. (2007). Low-operating-voltage organic transistors made of bifunctional self-assembled monolayers. *Adv. Funct. Mater.* 17 (4): 597.

45 Nielsen, C.B., Turbiez, M., and McCulloch, I. (2013). Recent advances in the development of semiconducting DPP-containing polymers for transistor applications. *Adv. Mater.* 25 (13): 1859.

46 Groves, C. (2017). Simulating charge transport in organic semiconductors and devices: a review. *Rep. Prog. Phys.* 80 (2): 026502.

47 Tessler, N., Preezant, Y., Rappaport, N., and Roichman, Y. (2009). Charge transport in disordered organic materials and its relevance to thin-film devices: a tutorial review. *Adv. Mater.* 21 (27): 2741.

48 Wang, L. and Beljonne, D. (2013). Flexible surface hopping approach to model the crossover from hopping to band-like transport in organic crystals. *J. Phys. Chem. Lett.* 4 (11): 1888.

49 Treat, N.D., Westacott, P., and Stingelin, N. (2015). The power of materials science tools for gaining insights into organic semiconductors. *Annu. Rev. Mater. Res.* 45 (1): 459.

50 Lee, W.H., Lim, J.A., Kwak, D. et al. (2009). Semiconductor-dielectric blends: a facile all solution route to flexible all-organic transistors. *Adv. Mater.* 21 (42): 4243.

51 Jo, P.S., Duong, D.T., Park, J. et al. (2015). Control of rubrene polymorphs via polymer binders: applications in organic field-effect transistors. *Chem. Mater.* 27 (11): 3979.

52 Prud'homme, R.E. (2016). Crystallization and morphology of ultrathin films of homopolymers and polymer blends. *Prog. Polym. Sci.* 54–55: 214.

53 Xue, L., Zhang, J., and Han, Y. (2012). Phase separation induced ordered patterns in thin polymer blend films. *Prog. Polym. Sci.* 37 (4): 564.

54 Anthopoulos, T.D., Anyfantis, G.C., Papavassiliou, G.C., and de Leeuw, D.M. (2007). Air-stable ambipolar organic transistors. *Appl. Phys. Lett.* 90 (12): 122105.

55 Wang, J., Wang, H.B., Yan, X.J. et al. (2006). Heterojunction ambipolar organic transistors fabricated by a two-step vacuum-deposition process. *Adv. Funct. Mater.* 16 (6): 824.

56 Yan, H., Kagata, T., and Okuzaki, H. (2009). Ambipolar pentacene/C60-based field-effect transistors with high hole and electron mobilities in ambient atmosphere. *Appl. Phys. Lett.* 94 (2): 023305.

57 Ljubic, D., Smithson, C.S., Wu, Y., and Zhu, S. (2016). Effect of polymer binders on UV-responsive organic thin-film phototransistors with benzothienobenzothiophene semiconductor. *ACS Appl. Mater. Interfaces* 8 (6): 3744.

58 Wu, Y.L., Ma, P., Wu, N.G. et al. (2016). Two-step solution-processed two-component bilayer phthalocyaninato copper-based heterojunctions with interesting Ambipolar organic transiting and ethanol-sensing properties. *Adv. Mater. Interfaces* 3 (16): 1600253.

59 Zhang, X., Mao, J., Deng, W. et al. (2018). Precise patterning of laterally stacked organic microbelt heterojunction arrays by surface-energy-controlled stepwise crystallization for ambipolar organic field-effect transistors. *Adv. Mater.* 30: 1800187.

60 Qin, Y., Zhang, J., Zheng, X. et al. (2014). Charge-transfer complex crystal based on extended-pi-conjugated acceptor and sulfur-bridged annulene: charge-transfer interaction and remarkable high ambipolar transport characteristics. *Adv. Mater.* 26 (24): 4093.

61 Zhu, W.G., Dong, H.L., Zhen, Y.G., and Hu, W.P. (2015). Challenges of organic "cocrystals". *Sci. China Mater.* 58 (11): 854.

62 Hasegawa, T., Mattenberger, K., Takeya, J., and Batlogg, B. (2004). Ambipolar field-effect carrier injections in organic Mott insulators. *Phys. Rev. B* 69 (24): 245115.

63 Kirtley, J.R. and Mannhart, J. (2008). Organic electronics: when TTF met TCNQ. *Nat. Mater.* 7 (7): 520.

64 Wen, S., Deng, W.Q., and Han, K.L. (2010). Ultra-low resistance at TTF-TCNQ organic interfaces. *Chem. Commun.* 46 (28): 5133.

65 Koch, N. (2014). *Supramolecular Materials for Opto-Electronics* (ed. N. Koch). RSC Publishing https://pubs.rsc.org/fa/content/ebook/978-1-84973-826-2.

66 Barnes, J.C., Dale, E.J., Prokofjevs, A. et al. (2015). Semiconducting single crystals comprising segregated arrays of complexes of C60. *J. Am. Chem. Soc.* 137 (6): 2392.

67 Kawase, T. and Kurata, H. (2006). Ball-, bowl-, and belt-shaped conjugated systems and their complexing abilities: exploration of the concave-convex pi-pi interaction. *Chem. Rev.* 106 (12): 5250.

68 Orgiu, E., Crivillers, N., Herder, M. et al. (2012). Optically switchable transistor via energy-level phototuning in a bicomponent organic semiconductor. *Nat. Chem.* 4 (8): 675.

69 Ishii, H., Sugiyama, K., Ito, E., and Seki, K. (1999). Energy level alignment and interfacial electronic structures at organic/metal and organic/organic interfaces. *Adv. Mater.* 11 (8): 605.

70 Sun, Z.Y., Shi, S.W., Bao, Q.Y. et al. (2015). Role of thick-lithium fluoride layer in energy level alignment at organic/metal interface: unifying effect on high metallic work functions. *Adv. Mater. Interfaces* 2 (4): 1400527.

71 Greczynski, G., Salaneck, W.R., and Fahlman, M. (2001). Hybrid interfaces in polymer-based electronics. *Synth. Met.* 121 (1–3): 1625.

72 Choong, V.E., Shi, S., Curless, J., and So, F. (2000). Bipolar transport organic light emitting diodes with enhanced reliability by LiF doping. *Appl. Phys. Lett.* 76 (8): 958.

73 Hajlaoui, R., Horowitz, G., Garnier, F. et al. (1997). Improved field-effect mobility in short oligothiophenes: quaterthiophene and quinquethiophene. *Adv. Mater.* 9 (5): 389.

74 Greiner, M.T. and Lu, Z.H. (2013). Thin-film metal oxides in organic semiconductor devices: their electronic structures, work functions and interfaces. *NPG Asia Mater.* 5: e55.

75 Meyer, J., Zilberberg, K., Riedl, T., and Kahn, A. (2011). Electronic structure of vanadium pentoxide: an efficient hole injector for organic electronic materials. *J. Appl. Phys.* 110 (3): 033710.

76 Greiner, M.T., Chai, L., Helander, M.G. et al. (2012). Transition metal oxide work functions: the influence of cation oxidation state and oxygen vacancies. *Adv. Funct. Mater.* 22 (21): 4557.

77 Yi, Y., Jeon, P.E., Lee, H. et al. (2009). The interface state assisted charge transport at the MoO_3/metal interface. *J. Chem. Phys.* 130 (9): 094704.

78 Cho, S., Seo, J.H., Lee, K., and Heeger, A.J. (2009). Enhanced performance of fullerene n-channel field-effect transistors with titanium sub-oxide injection layer. *Adv. Funct. Mater.* 19 (9): 1459.

79 Piromreun, P., Oh, H., Shen, Y.L. et al. (2000). Role of CsF on electron injection into a conjugated polymer. *Appl. Phys. Lett.* 77 (15): 2403.

80 Kim, J., Rim, Y.S., Liu, Y.S. et al. (2014). Interface control in organic electronics using mixed monolayers of carboranethiol isomers. *Nano Lett.* 14 (5): 2946.

81 Zhang, W.M., Han, Y., Zhu, X.X. et al. (2016). A novel alkylated indacenodithieno[3,2-b]thiophene-based polymer for high-performance field-effect transistors. *Adv. Mater.* 28 (20): 3922.

82 Fei, Z.P., Chen, L., Han, Y. et al. (2017). Alternating 5,5-dimethylcyclopentadiene and diketopyrrolopyrrole copolymer prepared at room temperature for high performance organic thin-film transistors. *J. Am. Chem. Soc.* 139 (24): 8094.

83 Khim, D., Baeg, K.J., Kim, J. et al. (2012). Electron injection enhancement by a Cs-salt interlayer in ambipolar organic field-effect transistors and complementary circuits. *J. Mater. Chem.* 22 (33): 16979.

84 Long, D.X., Baeg, K.J., Xu, Y. et al. (2014). Gradual controlling the work function of metal electrodes by solution-processed mixed interlayers for ambipolar polymer field-effect transistors and circuits. *Adv. Funct. Mater.* 24 (41): 6484.

85 Haas, S., Takahashi, Y., Takimiya, K., and Hasegawa, T. (2009). High-performance dinaphtho-thieno-thiophene single crystal field-effect transistors. *Appl. Phys. Lett.* 95 (2): 022111.

86 Lussem, B., Tietze, M.L., Kleemann, H. et al. (2013). Doped organic transistors operating in the inversion and depletion regime. *Nat. Commun.* 4: 2775.

87 Choi, S., Fuentes-Hernandez, C., Wang, C.Y. et al. (2016). A study on reducing contact resistance in solution-processed organic field-effect transistors. *ACS Appl. Mater. Interfaces* 8 (37): 24744.

88 Yan, X.J., Wang, J., Wang, H.B. et al. (2006). Improved n-type organic transistors by introducing organic heterojunction buffer layer under source/drain electrodes. *Appl. Phys. Lett.* 89 (5): 053510.

89 Li, Q., Yu, X.E., Shi, W., and Yu, J.S. (2013). Effect of organic buffer layers on the performance of n-type organic field-effect transistor based on C-60 active layer. *Synth. Met.* 163: 57.

90 Günther, A.A., Sawatzki, M., Formánek, P. et al. (2016). Contact doping for vertical organic field-effect transistors. *Adv. Funct. Mater.* 26 (5): 768.

91 Tietze, M.L., Pahner, P., Schmidt, K. et al. (2015). Doped organic semiconductors: trap-filling, impurity saturation, and reserve regimes. *Adv. Funct. Mater.* 25 (18): 2701.

92 Zhou, Y.H., Fuentes-Hernandez, C., Shim, J. et al. (2012). A universal method to produce low-work function electrodes for organic electronics. *Science* 336 (6079): 327.

93 Seah, W.L., Tang, C.G., Png, R.Q. et al. (2017). Interface doping for Ohmic organic semiconductor contacts using self-aligned polyelectrolyte counterion monolayer. *Adv. Funct. Mater.* 27 (18): 1606291.

94 Zhang, Y., Zalar, P., Kim, C. et al. (2012). DNA interlayers enhance charge injection in organic field-effect transistors. *Adv. Mater.* 24 (31): 4255.

95 Lee, C.G., Park, S., Ruoff, R.S., and Dodabalapur, A. (2009). Integration of reduced graphene oxide into organic field-effect transistors as conducting electrodes and as a metal modification layer. *Appl. Phys. Lett.* 95 (2): 023304.

96 Xu, Z.Y., Chen, X.S., Zhang, S.N. et al. (2016). Minimizing electrode edge in organic transistors with ultrathin reduced graphene oxide for improving charge injection efficiency. *Phys. Chem. Chem. Phys.* 18 (19): 13209.

97 Chen, X.S., Zhang, S.N., Wu, K.J. et al. (2016). Improving the charge injection in organic transistors by covalently linked graphene oxide/metal electrodes. *Adv. Electron. Mater.* 2 (4): 1500409.

98 Yang, S.Y., Shin, K., and Park, C.E. (2005). The effect of gate-dielectric surface energy on pentacene morphology and organic field-effect transistor characteristics. *Adv. Funct. Mater.* 15 (11): 1806.

99 Kim, S.H., Jang, M., Yang, H., and Park, C.E. (2010). Effect of pentacene-dielectric affinity on pentacene thin film growth morphology in organic field-effect transistors. *J. Mater. Chem.* 20 (27): 5612.

100 Sun, X.N., Liu, Y.Q., Di, C.A. et al. (2011). Interfacial heterogeneity of surface energy in organic field-effect transistors. *Adv. Mater.* 23 (8): 1009.

101 Gao, J., Asadi, K., Xu, J.B., and An, J. (2009). Controlling of the surface energy of the gate dielectric in organic field-effect transistors by polymer blend. *Appl. Phys. Lett.* 94 (9): 093302.

102 Jiang, Y.Y., Guo, Y.L., and Liu, Y.Q. (2017). Engineering of amorphous polymeric insulators for organic field-effect transistors. *Adv. Electron. Mater.* 3 (11): 1700161.

103 Sakai, H., Takahashi, Y., and Murata, H. (2007). Organic field effect transistors with dipole-polarized polymer gate dielectrics for control of threshold voltage. *Appl. Phys. Lett.* 91 (11): 113502.

104 Pierre, A., Sadeghi, M., Payne, M.M. et al. (2014). All-printed flexible organic transistors enabled by surface tension-guided blade coating. *Adv. Mater.* 26 (32): 5722.

105 Peng, X.Z., Horowitz, G., Fichou, D., and Garnier, F. (1990). All-organic thin-film transistors made of alpha-sexithienyl semiconducting and various polymeric insulating layers. *Appl. Phys. Lett.* 57 (19): 2013.

106 Khim, D., Xu, Y., Baeg, K.J. et al. (2016). Large enhancement of carrier transport in solution-processed field-effect transistors by fluorinated dielectric engineering. *Adv. Mater.* 28 (3): 518.

107 Yokota, T., Kajitani, T., Shidachi, R. et al. (2018). A few-layer molecular film on polymer substrates to enhance the performance of organic devices. *Nat. Nanotechnol.* 13 (2): 139–144.

108 Seiki, N., Shoji, Y., Kajitani, T. et al. (2015). Rational synthesis of organic thin films with exceptional long-range structural integrity. *Science* 348 (6239): 1122.

109 Campbell, I.H., Rubin, S., Zawodzinski, T.A. et al. (1996). Controlling Schottky energy barriers in organic electronic devices using self-assembled monolayers. *Phys. Rev. B* 54 (20): 14321.

110 Schreiber, F. (2000). Structure and growth of self-assembling monolayers. *Prog. Surf. Sci.* 65 (5–8): 151.

111 Queffelec, C., Petit, M., Janvier, P. et al. (2012). Surface modification using phosphonic acids and esters. *Chem. Rev.* 112 (7): 3777.

112 Kymissis, I., Dimitrakopoulos, C.D., and Purushothaman, S. (2001). High-performance bottom electrode organic thin-film transistors. *IEEE Trans. Electron Devices* 48 (6): 1060.

113 Chen, H.L. and Guo, X.F. (2013). Unique role of self-assembled monolayers in carbon nanomaterial-based field-effect transistors. *Small* 9 (8): 1144.

114 Hakkinen, H. (2012). The gold-sulfur interface at the nanoscale. *Nat. Chem.* 4 (6): 443.

115 Alt, M., Jesper, M., Schinke, J. et al. (2016). The Swiss-Army-Knife self-assembled monolayer: improving electron injection, stability, and wettability of metal electrodes with a one-minute process. *Adv. Funct. Mater.* 26 (18): 3172.

116 Tulevski, G.S., Miao, Q., Afzali, A. et al. (2006). Chemical complementarity in the contacts for nanoscale organic field-effect transistors. *J. Am. Chem. Soc.* 128 (6): 1788.

117 Zhao, Y., Perez-Segarra, W., Shi, Q.C., and Wei, A. (2005). Dithiocarbamate assembly on gold. *J. Am. Chem. Soc.* 127 (20): 7328.

118 Ford, W.E., Gao, D.Q., Knorr, N. et al. (2014). Organic dipole layers for ultralow work function electrodes. *ACS Nano* 8 (9): 9173.

119 Wang, G., Ruhling, A., Amirjalayer, S. et al. (2017). Ballbot-type motion of *N*-heterocyclic carbenes on gold surfaces. *Nat. Chem.* 9 (2): 152.

120 Engel, S., Fritz, E.C., and Ravoo, B.J. (2017). New trends in the functionalization of metallic gold: from organosulfur ligands to *N*-heterocyclic carbenes. *Chem. Soc. Rev.* 46 (8): 2057.

121 Pujari, S.P., Scheres, L., Marcelis, A.T.M., and Zuilhof, H. (2014). Covalent surface modification of oxide surfaces. *Angew. Chem. Int. Ed.* 53 (25): 6322.

122 Onclin, S., Ravoo, B.J., and Reinhoudt, D.N. (2005). Engineering silicon oxide surfaces using self-assembled monolayers. *Angew. Chem. Int. Ed.* 44 (39): 6282.

123 Hanson, E.L., Schwartz, J., Nickel, B. et al. (2003). Bonding self-assembled, compact organophosphonate monolayers to the native oxide surface of silicon. *J. Am. Chem. Soc.* 125 (51): 16074.

124 Haensch, C., Hoeppener, S., and Schubert, U.S. (2010). Chemical modification of self-assembled silane based monolayers by surface reactions. *Chem. Soc. Rev.* 39 (6): 2323.

125 Walter, S.R., Youn, J., Emery, J.D. et al. (2012). In-situ probe of gate dielectric-semiconductor interfacial order in organic transistors: origin and control of large performance sensitivities. *J. Am. Chem. Soc.* 134 (28): 11726.

126 Lee, H.S., Kim, D.H., Cho, J.H. et al. (2006). Enhancement of interconnectivity in the channels of pentacene thin-film transistors and its effect on field-effect mobility. *Adv. Funct. Mater.* 16 (14): 1859.

127 Briseno, A.L., Mannsfeld, S.C.B., Ling, M.M. et al. (2006). Patterning organic single-crystal transistor arrays. *Nature* 444: 913.

128 Todescato, F., Capelli, R., Dinelli, F. et al. (2008). Correlation between dielectric/organic interface properties and key electrical parameters in PPV-based OFETs. *J. Phys. Chem. B* 112 (33): 10130.

129 Lee, H.S., Kim, D.H., Cho, J.H. et al. (2008). Effect of the phase states of self-assembled monolayers on pentacene growth and thin-film transistor characteristics. *J. Am. Chem. Soc.* 130 (32): 10556.

130 Tello, M., Chiesa, M., Duffy, C.M., and Sirringhaus, H. (2008). Charge trapping in intergrain regions of pentacene thin film transistors. *Adv. Funct. Mater.* 18 (24): 3907.

131 Miyadera, T., Wang, S.D., Minari, T. et al. (2008). Charge trapping induced current instability in pentacene thin film transistors: trapping barrier and effect of surface treatment. *Appl. Phys. Lett.* 93 (3): 033304.

132 Minari, T., Kano, M., Miyadera, T. et al. (2008). Selective organization of solution-processed organic field-effect transistors. *Appl. Phys. Lett.* 92 (17): 173301.

133 Rawcliffe, R., Shkunov, M., Heeney, M. et al. (2008). Organic field-effect transistors of poly(2,5-bis(3-dodecylthiophen-2-yl)thieno [2,3-b]thiophene) deposited on five different silane self-assembled monolayers. *Chem. Commun.* 2008 (7): 871. https://doi.org/10.1039/b715536k.

134 Umeda, T., Kumaki, D., and Tokito, S. (2009). Surface-energy-dependent field-effect mobilities up to 1 $cm^2/V\,s$ for polymer thin-film transistor. *J. Appl. Phys.* 105 (2): 024516.

135 Yan, H., Uchida, T., Kagata, T. et al. (2009). Influence of surface energy and roughness of substrate on thin pentacene and C60 films prepared by vacuum deposition. *Chem. Lett.* 38 (6): 550.

136 Virkar, A., Mannsfeld, S., Oh, J.H. et al. (2009). The role of OTS density on pentacene and C_{60} nucleation, thin film growth, and transistor performance. *Adv. Funct. Mater.* 19 (12): 1962.

137 Celle, C., Suspène, C., Simonato, J.-P. et al. (2009). Self-assembled monolayers for electrode fabrication and efficient threshold voltage control of organic transistors with amorphous semiconductor layer. *Org. Electron.* 10 (1): 119.

138 Ito, Y., Virkar, A.A., Mannsfeld, S. et al. (2009). Crystalline ultrasmooth self-assembled monolayers of alkylsilanes for organic field-effect transistors. *J. Am. Chem. Soc.* 131 (26): 9396.

139 Liao, K.C., Ismail, A.G., Kreplak, L. et al. (2010). Designed organophosphonate self-assembled monolayers enhance device performance of pentacene-based organic thin-film transistors. *Adv. Mater.* 22 (28): 3081.

140 Mityashin, A., Roscioni, O.M., Muccioli, L. et al. (2014). Multiscale modeling of the electrostatic impact of self-assembled monolayers used as gate dielectric treatment in organic thin-film transistors. *ACS Appl. Mater. Interfaces* 6 (17): 15372.

141 Nakahara, K., Mitsui, C., Okamoto, T. et al. (2014). Furan fused V-shaped organic semiconducting materials with high emission and high mobility. *Chem. Commun.* 50 (40): 5342.

142 Kang, B., Park, N., Lee, J. et al. (2015). Surface-order mediated assembly of pi-conjugated molecules on self-assembled monolayers with controlled grain structures. *Chem. Mater.* 27 (13): 4669.

143 Nakano, M., Osaka, I., and Takimiya, K. (2017). Control of major carriers in an ambipolar polymer semiconductor by self-assembled monolayers. *Adv. Mater.* 29 (1): 1602893.

144 Dimitrakopoulos, C.D., Purushothaman, S., Kymissis, J. et al. (1999). Low-voltage organic transistors on plastic comprising high-dielectric constant gate insulators. *Science* 283 (5403): 822.

145 Dimitrakopoulos, C.D., Kymissis, I., Purushothaman, S. et al. (1999). Low-voltage, high-mobility pentacene transistors with solution-processed high dielectric constant insulators. *Adv. Mater.* 11 (16): 1372.

146 Baer, E. and Zhu, L. (2017). 50th anniversary perspective: dielectric phenomena in polymers and multilayered dielectric films. *Macromolecules* 50 (6): 2239.

147 Kim, S.H., Yang, S.Y., Shin, K. et al. (2006). Low-operating-voltage pentacene field-effect transistor with a high-dielectric-constant polymeric gate dielectric. *Appl. Phys. Lett.* 89 (18): 183516.

148 Ha, Y.G., Everaerts, K., Hersam, M.C., and Marks, T.J. (2014). Hybrid gate dielectric materials for unconventional electronic circuitry. *Acc. Chem. Res.* 47 (4): 1019.

149 Hanson, E.L., Guo, J., Koch, N. et al. (2005). Advanced surface modification of indium tin oxide for improved charge injection in organic devices. *J. Am. Chem. Soc.* 127 (28): 10058.

150 Ma, H., Acton, O., Hutchins, D.O. et al. (2012). Multifunctional phosphonic acid self-assembled monolayers on metal oxides as dielectrics, interface modification layers and semiconductors for low-voltage high-performance organic field-effect transistors. *Phys. Chem. Chem. Phys.* 14 (41): 14110.

151 Cattani-Scholz, A. (2017). Functional organophosphonate interfaces for nanotechnology: a review. *ACS Appl. Mater. Interfaces* 9 (31): 25643.

152 Hotchkiss, P.J., Jones, S.C., Paniagua, S.A. et al. (2012). The modification of indium tin oxide with phosphonic acids: mechanism of binding, tuning of surface properties, and potential for use in organic electronic applications. *Acc. Chem. Res.* 45 (3): 337.

153 Paniagua, S.A., Giordano, A.J., Smith, O.L. et al. (2016). Phosphonic acids for interfacial engineering of transparent conductive oxides. *Chem. Rev.* 116 (12): 7117.

154 Kim, C.S., Jo, S.J., Lee, S.W. et al. (2007). Surface-modified high-k oxide gate dielectrics for low-voltage high-performance pentacene thin-film transistors. *Adv. Funct. Mater.* 17 (6): 958.

155 Mühlenen, A.V., Castellani, M., Schaer, M., and Zuppiroli, L. (2008). Controlling charge-transfer at the gate interface of organic field-effect transistors. *Phys. Status Solidi B* 245 (6): 1170.
156 Acton, O., Ting, G., Ma, H. et al. (2008). Pi-sigma-phosphonic acid organic monolayer/sol-gel hafnium oxide hybrid dielectrics for low-voltage organic transistors. *Adv. Mater.* 20 (19): 3697.
157 Zschieschang, U., Halik, M., and Klauk, H. (2008). Microcontact-printed self-assembled monolayers as ultrathin gate dielectrics in organic thin-film transistors and complementary circuits. *Langmuir* 24 (5): 1665.
158 Acton, O., Osaka, I., Ting, G. et al. (2009). Phosphonic acid self-assembled monolayer and amorphous hafnium oxide hybrid dielectric for high performance polymer thin film transistors on plastic substrates. *Appl. Phys. Lett.* 95 (11): 113305.
159 Acton, O., Ting Ii, G.G., Ma, H. et al. (2009). π-σ-Phosphonic acid organic monolayer–amorphous sol–gel hafnium oxide hybrid dielectric for low-voltage organic transistors on plastic. *J. Mater. Chem.* 19 (42): 7929.
160 Acton, O., Ting, G.G., Shamberger, P.J. et al. (2010). Dielectric surface-controlled low-voltage organic transistors via n-alkyl phosphonic acid self-assembled monolayers on high-*k* metal oxide. *ACS Appl. Mater. Interfaces* 2 (2): 511.
161 Acton, O., Hutchins, D., Arnadottir, L. et al. (2011). Spin-cast and patterned organophosphonate self-assembled monolayer dielectrics on metal-oxide-activated Si. *Adv. Mater.* 23 (16): 1899.
162 Novak, M., Schmaltz, T., Faber, H., and Halik, M. (2011). Influence of self-assembled monolayer dielectrics on the morphology and performance of alpha, omega-dihexylquaterthiophene in thin film transistors. *Appl. Phys. Lett.* 98 (9): 093302.
163 Zschieschang, U., Yamamoto, T., Takimiya, K. et al. (2011). Organic electronics on banknotes. *Adv. Mater.* 23 (5): 654.
164 Salinas, M., Jager, C.M., Amin, A.Y. et al. (2012). The relationship between threshold voltage and dipolar character of self-assembled monolayers in organic thin-film transistors. *J. Am. Chem. Soc.* 134 (30): 12648.
165 Hutchins, D.O., Weidner, T., Baio, J. et al. (2013). Effects of self-assembled monolayer structural order, surface homogeneity and surface energy on pentacene morphology and thin film transistor device performance. *J. Mater. Chem. C* 1 (1): 101.
166 Kaltenbrunner, M., Sekitani, T., Reeder, J. et al. (2013). An ultra-lightweight design for imperceptible plastic electronics. *Nature* 499 (7459): 458.
167 Liu, D., He, Z., Su, Y. et al. (2014). Self-assembled monolayers of cyclohexyl-terminated phosphonic acids as a general dielectric surface for high-performance organic thin-film transistors. *Adv. Mater.* 26 (42): 7190.
168 Aghamohammadi, M., Rodel, R., Zschieschang, U. et al. (2015). Threshold-voltage shifts in organic transistors due to self-assembled monolayers at the dielectric: evidence for electronic coupling and dipolar effects. *ACS Appl. Mater. Interfaces* 7 (41): 22775.

169 Kraft, U., Sejfic, M., Kang, M.J. et al. (2015). Flexible low-voltage organic complementary circuits: finding the optimum combination of semiconductors and monolayer gate dielectrics. *Adv. Mater.* 27 (2): 207.

170 Guo, X.F., Myers, M., Xiao, S.X. et al. (2006). Chemoresponsive monolayer transistors. *Proc. Natl. Acad. Sci. U.S.A.* 103 (31): 11452.

171 Smits, E.C.P., Mathijssen, S.G.J., van Hal, P.A. et al. (2008). Bottom-up organic integrated circuits. *Nature* 455 (7215): 956.

172 Gholamrezaie, F., Mathijssen, S.G.J., Smits, E.C.P. et al. (2010). Ordered semiconducting self-assembled monolayers on polymeric surfaces utilized in organic integrated circuits. *Nano Lett.* 10 (6): 1998.

173 Novak, M., Ebel, A., Meyer-Friedrichsen, T. et al. (2011). Low-voltage p- and n-type organic self-assembled monolayer field effect transistors. *Nano Lett.* 11 (1): 156.

174 Hutchins, D.O., Acton, O., Weidner, T. et al. (2012). Spin cast self-assembled monolayer field effect transistors. *Org. Electron.* 13 (3): 464.

175 Schmaltz, T., Amin, A.Y., Khassanov, A. et al. (2013). Low-voltage self-assembled monolayer field-effect transistors on flexible substrates. *Adv. Mater.* 25 (32): 4511.

176 Ringk, A., Li, X.R., Gholamrezaie, F. et al. (2013). N-type self-assembled monolayer field-effect transistors and complementary inverters. *Adv. Funct. Mater.* 23 (16): 2016.

177 Ringk, A., Roelofs, W.S.C., Smits, E.C.P. et al. (2013). N-type self-assembled monolayer field-effect transistors for flexible organic electronics. *Org. Electron.* 14 (5): 1297.

178 Sizov, A.S., Anisimov, D.S., Agina, E.V. et al. (2014). Easily processable highly ordered Langmuir–Blodgett films of quaterthiophene disiloxane dimer for mono layer organic field-effect transistors. *Langmuir* 30 (50): 15327.

179 Parry, A.V.S., Lu, K.X., Tate, D.J. et al. (2014). Trichlorosilanes as anchoring groups for phenylene-thiophene molecular monolayer field effect transistors. *Adv. Funct. Mater.* 24 (42): 6677.

180 Cernetic, N., Weidner, T., Baio, J.E. et al. (2015). Enhanced performance of self-assembled monolayer field-effect transistors with top-contact geometry through molecular tailoring, heated assembly, and thermal annealing. *Adv. Funct. Mater.* 25 (33): 5376.

181 Gothe, B., de Roo, T., Will, J. et al. (2017). Self-assembled monolayer field-effect transistors based on oligo-9,9′-dioctylfluorene phosphonic acids. *Nanoscale* 9 (47): 18584.

182 Borshchev, O.V., Sizov, A.S., Agina, E.V. et al. (2017). Synthesis of organosilicon derivatives of [1]benzothieno[3,2-*b*][1]-benzothiophene for efficient monolayer Langmuir-Blodgett organic field effect transistors. *Chem. Commun.* 53 (5): 885.

183 Heinrich, C.D., Reichstein, P.M., and Thelakkat, M. (2018). Synthesis, surface grafting, and fabrication of ultrathin polymeric SAMFETs with high field-effect mobility. *ACS Appl. Mater. Interfaces* 10 (41): 35441.

184 Chen, H.L., Dong, S.H., Bai, M.L. et al. (2015). Solution-processable, low-voltage, and high-performance monolayer field-effect transistors with aqueous stability and high sensitivity. *Adv. Mater.* 27 (12): 2113.

185 Wang, C., Dong, H., Hu, W. et al. (2012). Semiconducting pi-conjugated systems in field-effect transistors: a material odyssey of organic electronics. *Chem. Rev.* 112 (4): 2208.

186 Gsanger, M., Bialas, D., Huang, L. et al. (2016). Organic semiconductors based on dyes and color pigments. *Adv. Mater.* 28 (19): 3615.

187 Sirringhaus, H. (2014). 25th anniversary article: organic field-effect transistors: the path beyond amorphous silicon. *Adv. Mater.* 26 (9): 1319.

188 Heeger, A.J. (2010). Semiconducting polymers: the third generation. *Chem. Soc. Rev.* 39 (7): 2354.

189 Chu, M., Fan, J.X., Yang, S. et al. (2018). Halogenated tetraazapentacenes with electron mobility as high as 27.8 $cm^2\ V^{-1}\ s^{-1}$ in solution-processed n-channel organic thin-film transistors. *Adv. Mater.* 30 (38): 1803467.

190 Takimiya, K., Osaka, I., Mori, T., and Nakano, M. (2014). Organic semiconductors based on [1]benzothieno[3,2-*b*][1]benzothiophene substructure. *Acc. Chem. Res.* 47 (5): 1493.

191 Li, Y.N., Sonar, P., Murphy, L., and Hong, W. (2013). High mobility diketopyrrolopyrrole (DPP)-based organic semiconductor materials for organic thin film transistors and photovoltaics. *Energy Environ. Sci.* 6 (6): 1684.

192 Kaji, T., Entani, S., Ikeda, S., and Saiki, K. (2008). Origin of carrier types in intrinsic organic semiconductors. *Adv. Mater.* 20 (11): 2084.

193 Mollinger, S.A., Salleo, A., and Spakowitz, A.J. (2016). Anomalous charge transport in conjugated polymers reveals underlying mechanisms of trapping and percolation. *ACS Cent. Sci.* 2 (12): 910.

194 Dongmin Kang, S. and Jeffrey Snyder, G. (2017). Charge-transport model for conducting polymers. *Nat. Mater.* 16 (2): 252.

195 Di Pietro, R., Nasrallah, I., Carpenter, J. et al. (2016). Coulomb enhanced charge transport in semicrystalline polymer semiconductors. *Adv. Funct. Mater.* 26 (44): 8011.

196 Liu, C., Huang, K.R., Park, W.T. et al. (2017). A unified understanding of charge transport in organic semiconductors: the importance of attenuated delocalization for the carriers. *Mater. Horiz.* 4 (4): 608.

197 Pinto, R.M., Macoas, E.M., Neves, A.I. et al. (2015). Effect of molecular stacking on exciton diffusion in crystalline organic semiconductors. *J. Am. Chem. Soc.* 137 (22): 7104.

198 Xu, Y., Sun, H., Li, W. et al. (2017). Exploring the charge transport in conjugated polymers. *Adv. Mater.* 29 (41): 1702729.

199 Fratini, S., Mayou, D., and Ciuchi, S. (2016). The transient localization scenario for charge transport in crystalline organic materials. *Adv. Funct. Mater.* 26 (14): 2292.

200 Coropceanu, V., Cornil, J., da Silva Filho, D.A. et al. (2007). Charge transport in organic semiconductors. *Chem. Rev.* 107 (4): 926.

201 Gershenson, M.E., Podzorov, V., and Morpurgo, A.F. (2006). Colloquium: electronic transport in single-crystal organic transistors. *Rev. Mod. Phys.* 78 (3): 973.

202 Cheung, D.L. and Troisi, A. (2008). Modelling charge transport in organic semiconductors: from quantum dynamics to soft matter. *Phys. Chem. Chem. Phys.* 10 (39): 5941.

203 Wang, L., Li, Q., Shuai, Z. et al. (2010). Multiscale study of charge mobility of organic semiconductor with dynamic disorders. *Phys. Chem. Chem. Phys.* 12 (13): 3309.

204 Troisi, A. (2011). Charge transport in high mobility molecular semiconductors: classical models and new theories. *Chem. Soc. Rev.* 40 (5): 2347.

205 Bassler, H. and Kohler, A. (2012). Charge transport in organic semiconductors. *Top. Curr. Chem.* 312: 1.

206 Troisi, A. (2010). Theories of the charge transport mechanism in ordered organic semiconductors. *Org. Electron.* 223: 259.

207 Bashir, A., Heck, A., Narita, A. et al. (2015). Charge carrier mobilities in organic semiconductors: crystal engineering and the importance of molecular contacts. *Phys. Chem. Chem. Phys.* 17 (34): 21988.

208 Yamashita, Y., Hinkel, F., Marszalek, T. et al. (2016). Mobility exceeding 10 $cm^2/(V{\cdot}s)$ in donor–acceptor polymer transistors with band-like charge transport. *Chem. Mater.* 28 (2): 420.

209 Mei, Y., Diemer, P.J., Niazi, M.R. et al. (2017). Crossover from band-like to thermally activated charge transport in organic transistors due to strain-induced traps. *Proc. Natl. Acad. Sci. U.S.A.* 114 (33): E6739.

210 Dong, H. and Hu, W. (2016). Multilevel investigation of charge transport in conjugated polymers. *Acc. Chem. Res.* 49 (11): 2435.

211 Dong, H., Jiang, L., and Hu, W. (2012). Interface engineering for high-performance organic field-effect transistors. *Phys. Chem. Chem. Phys.* 14 (41): 14165.

212 Yao, Y., Dong, H., and Hu, W. (2016). Charge transport in organic and polymeric semiconductors for flexible and stretchable devices. *Adv. Mater.* 28 (22): 4513.

213 Ma, Z., Geng, H., Wang, D., and Shuai, Z. (2016). Influence of alkyl side-chain length on the carrier mobility in organic semiconductors: herringbone vs. pi–pi stacking. *J. Mater. Chem. C* 4 (20): 4546.

214 Fazzi, D. and Caironi, M. (2015). Multi-length-scale relationships between the polymer molecular structure and charge transport: the case of poly-naphthalene diimide bithiophene. *Phys. Chem. Chem. Phys.* 17 (14): 8573.

215 Chung, H. and Diao, Y. (2016). Polymorphism as an emerging design strategy for high performance organic electronics. *J. Mater. Chem. C* 4 (18): 3915.

216 Becerril, H.A., Roberts, M.E., Liu, Z.H. et al. (2008). High-performance organic thin-film transistors through solution-sheared deposition of small-molecule organic semiconductors. *Adv. Mater.* 20 (13): 2588.

217 Giri, G., Verploegen, E., Mannsfeld, S.C. et al. (2011). Tuning charge transport in solution-sheared organic semiconductors using lattice strain. *Nature* 480 (7378): 504.

218 Park, S.K., Jackson, T.N., Anthony, J.E., and Mourey, D.A. (2007). High mobility solution processed 6,13-bis(triisopropyl-silylethynyl) pentacene organic thin film transistors. *Appl. Phys. Lett.* 91 (6): 063514.

219 Diao, Y., Tee, B.C., Giri, G. et al. (2013). Solution coating of large-area organic semiconductor thin films with aligned single-crystalline domains. *Nat. Mater.* 12 (7): 665.

220 Sirringhaus, H., Brown, P.J., Friend, R.H. et al. (1999). Two-dimensional charge transport in self-organized, high-mobility conjugated polymers. *Nature* 401 (6754): 685.

221 Noriega, R., Rivnay, J., Vandewal, K. et al. (2013). A general relationship between disorder, aggregation and charge transport in conjugated polymers. *Nat. Mater.* 12 (11): 1038.

222 Venkateshvaran, D., Nikolka, M., Sadhanala, A. et al. (2014). Approaching disorder-free transport in high-mobility conjugated polymers. *Nature* 515 (7527): 384.

223 Mas-Torrent, M. and Rovira, C. (2011). Role of molecular order and solid-state structure in organic field-effect transistors. *Chem. Rev.* 111 (8): 4833.

224 Qu, G., Kwok, J.J., and Diao, Y. (2016). Flow-directed crystallization for printed electronics. *Acc. Chem. Res.* 49 (12): 2756.

225 Treat, N.D., Nekuda Malik, J.A., Reid, O. et al. (2013). Microstructure formation in molecular and polymer semiconductors assisted by nucleation agents. *Nat. Mater.* 12 (7): 628.

226 Persson, N.E., Chu, P.H., McBride, M. et al. (2017). Nucleation, growth, and alignment of poly(3-hexylthiophene) nanofibers for high-performance OFETs. *Acc. Chem. Res.* 50 (4): 932.

227 Ji, S.X., Wan, L., Liu, C.C., and Nealey, P.F. (2016). Directed self-assembly of block copolymers on chemical patterns: a platform for nanofabrication. *Prog. Polym. Sci.* 54–55: 76.

228 Lou, S.J., Szarko, J.M., Xu, T. et al. (2011). Effects of additives on the morphology of solution phase aggregates formed by active layer components of high-efficiency organic solar cells. *J. Am. Chem. Soc.* 133 (51): 20661.

229 Liu, X., Huettner, S., Rong, Z. et al. (2012). Solvent additive control of morphology and crystallization in semiconducting polymer blends. *Adv. Mater.* 24 (5): 669.

230 Liao, H.C., Ho, C.C., Chang, C.Y. et al. (2013). Additives for morphology control in high-efficiency organic solar cells. *Mater. Today* 16 (9): 326.

231 Chang, J.F., Sun, B.Q., Breiby, D.W. et al. (2004). Enhanced mobility of poly(3-hexylthiophene) transistors by spin-coating from high-boiling-point solvents. *Chem. Mater.* 16 (23): 4772.

232 Peet, J., Cho, N.S., Lee, S.K., and Bazan, G.C. (2008). Transition from solution to the solid state in polymer solar cells cast from mixed solvents. *Macromolecules* 41 (22): 8655.

233 An, T.K., Kang, I., Yun, H.J. et al. (2013). Solvent additive to achieve highly ordered nanostructural semicrystalline DPP copolymers: toward a high charge carrier mobility. *Adv. Mater.* 25 (48): 7003.

234 Reichenberger, M., Baderschneider, S., Kroh, D. et al. (2016). Watching paint dry: the impact of diiodooctane on the kinetics of aggregate formation in thin films of poly(3-hexylthiophene). *Macromolecules* 49 (17): 6420.

235 Manley, E.F., Strzalka, J., Fauvell, T.J. et al. (2017). In situ GIWAXS analysis of solvent and additive effects on PTB7 thin film microstructure evolution during spin coating. *Adv. Mater.* 29 (43).

236 Ji, Y., Xiao, C., Wang, Q. et al. (2016). Asymmetric diketopyrrolopyrrole conjugated polymers for field-effect transistors and polymer solar cells processed from a nonchlorinated solvent. *Adv. Mater.* 28 (5): 943.

237 Li, M., An, C., Marszalek, T. et al. (2016). Controlling the surface organization of conjugated donor-acceptor polymers by their aggregation in solution. *Adv. Mater.* 28 (42): 9430.

238 Al-Mahboob, A., Fujikawa, Y., Sakurai, T., and Sadowski, J.T. (2013). Real-time microscopy of reorientation driven nucleation and growth in pentacene thin films on silicon dioxide. *Adv. Funct. Mater.* 23 (20): 2653.

239 Xu, W., Li, L., Tang, H. et al. (2011). Solvent-induced crystallization of poly(3-dodecylthiophene): morphology and kinetics. *J. Phys. Chem. B* 115 (20): 6412.

240 Chae, G.J., Jeong, S.H., Baek, J.H. et al. (2013). Improved performance in TIPS-pentacene field effect transistors using solvent additives. *J. Mater. Chem. C* 1 (27): 4216.

241 Park, Y.J., Seo, J.H., Elsawy, W. et al. (2015). Enhanced performance in isoindigo based organic small molecule field-effect transistors through solvent additives. *J. Mater. Chem. C* 3 (23): 5951.

242 Park, Y.D., Lee, H.S., Choi, Y.J. et al. (2009). Solubility-induced ordered polythiophene precursors for high-performance organic thin-film transistors. *Adv. Funct. Mater.* 19 (8): 1200.

243 Keum, J.K., Xiao, K., Ivanov, I.N. et al. (2013). Solvent quality-induced nucleation and growth of parallelepiped nanorods in dilute poly(3-hexylthiophene) (P3HT) solution and the impact on the crystalline morphology of solution-cast thin film. *Crystengcomm* 15 (6): 1114.

244 Aiyar, A.R., Hong, J.I., Nambiar, R. et al. (2011). Tunable crystallinity in regioregular poly(3-hexylthiophene) thin films and its impact on field effect mobility. *Adv. Funct. Mater.* 21 (14): 2652.

245 Choi, D., Chang, M., and Reichmanis, E. (2015). Controlled assembly of poly(3-hexylthiophene): managing the disorder to order transition on the nano- through meso-scales. *Adv. Funct. Mater.* 25 (6): 920.

246 Dimitrakopoulos, C.D. and Mascaro, D.J. (2001). Organic thin-film transistors: a review of recent advances. *IBM J. Res. Dev.* 45 (1): 11.

247 Street, R.A., Northrup, J.E., and Salleo, A. (2005). Transport in polycrystalline polymer thin-film transistors. *Phys. Rev. B* 71 (16): 165202.

248 Kline, R.J., McGehee, M.D., Kadnikova, E.N. et al. (2005). Dependence of regioregular poly(3-hexylthiophene) film morphology and field-effect mobility on molecular weight. *Macromolecules* 38 (8): 3312.

249 Rivnay, J., Jimison, L.H., Northrup, J.E. et al. (2009). Large modulation of carrier transport by grain-boundary molecular packing and microstructure in organic thin films. *Nat. Mater.* 8 (12): 952.

250 Kwon, S., Yu, K., Kweon, K. et al. (2014). Template-mediated nano-crystallite networks in semiconducting polymers. *Nat. Commun.* 5: 4183.

251 Kwon, S., Kim, J., Kim, G. et al. (2015). Organic single-crystal semiconductor films on a millimeter domain scale. *Adv. Mater.* 27 (43): 6870.

252 Smith, J., Hamilton, R., McCulloch, I. et al. (2010). Solution-processed organic transistors based on semiconducting blends. *J. Mater. Chem.* 20 (13): 2562.

253 Claudia Arias, A. (2006). Vertically segregated polymer blends: their use in organic electronics. *J. Macromol. Sci., Polym. Rev.* 46 (1): 103.

254 Brown, B.A., Veres, J., Anemian, R.M. et al. (2005). In International Application Published Under the Patent Cooperation Treaty, WO 2005/055248 A2.

255 Binder, K. (1995). Surface effects on polymer blends and block copolymer melts: theoretical concepts of surface enrichment, surface induced phase separation and ordering. *Acta Polym.* 46 (3): 204.

256 Babel, A. and Jenekhe, S.A. (2002). n-Channel field-effect transistors from blends of conjugated polymers. *J. Phys. Chem. B* 106 (24): 6129.

257 Babel, A. and Jenekhe, S.A. (2004). Morphology and field-effect mobility of charge carriers in binary blends of poly(3-hexylthiophene) with poly[2-methoxy-5-(2-ethylhexoxy)-1,4-phenylenevinylene] and polystyrene. *Macromolecules* 37 (26): 9835.

258 Borsenberger, P.M. and Bassler, H. (1991). Concerning the role of dipolar disorder on charge transport in molecularly doped polymers. *J. Chem. Phys.* 95 (7): 5327.

259 Goffri, S., Muller, C., Stingelin-Stutzmann, N. et al. (2006). Multicomponent semiconducting polymer systems with low crystallization-induced percolation threshold. *Nat. Mater.* 5 (12): 950.

260 Qiu, L., Lim, J.A., Wang, X. et al. (2008). Versatile use of vertical-phase-separation-induced bilayer structures in organic thin-film transistors. *Adv. Mater.* 20 (6): 1141.

261 Ford, M.J., Wang, M., Patel, S.N. et al. (2016). High mobility organic field-effect transistors from majority insulator blends. *Chem. Mater.* 28 (5): 1256.

262 Qiu, L.Z., Lee, W.H., Wang, X.H. et al. (2009). Organic thin-film transistors based on polythiophene nanowires embedded in insulating polymer. *Adv. Mater.* 21 (13): 1349.

263 Lin, J.C., Lee, W.Y., Wu, H.C. et al. (2012). Morphology and field-effect transistor characteristics of semicrystalline poly(3-hexylthiophene) and poly(stearyl acrylate) blend nanowires. *J. Mater. Chem.* 22 (29): 14682.

264 Lu, G.H., Chen, J.Y., Xu, W.T. et al. (2014). Aligned polythiophene and its blend film by direct-writing for anisotropic charge transport. *Adv. Funct. Mater.* 24 (31): 4959.

265 Lu, G., Blakesley, J., Himmelberger, S. et al. (2013). Moderate doping leads to high performance of semiconductor/insulator polymer blend transistors. *Nat. Commun.* 4: 1588.

266 Jacobs, I.E. and Moule, A.J. (2017). Controlling molecular doping in organic semiconductors. *Adv. Mater.* 29 (42): 1703063.

267 Smith, J., Heeney, M., McCulloch, I. et al. (2011). Percolation behaviour in high mobility p-channel polymer/small-molecule blend organic field-effect transistors. *Org. Electron.* 12 (1): 143.

268 Hamilton, R., Smith, J., Ogier, S. et al. (2009). High-performance polymer-Small molecule blend organic transistors. *Adv. Mater.* 21 (10–11): 1166.

269 Kang, B., Ge, F., Qiu, L.Z., and Cho, K. (2017). Effective use of electrically insulating units in organic semiconductor thin films for high-performance organic transistors. *Adv. Electron. Mater.* 3 (2): 1600240.

270 Chung, Y.S., Shin, N., Kang, J. et al. (2011). Zone-refinement effect in small molecule-polymer blend semiconductors for organic thin-film transistors. *J. Am. Chem. Soc.* 133 (3): 412.

271 Shin, N., Kang, J., Richter, L.J. et al. (2013). Vertically segregated structure and properties of small molecule-polymer blend semiconductors for organic thin-film transistors. *Adv. Funct. Mater.* 23 (3): 366.

272 Cho, S.Y., Ko, J.M., Lim, J. et al. (2013). Inkjet-printed organic thin film transistors based on TIPS pentacene with insulating polymers. *J. Mater. Chem. C* 1 (5): 914.

273 Kang, J., Shin, N., Jang, D.Y. et al. (2008). Structure and properties of small molecule-polymer blend semiconductors for organic thin film transistors. *J. Am. Chem. Soc.* 130 (37): 12273.

274 Yuan, Y., Giri, G., Ayzner, A.L. et al. (2014). Ultra-high mobility transparent organic thin film transistors grown by an off-Centre spin-coating method. *Nat. Commun.* 5: 3005.

275 Kimura, Y., Nagase, T., Kobayashi, T. et al. (2015). Soluble organic semiconductor precursor with specific phase separation for high-performance printed organic transistors. *Adv. Mater.* 27 (4): 727.

276 Niazi, M.R., Li, R., Qiang Li, E. et al. (2015). Solution-printed organic semiconductor blends exhibiting transport properties on par with single crystals. *Nat. Commun.* 6: 8598.

277 Smith, J., Zhang, W., Sougrat, R. et al. (2012). Solution-processed small molecule-polymer blend organic thin-film transistors with hole mobility greater than 5 cm^2/Vs. *Adv. Mater.* 24 (18): 2441.

278 Paterson, A.F., Treat, N.D., Zhang, W. et al. (2016). Small molecule/polymer blend organic transistors with hole mobility exceeding 13 $cm^2\ V^{-1}\ s^{-1}$. *Adv. Mater.* 28 (35): 7791.

279 Qian, Y., Zhang, X., Xie, L. et al. (2016). Stretchable organic semiconductor devices. *Adv. Mater.* 28 (42): 9243.

280 Trung, T.Q. and Lee, N.E. (2017). Recent progress on stretchable electronic devices with intrinsically stretchable components. *Adv. Mater.* 29 (3): 1603167.

281 Root, S.E., Savagatrup, S., Printz, A.D. et al. (2017). Mechanical properties of organic semiconductors for stretchable, highly flexible, and mechanically robust electronics. *Chem. Rev.* 117 (9): 6467.

282 Wang, S., Xu, J., Wang, W. et al. (2018). Skin electronics from scalable fabrication of an intrinsically stretchable transistor array. *Nature* 555 (7694): 83.

283 Chua, L.L., Ho, P.K.H., Sirringhaus, H., and Friend, R.H. (2004). Observation of field-effect transistor behavior at self-organized interfaces. *Adv. Mater.* 16 (18): 1609.

284 Lei, Y.L., Deng, P., Lin, M. et al. (2016). Enhancing crystalline structural orders of polymer semiconductors for efficient charge transport via polymer-matrix-mediated molecular self-assembly. *Adv. Mater.* 28 (31): 6687.

285 Lu, G.H., Blakesley, J., Himmelberger, S. et al. (2013). Moderate doping leads to high performance of semiconductor/insulator polymer blend transistors. *Nat. Commun.* 4: 1038.

286 Arias, A.C., Endicott, F., and Street, R.A. (2006). Surface-induced self-encapsulation of polymer thin-film transistors. *Adv. Mater.* 18 (21): 2900.

287 Qiu, L.Z., Xu, Q., Lee, W.H. et al. (2011). Organic thin-film transistors with a photo-patternable semiconducting polymer blend. *J. Mater. Chem.* 21 (39): 15637.

288 Gumyusenge, A., Tran, D.T., Luo, X.Y. et al. (2018). Semiconducting polymer blends that exhibit stable charge transport at high temperatures. *Science* 362 (6419): 1131–1134.

289 Sekitani, T. and Someya, T. (2011). Human-friendly organic integrated circuits. *Mater. Today* 14 (9): 398.

290 Yang, T.T., Xie, D., Li, Z.H., and Zhu, H.W. (2017). Recent advances in wearable tactile sensors: materials, sensing mechanisms, and device performance. *Mater. Sci. Eng. R.* 115: 1.

291 Savagatrup, S., Printz, A.D., O'Connor, T.F. et al. (2014). Molecularly stretchable electronics. *Chem. Mater.* 26 (10): 3028.

292 Lipomi, D.J. and Bao, Z.N. (2017). Stretchable and ultraflexible organic electronics. *MRS Bull.* 42 (2): 93.

293 O'Connor, B.T., Awartani, O.M., and Balar, N. (2017). Morphological considerations of organic electronic films for flexible and stretchable devices. *MRS Bull.* 42 (2): 108.

294 Song, E., Kang, B., Choi, H.H. et al. (2016). Stretchable and transparent organic semiconducting thin film with conjugated polymer nanowires embedded in an elastomeric matrix. *Adv. Electron. Mater.* 2 (1).

295 Shin, M., Oh, J.Y., Byun, K.E. et al. (2015). Polythiophene nanofibril bundles surface-embedded in elastomer: a route to a highly stretchable active channel layer. *Adv. Mater.* 27 (7): 1255.

296 Yu, K., Park, B., Kim, G. et al. (2016). Optically transparent semiconducting polymer nanonetwork for flexible and transparent electronics. *Proc. Natl. Acad. Sci. U.S.A.* 113 (50): 14261.

297 Kim, H.J., Sim, K., Thukral, A., and Yu, C. (2017). Rubbery electronics and sensors from intrinsically stretchable elastomeric composites of semiconductors and conductors. *Sci. Adv.* 3 (9): e1701114.

298 Meijer, E.J., de Leeuw, D.M., Setayesh, S. et al. (2003). Solution-processed ambipolar organic field-effect transistors and inverters. *Nat. Mater.* 2 (10): 678.

299 Chang, H.C., Shih, C.C., Liu, C.L., and Chen, W.C. (2015). A 1D electrospun nanofiber channel for organic field-effect transistors using a donor/acceptor planar heterojunction architecture. *Adv. Mater. Interfaces* 2 (6): 1500054.

300 Kim, F.S., Ahmed, E., Subramaniyan, S., and Jenekhe, S.A. (2010). Air-stable ambipolar field-effect transistors and complementary logic circuits from solution-processed n/p polymer heterojunctions. *ACS Appl. Mater. Interfaces* 2 (11): 2974.

301 Risteska, A., Chan, K.Y., Anthopoulos, T.D. et al. (2012). Designing organic and inorganic ambipolar thin-film transistors and inverters: theory and experiment. *Org. Electron.* 13 (12): 2816.

302 Baeg, K.J., Caironi, M., and Noh, Y.Y. (2013). Toward printed integrated circuits based on unipolar or ambipolar polymer semiconductors. *Adv. Mater.* 25 (31): 4210.

303 Bisri, S.Z., Piliego, C., Gao, J., and Loi, M.A. (2014). Outlook and emerging semiconducting materials for ambipolar transistors. *Adv. Mater.* 26 (8): 1176.

304 Zhu, W., Zhen, Y., Dong, H. et al. (2014). Organic cocrystal optoelectronic materials and devices. *Prog. Chem.* 26 (8): 1292.

305 Wang, H.B. and Yan, D.H. (2010). Organic heterostructures in organic field-effect transistors. *NPG Asia Mater.* 2 (2): 69.

306 Gao, D., Zhang, X., Kong, X. et al. (2015). (TFPP)Eu[Pc(OPh)$_8$]Eu[Pc(OPh)$_8$]/CuPc two-component bilayer heterojunction-based organic transistors with high ambipolar performance. *ACS Appl. Mater. Interfaces* 7 (4): 2486.

307 Shibao, M., Morita, T., Takashima, W., and Kaneto, K. (2007). Ambipolar transport in field-effect transistors based on composite films of poly(3-hexylthiophene) and fullerene derivative. *Jpn. J. Appl. Phys, Part 2* 46 (4–7): L123.

308 Cho, S.N., Yuen, J., Kim, J.Y. et al. (2006). Ambipolar organic field-effect transistors fabricated using a composite of semiconducting polymer and soluble fullerene. *Appl. Phys. Lett.* 89 (15): 153505.

309 Morita, T., Singh, V., Oku, S. et al. (2010). Ambipolar transport in bilayer organic field-effect transistor based on poly(3-hexylthiophene) and fullerene derivatives. *Jpn. J. Appl. Phys.* 49 (4): 041601.

310 Dodabalapur, A., Katz, H.E., Torsi, L., and Haddon, R.C. (1995). Organic heterostructure field-effect transistors. *Science* 269 (5230): 1560.

311 Tada, K., Harada, H., and Yoshino, K. (1996). Polymeric bipolar thin-film transistor utilizing conducting polymer containing electron transport dye. *Jpn. J. Appl. Phys.* 35 (7B): L944.

312 Rost, C., Gundlach, D.J., Karg, S., and Riess, W. (2004). Ambipolar organic field-effect transistor based on an organic heterostructure. *J. Appl. Phys.* 95 (10): 5782.

313 Wei, Z., Xu, W., Hu, W., and Zhu, D. (2009). Langmuir–Blodgett monolayer as an efficient p-conducting channel of ambipolar organic transistors and a template for n-type molecular alignment. *Langmuir* 25 (6): 3349.

314 Zhu, F., Wang, H.B., Song, D. et al. (2008). Charge transport in accumulation layers of organic heterojunctions. *Appl. Phys. Lett.* 93 (10): 103308.

315 Zhang, Y., Dong, H., Tang, Q. et al. (2010). Organic single-crystalline p–n junction nanoribbons. *J. Am. Chem. Soc.* 132 (33): 11580.

316 Wang, J., Wang, H.B., Yan, X.J. et al. (2005). Organic heterojunction and its application for double channel field-effect transistors. *Appl. Phys. Lett.* 87 (9): 093507.

317 Kang, S.J., Yi, Y., Kim, C.Y. et al. (2005). Ambipolar organic thin-film transistors using C60/pentacene structure: characterization of electronic structure and device property. *Appl. Phys. Lett.* 87 (23): 233502.

318 Noever, S.J., Fischer, S., and Nickel, B. (2013). Dual channel operation upon n-channel percolation in a pentacene-C60 ambipolar organic thin film transistor. *Adv. Mater.* 25 (15): 2147.

319 Sakamoto, Y., Suzuki, T., Kobayashi, M. et al. (2004). Perfluoropentacene: high-performance p–n junctions and complementary circuits with pentacene. *J. Am. Chem. Soc.* 126 (26): 8138.

320 Wang, H.B., Wang, X.J., Yu, B. et al. (2008). p–p isotype organic heterojunction and ambipolar field-effect transistors. *Appl. Phys. Lett.* 93 (11): 113303.

321 Yang, J., Yan, D., and Jones, T.S. (2015). Molecular template growth and its applications in organic electronics and optoelectronics. *Chem. Rev.* 115 (11): 5570.

322 Yang, J. and Yan, D. (2009). Weak epitaxy growth of organic semiconductor thin films. *Chem. Soc. Rev.* 38 (9): 2634.

323 Wang, H.B., Song, D., Yang, J.L. et al. (2007). High mobility vanadyl-phthalocyanine polycrystalline films for organic field-effect transistors. *Appl. Phys. Lett.* 90 (25): 253510.

324 Wang, H.B., Zhu, F., Yang, J.L. et al. (2007). Weak epitaxy growth affording high-mobility thin films of disk-like organic semiconductors. *Adv. Mater.* 19 (16): 2168.

325 Song, D., Wang, H.B., Zhu, F. et al. (2008). Phthalocyanato tin(IV) dichloride: an air-stable, high-performance, n-type organic semiconductor with a high field-effect electron mobility. *Adv. Mater.* 20 (11): 2142.

326 Pan, F., Tian, H.K., Qian, X.R. et al. (2011). High performance vanadyl phthalocyanine thin-film transistors based on fluorobenzene end-capped quaterthiophene as the inducing layer. *Org. Electron.* 12 (8): 1358.

327 Huang, L.Z., Zhu, F., Liu, C.F. et al. (2012). Crystalline organic heterostructures engineering based on vanadyl phthalocyanine and rod-like conjugated organic semiconductors with selected central groups. *Adv. Funct. Mater.* 22 (21): 4598.

328 Lunt, R.R., Benziger, J.B., and Forrest, S.R. (2007). Growth of an ordered crystalline organic heterojunction. *Adv. Mater.* 19 (23): 4229.

329 Zhu, F., Lou, K., Huang, L.Z. et al. (2009). Crystalline organic superlattice. *Appl. Phys. Lett.* 95 (20): 203106.

330 Wang, Z., Chang, H., Wang, T. et al. (2014). Alternate heteroepitaxial growth of highly oriented organic multilayer films. *J. Phys. Chem. B* 118 (15): 4212.

331 Kaneto, K., Yano, M., Shibao, M. et al. (2007). Ambipolar field-effect transistors based on poly(3-hexylthiophene)/fullerene derivative bilayer films. *Jpn. J. Appl. Phys, Part 1* 46 (4a): 1736.

332 Fan, C., Zoombelt, A.P., Jiang, H. et al. (2013). Solution-grown organic single-crystalline p–n junctions with ambipolar charge transport. *Adv. Mater.* 25 (40): 5762.

333 Wu, J., Fan, C., Xue, G. et al. (2015). Interfacing solution-grown C60 and (3-pyrrolinium)($CdCl_3$) single crystals for high-mobility transistor-based memory devices. *Adv. Mater.* 27: 4476.

334 Bender, C.J. (1986). Theoretical-models of charge-transfer complexes. *Chem. Soc. Rev.* 15 (4): 475.

335 Mori, T. (2011). Organic charge-transfer salts and the component molecules in organic transistors. *Chem. Lett.* 40 (5): 428.

336 Zhu, L., Kim, E.-G., Yi, Y., and Brédas, J.-L. (2011). Charge transfer in molecular complexes with 2,3,5,6-tetrafluoro-7,7,8,8-tetracyanoquinodimethane (F4-TCNQ): a density functional theory study. *Chem. Mater.* 23 (23): 5149.

337 Rückerl, F., Mahns, B., Dodbiba, E. et al. (2016). Electronic properties of the charge transfer material MnPc/F_4TCNQ. *J. Chem. Phys.* 145 (11): 114702.

338 Alves, H., Molinari, A.S., Xie, H., and Morpurgo, A.F. (2008). Metallic conduction at organic charge-transfer interfaces. *Nat. Mater.* 7 (7): 574.

339 Hayashi, H., Nihashi, W., Umeyama, T. et al. (2011). Segregated donor–acceptor columns in liquid crystals that exhibit highly efficient ambipolar charge transport. *J. Am. Chem. Soc.* 133 (28): 10736.

340 Zhang, J., Tan, J., Ma, Z. et al. (2013). Fullerene/sulfur-bridged annulene cocrystals: two-dimensional segregated heterojunctions with ambipolar transport properties and photoresponsivity. *J. Am. Chem. Soc.* 135 (2): 558.

341 Krupskaya, Y., Gibertini, M., Marzari, N., and Morpurgo, A.F. (2015). Band-like electron transport with record-high mobility in the TCNQ family. *Adv. Mater.* 27 (15): 2453.

342 Geng, H., Zheng, X., Shuai, Z. et al. (2015). Understanding the charge transport and polarities in organic donor-acceptor mixed-stack crystals: molecular insights from the super-exchange couplings. *Adv. Mater.* 27 (8): 1443.

343 Goetz, K.P., Vermeulen, D., Payne, M.E. et al. (2014). Charge-transfer complexes: new perspectives on an old class of compounds. *J. Mater. Chem. C* 2 (17): 3065.

344 Zhu, W., Zheng, R., Zhen, Y. et al. (2015). Rational design of charge-transfer interactions in halogen-bonded co-crystals toward versatile solid-state optoelectronics. *J. Am. Chem. Soc.* 137 (34): 11038.

345 Zhang, J., Xu, W., Sheng, P. et al. (2017). Organic donor-acceptor complexes as novel organic semiconductors. *Acc. Chem. Res.* 50 (7): 1654.

346 Zhang, J., Geng, H., Virk, T.S. et al. (2012). Sulfur-bridged annulene-TCNQ co-crystal: a self-assembled "molecular level heterojunction" with air stable ambipolar charge transport behavior. *Adv. Mater.* 24 (19): 2603.

347 Hinderhofer, A. and Schreiber, F. (2012). Organic–organic heterostructures: concepts and applications. *ChemPhysChem* 13 (3): 628.

348 Brooks, J.S. (1998). Towards the application of charge transfer salts and conducting polymer wires to electronic devices and sensors. *Adv. Mater. Opt. Electron.* 8 (5): 269.

349 Jerome, D. (2004). Organic conductors: from charge density wave TTF-TCNQ to superconducting $(TMTSF)_2PF_6$. *Chem. Rev.* 104 (11): 5565.

350 Wang, Y., Zhu, W., Dong, H. et al. (2016). Organic cocrystals: new strategy for molecular collaborative innovation. *Top. Curr. Chem.* 374 (6): 83.

351 Bond, A.D. (2007). What is a co-crystal? *Crystengcomm* 9 (9): 833.

352 Friščić, T. and Jones, W. (2009). Recent advances in understanding the mechanism of cocrystal formation via grinding. *Cryst. Growth Des.* 9 (3): 1621.

353 Aitipamula, S., Banerjee, R., Bansal, A.K. et al. (2012). Polymorphs, salts, and cocrystals: what's in a name? *Cryst. Growth Des.* 12 (5): 2147.

354 Munnoch, P.J. and Wright, J.D. (1974). Crystal structure of the 1 : 1 molecular complex of chrysene and 7,7,8,8-tetracyanoquinodimethane. *J. Chem. Soc., Perk. Trans. 2* 11: 1397. https://doi.org/10.1039/p29740001397.

355 Ferraris, J., Walatka, V., Perlstei, J., and Cowan, D.O. (1973). Electron-transfer in a new highly conducting donor-acceptor complex. *J. Am. Chem. Soc.* 95 (3): 948.

356 Tsutsumi, J.Y., Matsuoka, S., Inoue, S. et al. (2015). N-type field-effect transistors based on layered crystalline donor–acceptor semiconductors with dialkylated benzothienobenzothiophenes as electron donors. *J. Mater. Chem. C* 3 (9): 1976.

357 Zhu, L., Yi, Y., Li, Y. et al. (2012). Prediction of remarkable ambipolar charge-transport characteristics in organic mixed-stack charge-transfer crystals. *J. Am. Chem. Soc.* 134 (4): 2340.

358 Qin, Y., Cheng, C., Geng, H. et al. (2016). Efficient ambipolar transport properties in alternate stacking donor–acceptor complexes: from experiment to theory. *Phys. Chem. Chem. Phys.* 18 (20): 14094.

359 Pfattner, R., Bromley, S.T., Rovira, C., and Mas-Torrent, M. (2016). Tuning crystal ordering, electronic structure, and morphology in organic semiconductors: tetrathiafulvalenes as a model case. *Adv. Funct. Mater.* 26 (14): 2256.

360 Zhang, J., Zhao, G., Qin, Y. et al. (2014). Enhancement of the p-channel performance of sulfur-bridged annulene through a donor–acceptor co-crystal approach. *J. Mater. Chem. C* 2 (42): 8886.

361 Shibata, Y., Tsutsumi, J.Y., Matsuoka, S. et al. (2015). Uniaxially oriented polycrystalline thin films and air-stable n-type transistors based on donor-acceptor semiconductor $(diC_8BTBT)(F_nTCNQ)$ [$n = 0, 2, 4$]. *Appl. Phys. Lett.* 106 (14): 143303.

362 Park, S.K., Varghese, S., Kim, J.H. et al. (2013). Tailor-made highly luminescent and ambipolar transporting organic mixed stacked charge-transfer crystals: an isometric donor-acceptor approach. *J. Am. Chem. Soc.* 135 (12): 4757.

363 Park, S.K., Kim, J.H., Ohto, T. et al. (2017). Highly luminescent 2D-type slab crystals based on a molecular charge-transfer complex as promising organic light-emitting transistor materials. *Adv. Mater.* 29 (36): 1701346.

364 Vermeulen, D., Zhu, L.Y., Goetz, K.P. et al. (2014). Charge transport properties of perylene–TCNQ crystals: the effect of stoichiometry. *J. Phys. Chem. C* 118 (42): 24688.

365 Hu, P., Ma, L., Tan, K.J. et al. (2014). Solvent-dependent stoichiometry in perylene–7,7,8,8-tetracyanoquinodimethane charge transfer compound single crystals. *Cryst. Growth Des.* 14 (12): 6376.

366 Zhang, J., Gu, P.Y., Long, G.K. et al. (2016). Switching charge-transfer characteristics from p-type to n-type through molecular "doping" (co-crystallization). *Chem. Sci.* 7 (6): 3851.

367 Xu, X., Xiao, T., Gu, X. et al. (2015). Solution-processed ambipolar organic thin-film transistors by blending p- and n-type semiconductors: solid solution versus microphase separation. *ACS Appl. Mater. Interfaces* 7 (51): 28019.

368 Moulin, E., Cid, J.J., and Giuseppone, N. (2013). Advances in supramolecular electronics – from randomly self-assembled nanostructures to addressable self-organized interconnects. *Adv. Mater.* 25 (3): 477.

369 Kumar, M., Rao, K.V., and George, S.J. (2014). Supramolecular charge transfer nanostructures. *Phys. Chem. Chem. Phys.* 16 (4): 1300.

370 Segura, J.L. and Martin, N. (2001). New concepts in tetrathiafulvalene chemistry. *Angew. Chem. Int. Ed.* 40 (8): 1372.

371 Fourmigue, M. and Batail, P. (2004). Activation of hydrogen- and halogen-bonding interactions in tetrathiafulvalene-based crystalline molecular conductors. *Chem. Rev.* 104 (11): 5379.

372 Kato, R. (2004). Conducting metal dithiolene complexes: structural and electronic properties. *Chem. Rev.* 104 (11): 5319.

373 Berger, G., Soubhye, J., and Meyer, F. (2015). Halogen bonding in polymer science: from crystal engineering to functional supramolecular polymers and materials. *Polym. Chem.* 6 (19): 3559.

374 Yamamoto, H.M., Yamaura, J.-I., and Kato, R. (1998). Structural and electrical properties of $(BEDT\text{-}TTF)_2X$(diiodoacetylene) (X=Cl, Br): the novel self-assembly of neutral Lewis-acidic molecules and halide anions in a molecular metal. *J. Mater. Chem.* 8 (1): 15.

375 Yamamoto, H.M., Yamaura, J.-I., and Kato, R. (1998). Multicomponent molecular conductors with supramolecular assembly: iodine-containing neutral molecules as building blocks. *J. Am. Chem. Soc.* 120 (24): 5905.

376 Yamamoto, H.M., Kosaka, Y., Maeda, R. et al. (2008). Supramolecular insulating networks sheathing conducting nanowires based on organic radical cations. *ACS Nano* 2 (1): 143.

377 Rao, K.V., Jayaramulu, K., Maji, T.K., and George, S.J. (2010). Supramolecular hydrogels and high-aspect-ratio nanofibers through charge-transfer-induced alternate coassembly. *Angew. Chem. Int. Ed.* 122 (25): 4314.

378 Fahrenbach, A.C., Sampath, S., Late, D.J. et al. (2012). A semiconducting organic radical cationic host-guest complex. *ACS Nano* 6 (11): 9964.

379 Das, A. and Ghosh, S. (2014). Supramolecular assemblies by charge-transfer interactions between donor and acceptor chromophores. *Angew. Chem. Int. Ed.* 53 (8): 2038.

380 Sagade, A.A., Rao, K.V., Mogera, U. et al. (2013). High-mobility field effect transistors based on supramolecular charge transfer nanofibres. *Adv. Mater.* 25 (4): 559.

381 Black, H.T. and Perepichka, D.F. (2014). Crystal engineering of dual channel p/n organic semiconductors by complementary hydrogen bonding. *Angew. Chem. Int. Ed.* 53 (8): 2138.

382 Black, H.T., Lin, H., Belanger-Gariepy, F., and Perepichka, D.F. (2014). Supramolecular control of organic p/n-heterojunctions by complementary hydrogen bonding. *Faraday Discuss.* 174: 297.

383 Black, H.T., Yee, N., Zems, Y., and Perepichka, D.F. (2016). Complementary hydrogen bonding modulates electronic properties and controls self-assembly of donor/acceptor semiconductors. *Chemistry* 22 (48): 17251.

384 Oh, J.Y., Rondeau-Gagne, S., Chiu, Y.C. et al. (2016). Intrinsically stretchable and healable semiconducting polymer for organic transistors. *Nature* 539 (7629): 411.

385 Wang, Y., Li, Y., Zhu, W. et al. (2016). Co-crystal engineering: a novel method to obtain one-dimensional (1D) carbon nanocrystals of corannulene-fullerene by a solution process. *Nanoscale* 8 (32): 14920.

386 Wu, G., Huang, J., Zang, Y. et al. (2017). Porous field-effect transistors based on a semiconductive metal-organic framework. *J. Am. Chem. Soc.* 139 (4): 1360.

387 Dong, R., Han, P., Arora, H. et al. (2018). High-mobility band-like charge transport in a semiconducting two-dimensional metal-organic framework. *Nat. Mater.* 17 (11): 1027.

388 Kawase, T., Tanaka, K., Fujiwara, N. et al. (2003). Complexation of a carbon nanoring with fullerenes. *Angew. Chem. Int. Ed.* 42 (14): 1624.

389 Barnes, J.C., Juricek, M., Strutt, N.L. et al. (2013). ExBox: a polycyclic aromatic hydrocarbon scavenger. *J. Am. Chem. Soc.* 135 (1): 183.

390 Hausermann, R. and Batlogg, B. (2011). Gate bias stress in pentacene field-effect-transistors: charge trapping in the dielectric or semiconductor. *Appl. Phys. Lett.* 99 (8): 083303.

391 Sharma, A., Mathijssen, S.G.J., Smits, E.C.P. et al. (2010). Proton migration mechanism for operational instabilities in organic field-effect transistors. *Phys. Rev. B* 82 (7): 075322.

392 Egginger, M., Bauer, S., Schwodiauer, R. et al. (2009). Current versus gate voltage hysteresis in organic field effect transistors. *Monatsh. Chem.* 140 (7): 735.

393 Lee, B.H., Bazan, G.C., and Heeger, A.J. (2016). Doping-induced carrier density modulation in polymer field-effect transistors. *Adv. Mater.* 28 (1): 57.

394 Zhou, Y., Han, S.T., Chen, X. et al. (2014). An upconverted photonic nonvolatile memory. *Nat. Commun.* 5: 4720.

395 Smithson, C.S., Wu, Y., Wigglesworth, T., and Zhu, S. (2015). A more than six orders of magnitude UV-responsive organic field-effect transistor utilizing a benzothiophene semiconductor and disperse red 1 for enhanced charge separation. *Adv. Mater.* 27 (2): 228.

396 Sirringhaus, H. (2005). Device physics of solution-processed organic field-effect transistors. *Adv. Mater.* 17 (20): 2411.

397 Walzer, K., Maennig, B., Pfeiffer, M., and Leo, K. (2007). Highly efficient organic devices based on electrically doped transport layers. *Chem. Rev.* 107 (4): 1233.

398 Pfister, G. (1977). Hopping transport in a molecularly doped organic polymer. *Phys. Rev. B* 16 (8): 3676.

399 Kim, J.H., Yun, S.W., An, B.K. et al. (2013). Remarkable mobility increase and threshold voltage reduction in organic field-effect transistors by overlaying discontinuous nano-patches of charge-transfer doping layer on top of semiconducting film. *Adv. Mater.* 25 (5): 719.

400 Karimi-Alavijeh, H., Panahi, F., and Gharavi, A. (2014). Photo-switching effect in stilbene organic field effect transistors. *J. Appl. Phys.* 115 (9): 093706.

401 Lim, S.L., Li, N.J., Lu, J.M. et al. (2009). Conductivity switching and electronic memory effect in polymers with pendant azobenzene chromophores. *ACS Appl. Mater. Interfaces* 1 (1): 60.

402 Zhou, L., Mao, J., Ren, Y. et al. (2018). Recent advances of flexible data storage devices based on organic nanoscaled materials. *Small* 14 (10).

403 Ljubic, D., Smithson, C.S., Wu, Y.L., and Zhu, S.P. (2015). Highly UV-sensitive and responsive benzothiophene/dielectric polymer blend-based organic thin-film phototransistor. *Adv. Electron. Mater.* 1 (8): 1500119.

404 Smithson, C.S., Ljubic, D., Wu, Y., and Zhu, S. (2015). The effect of azobenzene derivatives on UV-responsive organic thin-film transistors with a 2,7-dipentylbenzo[*b*]benzo[4,5]thieno[2,3-*d*]thiophene semiconductor. *J. Mater. Chem. C* 3 (31): 8090.

405 Orgiu, E. and Samori, P. (2014). 25th anniversary article: organic electronics marries photochromism: generation of multifunctional interfaces, materials, and devices. *Adv. Mater.* 26 (12): 1827.

406 Irie, M., Fukaminato, T., Matsuda, K., and Kobatake, S. (2014). Photochromism of diarylethene molecules and crystals: memories, switches, and actuators. *Chem. Rev.* 114 (24): 12174.

407 Fleming, C.L., Li, S., Grotli, M., and Andreasson, J. (2018). Shining new light on the spiropyran photoswitch: a photocage decides between cis- trans or spiro-merocyanine isomerization. *J. Am. Chem. Soc.* 140 (43): 14069.

408 Hayakawa, R., Petit, M., Higashiguchi, K. et al. (2015). Interface engineering for improving optical switching in a diarylethene-channel transistor. *Org. Electron.* 21: 149.

409 Hayakawa, R., Higashiguchi, K., Matsuda, K. et al. (2013). Optically and electrically driven organic thin film transistors with diarylethene photochromic channel layers. *ACS Appl. Mater. Interfaces* 5 (9): 3625.

410 Li, Y., Zhang, H., Qi, C., and Guo, X. (2012). Light-driven photochromism-induced reversible switching in P3HT–spiropyran hybrid transistors. *J. Mater. Chem.* 22 (10): 4261.

411 Ishiguro, Y., Hayakawa, R., Chikyow, T., and Wakayama, Y. (2013). Optical switching of carrier transport in polymeric transistors with photochromic spiropyran molecules. *J. Mater. Chem. C* 1 (17): 3012.

412 Ishiguro, Y., Hayakawa, R., Yasuda, T. et al. (2013). Unique device operations by combining optical-memory effect and electrical-gate modulation in a photochromism-based dual-gate transistor. *ACS Appl. Mater. Interfaces* 5 (19): 9726.

413 Ishiguro, Y., Hayakawa, R., Chikyow, T., and Wakayama, Y. (2014). Optically controllable dual-gate organic transistor produced via phase separation between polymer semiconductor and photochromic spiropyran molecules. *ACS Appl. Mater. Interfaces* 6 (13): 10415.

414 Gemayel, M.E., Borjesson, K., Herder, M. et al. (2015). Optically switchable transistors by simple incorporation of photochromic systems into small-molecule semiconducting matrices. *Nat. Commun.* 6: 6330.

415 Leydecker, T., Herder, M., Pavlica, E. et al. (2016). Flexible non-volatile optical memory thin-film transistor device with over 256 distinct levels based on an organic bicomponent blend. *Nat. Nanotechnol.* 11 (9): 769.

416 Sze, S.M. and Ng, K.K. (2006). *Physics of Semiconductor Devices*. Wiley. https://onlinelibrary.wiley.com/doi/book/10.1002/0470068329.

417 Shen, Y., Hosseini, A.R., Wong, M.H., and Malliaras, G.G. (2004). How to make ohmic contacts to organic semiconductors. *ChemPhysChem* 5 (1): 16.

418 Xu, Y., Sun, H., and Noh, Y.Y. (2017). Schottky barrier in organic transistors. *IEEE Trans. Electron Devices* 64 (5): 1932.

419 Gorgoi, M. and Zahn, D.R.T. (2005). "Band bending" in copper phthalocyanine on hydrogen-passivated Si(111). *Org. Electron.* 6 (4): 168.

420 Kampen, T.U. (2005). Electronic structure of organic interfaces – a case study on perylene derivatives. *Appl. Phys. A* 82 (3): 457.

421 Limketkai, B.N. and Baldo, M.A. (2005). Charge injection into cathode-doped amorphous organic semiconductors. *Phys. Rev. B* 71 (8): 085207.

422 Liu, Z.H., Kobayashi, M., Paul, B.C. et al. (2010). Contact engineering for organic semiconductor devices via Fermi level depinning at the metal-organic interface. *Phys. Rev. B* 82 (3): 035311.

423 von Wrochem, F., Gao, D., Scholz, F. et al. (2010). Efficient electronic coupling and improved stability with dithiocarbamate-based molecular junctions. *Nat. Nanotechnol.* 5 (8): 618.

424 Morikawa, Y., Toyoda, K., Hamada, I. et al. (2012). First-principles theoretical study of organic/metal interfaces: vacuum level shifts and interface dipoles. *Curr. Appl Phys.* 12: S2.

425 Natali, D. and Caironi, M. (2012). Charge injection in solution-processed organic field-effect transistors: physics, models and characterization methods. *Adv. Mater.* 24 (11): 1357.

426 Wang, H., Amsalem, P., Heimel, G. et al. (2014). Band-bending in organic semiconductors: the role of alkali-halide interlayers. *Adv. Mater.* 26 (6): 925.

427 Zhang, P.L., Zhao, S., Wang, H.B. et al. (2017). Relation between interfacial band-bending and electronic properties in organic semiconductor pentacene. *Adv. Electron. Mater.* 3 (11): 1700136.

428 Zhang, Z. and Yates, J.T., Jr. (2012). Band bending in semiconductors: chemical and physical consequences at surfaces and interfaces. *Chem. Rev.* 112 (10): 5520.

429 Tung, R.T. (2001). Recent advances in Schottky barrier concepts. *Mater. Sci. Eng. R* 35 (1–3): 1.

430 Hwang, J., Wan, A., and Kahn, A. (2009). Energetics of metal–organic interfaces: new experiments and assessment of the field. *Mater. Sci. Eng.: R: Rep.* 64 (1–2): 1.

431 Flores, F., Ortega, J., and Vazquez, H. (2009). Modelling energy level alignment at organic interfaces and density functional theory. *Phys. Chem. Chem. Phys.* 11 (39): 8658.

432 Tung, R.T. (2014). The physics and chemistry of the Schottky barrier height. *Appl. Phys. Rev.* 1 (1): 011304.

433 Li, L., Meise-Gresch, K., Jiang, L. et al. (2012). The electrode's effect on the stability of organic transistors and circuits. *Adv. Mater.* 24 (22): 3053.

434 Liu, C., Huseynova, G., Xu, Y. et al. (2016). Universal diffusion-limited injection and the hook effect in organic thin-film transistors. *Sci. Rep.* 6: 29811.

435 Jung, S., Kim, C.H., Bonnassieux, Y., and Horowitz, G. (2015). Injection barrier at metal/organic semiconductor junctions with a Gaussian density-of-states. *J. Phys. D: Appl. Phys.* 48 (39): 395103.

436 Bao, Z. and Locklin, J. (2007). *Organic Field-Effect Transistors*. CRC Press.

437 Parthasarathy, G., Shen, C., Kahn, A., and Forrest, S.R. (2001). Lithium doping of semiconducting organic charge transport materials. *J. Appl. Phys.* 89 (9): 4986.

438 Watkins, N.J., Yan, L., and Gao, Y.L. (2002). Electronic structure symmetry of interfaces between pentacene and metals. *Appl. Phys. Lett.* 80 (23): 4384.

439 Shen, C., Kahn, A., and Schwartz, J. (2001). Chemical and electrical properties of interfaces between magnesium and aluminum and tris-(8-hydroxy quinoline) aluminum. *J. Appl. Phys.* 89 (1): 449.

440 Kafer, D., Ruppel, L., and Witte, G. (2007). Growth of pentacene on clean and modified gold surfaces. *Phys. Rev. B* 75 (8): 085309.

441 Rentenberger, S., Vollmer, A., Zojer, E. et al. (2006). UV/ozone treated au for air-stable, low hole injection barrier electrodes in organic electronics. *J. Appl. Phys.* 100 (5): 053701.

442 Koch, N. (2007). Organic electronic devices and their functional interfaces. *ChemPhysChem* 8 (10): 1438.

443 Braun, S., Salaneck, W.R., and Fahlman, M. (2009). Energy-level alignment at organic/metal and organic/organic interfaces. *Adv. Mater.* 21 (14–15): 1450.

444 Wan, A., Hwang, J., Amy, F., and Kahn, A. (2005). Impact of electrode contamination on the alpha-NPD/Au hole injection barrier. *Org. Electron.* 6 (1): 47.

445 Smith, R.K., Lewis, P.A., and Weiss, P.S. (2004). Patterning self-assembled monolayers. *Prog. Surf. Sci.* 75 (1–2): 1.

446 Ulman, A. (1996). Formation and structure of self-assembled monolayers. *Chem. Rev.* 96 (4): 1533.

447 Schulz, P., Schafer, T., Zangmeister, C.D. et al. (2014). A new route to low resistance contacts for performance-enhanced organic electronic devices. *Adv. Mater. Interfaces* 1 (5): 1300130.

448 Gao, D., Scholz, F., Nothofer, H.G. et al. (2011). Fabrication of asymmetric molecular junctions by the oriented assembly of dithiocarbamate rectifiers. *J. Am. Chem. Soc.* 133 (15): 5921.

449 Meyer, D., Schafer, T., Schulz, P. et al. (2016). Dithiocarbamate self-assembled monolayers as efficient surface modifiers for low work function noble metals. *Langmuir* 32 (35): 8812.

450 Crispin, X., Geskin, V., Crispin, A. et al. (2002). Characterization of the interface dipole at organic/metal interfaces. *J. Am. Chem. Soc.* 124 (27): 8131.

451 Heimel, G., Rissner, F., and Zojer, E. (2010). Modeling the electronic properties of pi-conjugated self-assembled monolayers. *Adv. Mater.* 22 (23): 2494.

452 Tao, F. and Bernasek, S.L. (2007). Understanding odd-even effects in organic self-assembled monolayers. *Chem. Rev.* 107 (5): 1408.

453 Thuo, M.M., Reus, W.F., Nijhuis, C.A. et al. (2011). Odd-even effects in charge transport across self-assembled monolayers. *J. Am. Chem. Soc.* 133 (9): 2962.

454 Osella, S., Cornil, D., and Cornil, J. (2014). Work function modification of the (111) gold surface covered by long alkanethiol-based self assembled monolayers. *Phys. Chem. Chem. Phys.* 16 (7): 2866.

455 Lee, H.J., Jamison, A.C., and Lee, T.R. (2015). Surface dipoles: a growing body of evidence supports their impact and importance. *Acc. Chem. Res.* 48 (12): 3007.

456 Alloway, D.M., Graham, A.L., Yang, X. et al. (2009). Tuning the effective work function of gold and silver using omega-functionalized alkanethiols: varying surface composition through dilution and choice of terminal groups. *J. Phys. Chem. C* 113 (47): 20328.

457 Kuzumoto, Y. and Kitamura, M. (2014). Work function of gold surfaces modified using substituted benzenethiols: reaction time dependence and thermal stability. *Appl. Phys Express* 7 (3): 035701.

458 Zehner, R.W., Parsons, B.F., Hsung, R.P., and Sita, L.R. (1999). Tuning the work function of gold with self-assembled monolayers derived from $X{-}[C_6H_4{-}C{\equiv}C{-}]nC_6H_4{-}SH$ ($n = 0, 1, 2$; X = H, F, CH_3, CF_3, and OCH_3). *Langmuir* 15 (4): 1121.

459 Wang, L.J., Rangger, G.M., Romaner, L. et al. (2009). Electronic structure of self-assembled monolayers on Au(111) surfaces: the impact of backbone polarizability. *Adv. Funct. Mater.* 19 (23): 3766.

460 Masillamani, A.M., Crivillers, N., Orgiu, E. et al. (2012). Multiscale charge injection and transport properties in self-assembled monolayers of biphenyl thiols with varying torsion angles. *Chemistry* 18 (33): 10335.

461 Xie, Z., Baldea, I., Oram, S. et al. (2017). Effect of heteroatom substitution on transport in alkanedithiol-based molecular tunnel junctions: evidence for universal behavior. *ACS Nano* 11 (1): 569.

462 Kovalchuk, A., Abu-Husein, T., Fracasso, D. et al. (2016). Transition voltages respond to synthetic reorientation of embedded dipoles in self-assembled monolayers. *Chem. Sci.* 7 (1): 781.

463 Briseno, A.L., Aizenberg, J., Han, Y.J. et al. (2005). Patterned growth of large oriented organic semiconductor single crystals on self-assembled monolayer templates. *J. Am. Chem. Soc.* 127 (35): 12164.

464 Park, S., Giri, G., Shaw, L. et al. (2015). Large-area formation of self-aligned crystalline domains of organic semiconductors on transistor channels using CONNECT. *Proc. Natl. Acad. Sci. U.S.A.* 112 (18): 5561.

465 Heimel, G., Romaner, L., Zojer, E., and Bredas, J.L. (2008). The interface energetics of self-assembled monolayers on metals. *Acc. Chem. Res.* 41 (6): 721.

466 DiBenedetto, S.A., Facchetti, A., Ratner, M.A., and Marks, T.J. (2009). Molecular self-assembled monolayers and multilayers for organic and unconventional inorganic thin-film transistor applications. *Adv. Mater.* 21 (14–15): 1407.

467 Katz, H.E., Johnson, J., Lovinger, A.J., and Li, W. (2000). Naphthalenetetracarboxylic diimide-based n-channel transistor semiconductors: structural variation and thiol-enhanced gold contacts. *J. Am. Chem. Soc.* 122 (32): 7787.

468 Gundlach, D.J., Royer, J.E., Park, S.K. et al. (2008). Contact-induced crystallinity for high-performance soluble acene-based transistors and circuits. *Nat. Mater.* 7 (3): 216.

469 Niazi, M.R., Li, R.P., Abdelsamie, M. et al. (2016). Contact-induced nucleation in high-performance bottom-contact organic thin film transistors manufactured by large-area compatible solution processing. *Adv. Funct. Mater.* 26 (14): 2371.

470 Ho, P.K., Kim, J.S., Burroughes, J.H. et al. (2000). Molecular-scale interface engineering for polymer light-emitting diodes. *Nature* 404 (6777): 481.

471 Giordano, L., Cinquini, F., and Pacchioni, G. (2006). Tuning the surface metal work function by deposition of ultrathin oxide films: density functional calculations. *Phys. Rev. B* 73 (4): 045414.

472 Prada, S., Martinez, U., and Pacchioni, G. (2008). Work function changes induced by deposition of ultrathin dielectric films on metals: a theoretical analysis. *Phys. Rev. B* 78 (23): 235423.

473 Giordano, L. and Pacchioni, G. (2011). Oxide films at the nanoscale: new structures, new functions, and new materials. *Acc. Chem. Res.* 44 (11): 1244.

474 Hu, Z.H., Zhong, Z.M., Zhang, K. et al. (2017). Dipole formation at organic/metal interfaces with pre-deposited and post-deposited metal. *NPG Asia Mater.* 9 (5): e379.

475 Park, Y., Choong, V., Ettedgui, E. et al. (1996). Energy level bending and alignment at the interface between Ca and a phenylene vinylene oligomer. *Appl. Phys. Lett.* 69 (8): 1080.

476 Chu, C.W., Li, S.H., Chen, C.W. et al. (2005). High-performance organic thin-film transistors with metal oxide/metal bilayer electrode. *Appl. Phys. Lett.* 87 (19): 193508.

477 Kumaki, D., Umeda, T., and Tokito, S. (2008). Reducing the contact resistance of bottom-contact pentacene thin-film transistors by employing a MoO_x carrier injection layer. *Appl. Phys. Lett.* 92 (1): 013301.

478 Minagawa, M., Kim, Y., Claus, M., and Bao, Z. (2017). Reducing the contact resistance in bottom-contact-type organic field-effect transitors using an AgO_x interface layer. *Appl. Phys Express* 10 (9): 091601.

479 Gao, Y.H., Shao, Y.D., Yan, L.J. et al. (2016). Efficient charge injection in organic field-effect transistors enabled by low-temperature atomic layer deposition of ultrathin VOx interlayer. *Adv. Funct. Mater.* 26 (25): 4456.

480 Gui, H., Wei, B., and Wang, J. (2014). Inserting a Mn-doped TiO_2 layer for improving performance of pentacene organic thin-film transistors. *Org. Electron.* 15 (11): 3349.

481 Kumatani, A., Li, Y., Darmawan, P. et al. (2013). On practical charge injection at the metal/organic semiconductor interface. *Sci. Rep.* 3: 1026.

482 Khim, D., Han, H., Baeg, K.J. et al. (2013). Simple bar-coating process for large-area, high-performance organic field-effect transistors and ambipolar complementary integrated circuits. *Adv. Mater.* 25 (31): 4302.

483 Gwinner, M.C., Vaynzof, Y., Banger, K.K. et al. (2010). Solution-processed zinc oxide as high-performance air-stable electron injector in organic ambipolar light emitting field-effect transistors. *Adv. Funct. Mater.* 20 (20): 3457.

484 Chu, C.W., Sung, C.F., Lee, Y.Z., and Cheng, K. (2008). Improved performance in n-channel organic thin film transistors by nanoscale interface modification. *Org. Electron.* 9 (2): 262.

485 Waqas Alam, M., Wang, Z., Naka, S., and Okada, H. (2013). Mobility enhancement of top contact pentacene based organic thin film transistor with bi-layer GeO/Au electrodes. *Appl. Phys. Lett.* 102 (6): 061105.

486 Kano, M., Minari, T., and Tsukagoshi, K. (2009). Improvement of subthreshold current transport by contact interface modification in p-type organic field-effect transistors. *Appl. Phys. Lett.* 94 (14): 143304.

487 Wang, Z.K., Alam, M.W., Lou, Y.H. et al. (2012). Enhanced carrier injection in pentacene thin-film transistors by inserting a MoO_3-doped pentacene layer. *Appl. Phys. Lett.* 100 (4): 043302.

488 Minari, T., Darmawan, P., Liu, C. et al. (2012). Highly enhanced charge injection in thienoacene-based organic field-effect transistors with chemically doped contact. *Appl. Phys. Lett.* 100 (9): 093303.

489 Baeg, K.J., Kim, J., Khim, D. et al. (2011). Charge injection engineering of ambipolar field-effect transistors for high-performance organic complementary circuits. *ACS Appl. Mater. Interfaces* 3 (8): 3205.

490 Kim, N.K., Khim, D., Xu, Y. et al. (2014). Solution-processed barium salts as charge injection layers for high performance N-channel organic field-effect transistors. *ACS Appl. Mater. Interfaces* 6 (12): 9614.

491 Winter, R., Hammer, M.S., Deibel, C., and Pflaum, J. (2009). Improvement of the poly-3-hexylthiophene transistor performance using small molecule contact functionalization. *Appl. Phys. Lett.* 95 (26): 263313.

492 Wang, J., Wang, H.B., Zhang, J. et al. (2005). Organic thin-film transistors with improved characteristics using lutetium bisphthalocyanine as a buffer layer. *J. Appl. Phys.* 97 (2): 026106.

493 Ante, F., Kalblein, D., Zschieschang, U. et al. (2011). Contact doping and ultrathin gate dielectrics for nanoscale organic thin-film transistors. *Small* 7 (9): 1186.

494 Zschieschang, U., Kang, M.J., Takimiya, K. et al. (2012). Flexible low-voltage organic thin-film transistors and circuits based on C10-DNTT. *J. Mater. Chem.* 22 (10): 4273.

495 Hou, J.L., Kasemann, D., Widmer, J. et al. (2016). Reduced contact resistance in top-contact organic field-effect transistors by interface contact doping. *Appl. Phys. Lett.* 108 (10): 103303.

496 Singh, S., Mohapatra, S.K., Sharma, A. et al. (2013). Reduction of contact resistance by selective contact doping in fullerene n-channel organic field-effect transistors. *Appl. Phys. Lett.* 102 (15): 153303.

497 Li, J., Shi, W., Shu, L.F., and Yu, J.S. (2015). Performance enhancement of p-type organic field-effect transistor through introducing organic buffer layers. *J. Mater. Sci. – Mater. Electron.* 26 (11): 8301.

498 Li, Q., Yu, X., Shi, W., and Yu, J. (2013). Effect of organic buffer layers on the performance of n-type organic field-effect transistor based on C60 active layer. *Synth. Met.* 163: 57.

499 Ho, P.K.H., Granstrom, M., Friend, R.H., and Greenham, N.C. (1998). Ultra-thin self-assembled layers at the ITO interface to control charge injection and electroluminescence efficiency in polymer light-emitting diodes. *Adv. Mater.* 10 (10): 769.

500 Brown, T.M., Kim, J.S., Friend, R.H. et al. (1999). Built-in field electroabsorption spectroscopy of polymer light-emitting diodes incorporating a doped poly(3,4-ethylene dioxythiophene) hole injection layer. *Appl. Phys. Lett.* 75 (12): 1679.

501 Shi, W., Yu, J.S., Huang, W. et al. (2013). Performance enhancement of poly(3-hexylthiophene) organic field-effect transistor by inserting poly(methylmethacrylate) buffer layer. *Appl. Phys. Lett.* 102 (11): 111607.

502 Sun, B., Hong, W., Thibau, E.S. et al. (2015). Polyethylenimine (PEI) as an effective dopant To conveniently convert Ambipolar and p-type polymers into unipolar n-type polymers. *ACS Appl. Mater. Interfaces* 7 (33): 18662.

503 Seo, J.H., Gutacker, A., Walker, B. et al. (2009). Improved injection in n-type organic transistors with conjugated polyelectrolytes. *J. Am. Chem. Soc.* 131 (51): 18220.

504 Gim, Y., Kang, B., Kim, B. et al. (2015). Atomically-thin molecular layers for electrode modification of organic transistors. *Nanoscale* 7 (33): 14100.

505 Lefenfeld, M., Blanchet, G., and Rogers, J.A. (2003). High-performance contacts in plastic transistors and logic gates that use printed electrodes of DNNSA-PANI doped with single-walled carbon nanotubes. *Adv. Mater.* 15 (14): 1188.

506 Choi, M.C., Kim, Y., and Ha, C.S. (2008). Polymers for flexible displays: from material selection to device applications. *Prog. Polym. Sci.* 33 (6): 581.

507 Lee, M.W., Lee, M.Y., Choi, J.C. et al. (2010). Fine patterning of glycerol-doped PEDOT:PSS on hydrophobic PVP dielectric with ink jet for source and drain electrode of OTFTs. *Org. Electron.* 11 (5): 854.

508 Li, L., Hirtz, M., Wang, W. et al. (2010). Patterning of polymer electrodes by nanoscratching. *Adv. Mater.* 22 (12): 1374.

509 Heeger, A.J. (2001). Nobel lecture: semiconducting and metallic polymers: the fourth generation of polymeric materials. *Rev. Mod. Phys.* 73 (3): 681.

510 Lee, M.S., Kang, H.S., Kang, H.S. et al. (2005). Flexible all-polymer field effect transistors with optical transparency using electrically conducting polymers. *Thin Solid Films* 477 (1–2): 169.

511 Li, L., Jiang, L., Wang, W. et al. (2012). High-performance and stable organic transistors and circuits with patterned polypyrrole electrodes. *Adv. Mater.* 24 (16): 2159.

512 Zhang, J.J., Wang, C., Chen, J. et al. (2016). The highly conducting carbon electrodes derived from spin-coated polyacrylonitrile films. *Sci. China Chem.* 59 (6): 672.

513 Burns, S.E., Cain, P., Mills, J. et al. (2011). Inkjet printing of polymer thin-film transistor circuits. *MRS Bull.* 28 (11): 829.

514 Schmidt, G.C., Hoft, D., Bhuie, M. et al. (2013). Modified poly(3,4-ethylenedioxythiophene):poly(styrenesulfonate) source/drain electrodes for fully printed organic field-effect transistors consisting of a semiconductor blend. *Appl. Phys. Lett.* 103 (11): 113302.

515 Kim, S.H., Kim, J., Nam, S. et al. (2017). Tuning the work function of printed polymer electrodes by introducing a fluorinated polymer to enhance the operational stability in bottom-contact organic field-effect transistors. *ACS Appl. Mater. Interfaces* 9 (14): 12637.

516 Cosseddu, P., Bonfiglio, A., Salzmann, I. et al. (2008). Ambipolar transport in transparent and flexible all-organic heterojunction field effect transistors at ambient conditions. *Org. Electron.* 9 (2): 191.

517 Cosseddu, P. and Bonfiglio, A. (2010). Influence of device geometry in the electrical behavior of all organic ambipolar field effect transistors. *Appl. Phys. Lett.* 97 (20): 203305.

518 Koezuka, H., Tsumura, A., Fuchigami, H., and Kuramoto, K. (1993). Polythiophene field-effect transistor with polypyrrole worked as source and drain electrodes. *Appl. Phys. Lett.* 62 (15): 1794.

519 Jiang, L., Sun, Y., Peng, H. et al. (2011). Enhanced electrical conductivity of individual conducting polymer nanobelts. *Small* 7 (14): 1949.

520 Drury, C.J., Mutsaers, C.M.J., Hart, C.M. et al. (1998). Low-cost all-polymer integrated circuits. *Appl. Phys. Lett.* 73 (1): 108.

521 Matters, M., de Leeuw, D.M., Vissenberg, M.J.C.M. et al. (1999). Organic field-effect transistors and all-polymer integrated circuits. *Opt. Mater.* 12 (2–3): 189.

522 Gelinck, G.H., Geuns, T.C.T., and de Leeuw, D.M. (2000). High-performance all-polymer integrated circuits. *Appl. Phys. Lett.* 77 (10): 1487.

523 Blanchet, G.B., Loo, Y.L., Rogers, J.A. et al. (2003). Large area, high resolution, dry printing of conducting polymers for organic electronics. *Appl. Phys. Lett.* 82 (3): 463.

524 Hellstrom, S.L., Jin, R.Z., Stoltenberg, R.M., and Bao, Z. (2010). Driving high-performance n- and p-type organic transistors with carbon nanotube/conjugated polymer composite electrodes patterned directly from solution. *Adv. Mater.* 22 (37): 4204.

525 Schroeder, R., Majewski, L.A., Grell, M. et al. (2005). Electrode specific electropolymerization of ethylenedioxythiophene: injection enhancement in organic transistors. *Appl. Phys. Lett.* 87 (11): 113501.

526 Mori, T. (2008). Molecular materials for organic field-effect transistors. *J. Phys.: Condens. Mater.* 20 (18): 184010.

527 Mir, I.A., Ahuja, T., Kumar, D. et al. (2010). Carrier injection from polypyrrole coated gold electrode in pentacene field effect transistors. *Synth. Met.* 160 (19–20): 2116.

528 Hong, K., Yang, S.Y., Yang, C. et al. (2008). Reducing the contact resistance in organic thin-film transistors by introducing a PEDOT : PSS hole-injection layer. *Org. Electron.* 9 (5): 864.

529 Hong, K., Kim, S.H., Yang, C. et al. (2010). Improved n-type bottom-contact organic transistors by introducing a poly(3,4-ethylenedioxythiophene):poly(4-styrene sulfonate) coating on the source/drain electrodes. *Appl. Phys. Lett.* 97 (10): 103304.

530 Feldman, A.K., Steigerwald, M.L., Guo, X., and Nuckolls, C. (2008). Molecular electronic devices based on single-walled carbon nanotube electrodes. *Acc. Chem. Res.* 41 (12): 1731.

531 Valitova, I., Amato, M., Mahvash, F. et al. (2013). Carbon nanotube electrodes in organic transistors. *Nanoscale* 5 (11): 4638.

532 Jia, C., Ma, B., Xin, N., and Guo, X. (2015). Carbon electrode-molecule junctions: a reliable platform for molecular electronics. *Acc. Chem. Res.* 48 (9): 2565.

533 Cao, Y., Steigerwald, M.L., Nuckolls, C., and Guo, X. (2010). Current trends in shrinking the channel length of organic transistors down to the nanoscale. *Adv. Mater.* 22 (1): 20.

534 Jia, C. and Guo, X. (2013). Molecule-electrode interfaces in molecular electronic devices. *Chem. Soc. Rev.* 42 (13): 5642.

535 Suzuki, S., Watanabe, Y., Homma, Y. et al. (2004). Work functions of individual single-walled carbon nanotubes. *Appl. Phys. Lett.* 85 (1): 127.

536 Tsukagoshi, K., Yagi, I., and Aoyagi, Y. (2004). Pentacene nanotransistor with carbon nanotube electrodes. *Appl. Phys. Lett.* 85 (6): 1021.

537 Qi, P., Javey, A., Rolandi, M. et al. (2004). Miniature organic transistors with carbon nanotubes as quasi-one-dimensional electrodes. *J. Am. Chem. Soc.* 126 (38): 11774.

538 Aguirre, C.M., Ternon, C., Paillet, M. et al. (2009). Carbon nanotubes as injection electrodes for organic thin film transistors. *Nano Lett.* 9 (4): 1457.

539 Guo, X., Xiao, S., Myers, M. et al. (2009). Photoresponsive nanoscale columnar transistors. *Proc. Natl. Acad. Sci. U.S.A.* 106 (3): 691.

540 Cicoira, F., Aguirre, C.M., and Martel, R. (2011). Making contacts to n-type organic transistors using carbon nanotube arrays. *ACS Nano* 5 (1): 283.

541 Cicoira, F., Coppede, N., Iannotta, S., and Martel, R. (2011). Ambipolar copper phthalocyanine transistors with carbon nanotube array electrodes. *Appl. Phys. Lett.* 98 (18): 183303.

542 Sarker, B.K., Liu, J., Zhai, L., and Khondaker, S.I. (2011). Fabrication of organic field effect transistor by directly grown poly(3 hexylthiophene) crystalline nanowires on carbon nanotube aligned array electrode. *ACS Appl. Mater. Interfaces* 3 (4): 1180.

543 Marzouk, J., Lucas, B., Trigaud, T. et al. (2014). Simple strategy to tune the charge transport properties of conjugated polymer/carbon nanotube composites using an electric field assisted deposition technique. *Polym. Int.* 63 (8): 1378.

544 Sarker, B.K. and Khondaker, S.I. (2012). High-performance short channel organic transistors using densely aligned carbon nanotube array electrodes. *Appl. Phys. Lett.* 100 (2): 023301.

545 Kang, N., Sarker, B.K., and Khondaker, S.I. (2012). The effect of carbon nanotube/organic semiconductor interfacial area on the performance of organic transistors. *Appl. Phys. Lett.* 101 (23): 233302.

546 Sarker, B.K. and Khondaker, S.I. (2012). Thermionic emission and tunneling at carbon nanotube-organic semiconductor interface. *ACS Nano* 6 (6): 4993.

547 Liu, J., Rinzler, A.G., Dai, H. et al. (1998). Fullerene pipes. *Science* 280 (5367): 1253.

548 Tenent, R.C., Barnes, T.M., Bergeson, J.D. et al. (2009). Ultrasmooth, large-area, high-uniformity, conductive transparent single-walled-carbon-nanotube films for photovoltaics produced by ultrasonic spraying. *Adv. Mater.* 21 (31): 3210.

549 Liu, A.H., Honma, I., Ichihara, M., and Zhou, H.S. (2006). Poly(acrylic acid)-wrapped multi-walled carbon nanotubes composite solubilization in water: definitive spectroscopic properties. *Nanotechnology* 17 (12): 2845.

550 Hong, K., Nam, S., Yang, C. et al. (2009). Solution-processed organic field-effect transistors composed of poly(4-styrene sulfonate) wrapped multiwalled carbon nanotube source/drain electrodes. *Org. Electron.* 10 (2): 363.

551 Zhang, Y.Y., Shi, Y., Chen, F. et al. (2007). Poly(3,3‴-didodecylquarterthiophene) field effect transistors with single-walled carbon nanotube based source and drain electrodes. *Appl. Phys. Lett.* 91 (22): 223512.

552 Hong, K., Yang, C., Kim, S.H. et al. (2009). Photopatternable source/drain electrodes using multiwalled carbon nanotube/polymer nanocomposites for organic field-effect transistors. *ACS Appl. Mater. Interfaces* 1 (10): 2332.

553 Hong, K., Kim, S.H., Yang, C. et al. (2011). Photopatternable poly(4-styrene sulfonic acid)-wrapped MWNT thin-film source/drain electrodes for use in organic field-effect transistors. *ACS Appl. Mater. Interfaces* 3 (1): 74.

554 Jeong, Y.J., Lee, X., Bae, J. et al. (2016). Direct patterning of conductive carbon nanotube/polystyrene sulfonate composites via electrohydrodynamic jet printing for use in organic field-effect transistors. *J. Mater. Chem. C* 4 (22): 4912.

555 Pang, S., Hernandez, Y., Feng, X., and Mullen, K. (2011). Graphene as transparent electrode material for organic electronics. *Adv. Mater.* 23 (25): 2779.

556 Lee, S., Jo, G., Kang, S.J. et al. (2011). Enhanced charge injection in pentacene field-effect transistors with graphene electrodes. *Adv. Mater.* 23 (1): 100.

557 Lee, W.H., Park, J., Sim, S.H. et al. (2011). Transparent flexible organic transistors based on monolayer graphene electrodes on plastic. *Adv. Mater.* 23 (15): 1752.

558 Jeong, Y.J., Jang, J., Nam, S. et al. (2014). High-performance organic complementary inverters using monolayer graphene electrodes. *ACS Appl. Mater. Interfaces* 6 (9): 6816.

559 Lin, Y. and Guo, X. (2015). Ultrahigh photogain nanoscale hybrid photodetectors. *Small* 11 (24): 2856.

560 Liu, S., Wei, Z.M., Cao, Y. et al. (2011). Ultrasensitive water-processed monolayer photodetectors. *Chem. Sci.* 2 (4): 796.

561 Liu, S. and Guo, X.F. (2012). Carbon nanomaterials field-effect-transistor-based biosensors. *NPG Asia Mater.* 4 (8): 1.

562 Zhang, D., Gan, L., Cao, Y. et al. (2012). Understanding charge transfer at PbS-decorated graphene surfaces toward a tunable photosensor. *Adv. Mater.* 24 (20): 2715.

563 Yu, Y.J., Zhao, Y., Ryu, S. et al. (2009). Tuning the graphene work function by electric field effect. *Nano Lett.* 9 (10): 3430.

564 Shi, Y., Kim, K.K., Reina, A. et al. (2010). Work function engineering of graphene electrode via chemical doping. *ACS Nano* 4 (5): 2689.

565 Cao, Y., Liu, S., Shen, Q. et al. (2009). High-performance Photoresponsive organic nanotransistors with single-layer graphenes as two-dimensional electrodes. *Adv. Funct. Mater.* 19 (17): 2743.

566 Cao, Y., Wei, Z., Liu, S. et al. (2010). High-performance Langmuir–Blodgett monolayer transistors with high responsivity. *Angew. Chem. Int. Ed.* 49 (36): 6319.

567 Lee, W.H., Park, J., Sim, S.H. et al. (2011). Surface-directed molecular assembly of pentacene on monolayer graphene for high-performance organic transistors. *J. Am. Chem. Soc.* 133 (12): 4447.

568 Di, C.A., Wei, D.C., Yu, G. et al. (2008). Patterned graphene as source/drain electrodes for bottom-contact organic field-effect transistors. *Adv. Mater.* 20 (17): 3289.

569 Lee, C.H., Schiros, T., Santos, E.J. et al. (2014). Epitaxial growth of molecular crystals on van der waals substrates for high-performance organic electronics. *Adv. Mater.* 26 (18): 2812.

570 Kang, S.J., Lee, G.H., Yu, Y.J. et al. (2014). Organic field effect transistors based on graphene and hexagonal boron nitride heterostructures. *Adv. Funct. Mater.* 24 (32): 5157.

571 He, D., Zhang, Y., Wu, Q. et al. (2014). Two-dimensional quasi-freestanding molecular crystals for high-performance organic field-effect transistors. *Nat. Commun.* 5: 5162.

572 Zhang, Y., Qiao, J., Gao, S. et al. (2016). Probing carrier transport and structure-property relationship of highly ordered organic semiconductors at the two-dimensional limit. *Phys. Rev. Lett.* 116 (1): 016602.

573 He, D., Qiao, J., Zhang, L. et al. (2017). Ultrahigh mobility and efficient charge injection in monolayer organic thin-film transistors on boron nitride. *Sci. Adv.* 3 (9): e1701186.

574 Liu, X., Luo, X., Nan, H. et al. (2016). Epitaxial ultrathin organic crystals on graphene for high-efficiency phototransistors. *Adv. Mater.* 28 (26): 5200.

575 Liu, J., Yin, Z., Cao, X. et al. (2010). Bulk heterojunction polymer memory devices with reduced graphene oxide as electrodes. *ACS Nano* 4 (7): 3987.

576 Kong, B.S., Geng, J., and Jung, H.T. (2009). Layer-by-layer assembly of graphene and gold nanoparticles by vacuum filtration and spontaneous reduction of gold ions. *Chem. Commun.* 2174.

577 Sygellou, L., Paterakis, G., Galiotis, C., and Tasis, D. (2016). Work function tuning of reduced graphene oxide thin films. *J. Phys. Chem. C* 120 (1): 281.

578 Pang, S.P., Tsao, H.N., Feng, X.L., and Mullen, K. (2009). Patterned graphene electrodes from solution-processed graphite oxide films for organic field-effect transistors. *Adv. Mater.* 21 (34): 3488.

579 Becerril, H.A., Stoltenberg, R.M., Tang, M.L. et al. (2010). Fabrication and evaluation of solution-processed reduced graphene oxide electrodes for p- and n-channel bottom-contact organic thin-film transistors. *ACS Nano* 4 (11): 6343.

580 Kang, B., Lim, S., Lee, W.H. et al. (2013). Work-function-tuned reduced graphene oxide via direct surface functionalization as source/drain electrodes in bottom-contact organic transistors. *Adv. Mater.* 25 (41): 5856.

581 Xiang, D., Wang, X., Jia, C. et al. (2016). Molecular-scale electronics: from concept to function. *Chem. Rev.* 116 (7): 4318.

582 Lortscher, E. (2013). Wiring molecules into circuits. *Nat. Nanotechnol.* 8 (6): 381.

583 Sun, L., Diaz-Fernandez, Y.A., Gschneidtner, T.A. et al. (2014). Single-molecule electronics: from chemical design to functional devices. *Chem. Soc. Rev.* 43 (21): 7378.

584 Sadeghi, H., Sangtarash, S., and Lambert, C. (2017). Robust molecular anchoring to graphene electrodes. *Nano Lett.* 17 (8): 4611.

585 Guo, X., Small, J.P., Klare, J.E. et al. (2006). Covalently bridging gaps in single-walled carbon nanotubes with conducting molecules. *Science* 311 (5759): 356.

586 Guo, X., Gorodetsky, A.A., Hone, J. et al. (2008). Conductivity of a single DNA duplex bridging a carbon nanotube gap. *Nat. Nanotechnol.* 3 (3): 163.

587 Zhu, J., McMorrow, J., Crespo-Otero, R. et al. (2016). Solution-processable carbon nanoelectrodes for single-molecule investigations. *J. Am. Chem. Soc.* 138 (9): 2905.

588 Jia, C., Migliore, A., Xin, N. et al. (2016). Covalently bonded single-molecule junctions with stable and reversible photoswitched conductivity. *Science* 352 (6292): 1443.

589 Cao, Y., Dong, S., Liu, S. et al. (2012). Building high-throughput molecular junctions using indented graphene point contacts. *Angew. Chem. Int. Ed.* 51 (49): 12228.

590 Cao, Y., Dong, S., Liu, S. et al. (2013). Toward functional molecular devices based on graphene-molecule junctions. *Angew. Chem. Int. Ed.* 52 (14): 3906.

591 Wen, H., Li, W., Chen, J. et al. (2016). Complex formation dynamics in a single-molecule electronic device. *Sci. Adv.* 2 (11): e1601113.

592 Xin, N., Wang, J., Jia, C. et al. (2017). Stereoelectronic effect-induced conductance switching in aromatic chain single-molecule junctions. *Nano Lett.* 17 (2): 856.

593 Xu, Q., Scuri, G., Mathewson, C. et al. (2017). Single electron transistor with single aromatic ring molecule covalently connected to graphene nanogaps. *Nano Lett.* 17 (9): 5335.

594 Ji, Y., Kim, J., Cha, A.N. et al. (2016). Graphene quantum dots as a highly efficient solution-processed charge trapping medium for organic nano-floating gate memory. *Nanotechnology* 27 (14): 145204.

595 Xin, N. and Guo, X. (2017). Catalyst: the renaissance of molecular electronics. *Chemistry* 3 (3): 373.

596 Nouchi, R., Shigeno, M., Yamada, N. et al. (2014). Reversible switching of charge injection barriers at metal/organic-semiconductor contacts modified with structurally disordered molecular monolayers. *Appl. Phys. Lett.* 104 (1): 013308.

597 Nouchi, R. and Tanimoto, T. (2015). Substituent-controlled reversible switching of charge injection barrier heights at metal/organic semiconductor contacts modified with disordered molecular monolayers. *ACS Nano* 9 (7): 7429.

598 Zhang, H.T., Chen, H.L., Ma, W. et al. (2016). Photocontrol of charge injection/extraction at electrode/semiconductor interfaces for high-photoresponsivity organic transistors. *J. Mater. Chem. C* 4 (23): 5289.

599 Crivillers, N., Osella, S., Van Dyck, C. et al. (2013). Large work function shift of gold induced by a novel perfluorinated azobenzene-based self-assembled monolayer. *Adv. Mater.* 25 (3): 432.

600 Ah Qune, L.F.N., Akiyama, H., Nagahiro, T. et al. (2008). Reversible work function changes induced by photoisomerization of asymmetric azobenzene dithiol self-assembled monolayers on gold. *Appl. Phys. Lett.* 93 (8): 083109.

601 Crivillers, N., Orgiu, E., Reinders, F. et al. (2011). Optical modulation of the charge injection in an organic field-effect transistor based on photochromic self-assembled-monolayer-functionalized electrodes. *Adv. Mater.* 23 (12): 1447.

602 Scott, J.C. (2003). Metal–organic interface and charge injection in organic electronic devices. *J. Vac. Sci. Technol., A* 21 (3): 521.

603 Martinez Hardigree, J.F. and Katz, H.E. (2014). Through thick and thin: tuning the threshold voltage in organic field-effect transistors. *Acc. Chem. Res.* 47 (4): 1369.

604 Yoneya, N., Noda, M., Hirai, N. et al. (2004). Reduction of contact resistance in pentacene thin-film transistors by direct carrier injection into a-few-molecular-layer channel. *Appl. Phys. Lett.* 85 (20): 4663.

605 Kronemeijer, A.J., Pecunia, V., Venkateshvaran, D. et al. (2014). Two-dimensional carrier distribution in top-gate polymer field-effect transistors: correlation between width of density of localized states and Urbach energy. *Adv. Mater.* 26 (5): 728.

606 Veres, J., Ogier, S., Lloyd, G., and de Leeuw, D. (2004). Gate insulators in organic field-effect transistors. *Chem. Mater.* 16 (23): 4543.

607 Facchetti, A., Yoon, M.H., and Marks, T.J. (2005). Gate dielectrics for organic field-effect transistors: new opportunities for organic electronics. *Adv. Mater.* 17 (14): 1705.

608 Kim, C., Facchetti, A., and Marks, T.J. (2007). Polymer gate dielectric surface viscoelasticity modulates pentacene transistor performance. *Science* 318 (5847): 76.

609 Yang, S.Y., Kim, S.H., Shin, K. et al. (2006). Low-voltage pentacene field-effect transistors with ultrathin polymer gate dielectrics. *Appl. Phys. Lett.* 88 (17): 173507.

610 Kalb, W.L., Mathis, T., Haas, S. et al. (2007). Organic small molecule field-effect transistors with Cytop™ gate dielectric: eliminating gate bias stress effects. *Appl. Phys. Lett.* 90 (9): 092104.

611 Volksen, W., Miller, R.D., and Dubois, G. (2010). Low dielectric constant materials. *Chem. Rev.* 110 (1): 56.

612 Jiang, Y.Y., Guo, Y.L., and Liu, Y.Q. (2017). Engineering of amorphous polymeric insulators for organic field-effect transistors. *Adv. Electron. Mater.* 3 (11): 1700157.

613 Mao, L.-K., Hwang, J.-C., and Chueh, Y.-L. (2014). Materials and interfaces issues in pentacene/PTCDI-C8 ambipolar organic field-effect transistors with solution-based gelatin dielectric. *Org. Electron.* 15 (10): 2400.

614 Ko, J., Nguyen, L.T.H., Surendran, A. et al. (2017). Human hair keratin for biocompatible flexible and transient electronic devices. *ACS Appl. Mater. Interfaces* 9 (49): 43004.

615 Yumusak, C., Singh, T.B., Sariciftci, N.S., and Grote, J.G. (2009). Bio-organic field effect transistors based on crosslinked deoxyribonucleic acid (DNA) gate dielectric. *Appl. Phys. Lett.* 95 (26): 263304.

616 Stadler, P., Oppelt, K., Singh, T.B. et al. (2007). Organic field-effect transistors and memory elements using deoxyribonucleic acid (DNA) gate dielectric. *Org. Electron.* 8 (6): 648.

617 Sun, X.; Di, C.-a.; Liu, Y. Engineering of the dielectric–semiconductor interface in organic field-effect transistors. *J. Mater. Chem.* 2010, 20 (13), 2599.

618 Halik, M. and Hirsch, A. (2011). The potential of molecular self-assembled monolayers in organic electronic devices. *Adv. Mater.* 23 (22–23): 2689.

619 Lanza, M. (2014). A review on resistive switching in high-*k* dielectrics: a nanoscale point of view using conductive atomic force microscope. *Materials* 7 (3): 2155.

620 Robertson, J. (2004). High dielectric constant oxides. *Eur. Phys. J. Appl. Phys.* 28 (3): 265.

621 He, G., Zhu, L.Q., Sun, Z.Q. et al. (2011). Integrations and challenges of novel high-*k* gate stacks in advanced CMOS technology. *Prog. Mater Sci.* 56 (5): 475.

622 Shao, W., Dong, H.L., Jiang, L., and Hu, W.P. (2011). Morphology control for high performance organic thin film transistors. *Chem. Sci.* 2 (4): 590.

623 Jones, A.O.F., Chattopadhyay, B., Geerts, Y.H., and Resel, R. (2016). Substrate-induced and thin-film phases: polymorphism of organic materials on surfaces. *Adv. Funct. Mater.* 26 (14): 2233.

624 Prisawong, P., Zalar, P., Reuveny, A. et al. (2016). Vacuum ultraviolet treatment of self-assembled monolayers: a tool for understanding growth and tuning charge transport in organic field-effect transistors. *Adv. Mater.* 28 (10): 2049.

625 Yang, H., Kim, S.H., Yang, L. et al. (2007). Pentacene nanostructures on surface-hydrophobicity-controlled polymer/SiO_2 bilayer gate-dielectrics. *Adv. Mater.* 19 (19): 2868.

626 Kwak, S.-Y., Choi, C.G., and Bae, B.-S. (2009). Effect of surface energy on pentacene growth and characteristics of organic thin-film transistors. *Electrochem. Solid-State Lett.* 12 (8): G37.

627 Chou, W.Y., Kuo, C.W., Cheng, H.L. et al. (2006). Effect of surface free energy in gate dielectric in pentacene thin-film transistors. *Appl. Phys. Lett.* 89 (11): 112126.

628 Chou, W.Y., Kuo, C.W., Chang, C.W. et al. (2010). Tuning surface properties in photosensitive polyimide. Material design for high performance organic thin-film transistors. *J. Mater. Chem.* 20 (26): 5474.

629 Liu, C.H., Zhu, Q.P., Jin, W.P. et al. (2011). The ultraviolet-ozone effects on organic thin-film transistors with double polymeric dielectric layers. *Synth. Met.* 161 (15–16): 1635.

630 Dimitrakopoulos, C.D., Brown, A.R., and Pomp, A. (1996). Molecular beam deposited thin films of pentacene for organic field effect transistor applications. *J. Appl. Phys.* 80 (4): 2501.

631 Gundlach, D.J., Lin, Y.Y., Jackson, T.N. et al. (1997). Pentacene organic thin-film transistors - molecular ordering and mobility. *IEEE Electr. Device Lett.* 18 (3): 87.

632 Daraktchiev, M., von Muhlenen, A., Nuesch, F. et al. (2005). Ultrathin organic transistors on oxide surfaces. *New J. Phys.* 7: 133.

633 Deman, A.L., Erouel, M., Lallemand, D. et al. (2008). Growth related properties of pentacene thin film transistors with different gate dielectrics. *J. Non-Cryst. Solids* 354 (15–16): 1598.

634 Meredig, B., Salleo, A., and Gee, R. (2009). Ordering of poly(3-hexylthiophene) nanocrystallites on the basis of substrate surface energy. *ACS Nano* 3 (10): 2881.

635 Yen-Yi, L., Gundlach, D.I., Nelson, S.F., and Jackson, T.N. (1997). Pentacene-based organic thin-film transistors. *IEEE Trans. Electron Devices* 44 (8): 1325.

636 Don Park, Y., Lim, J.A., Lee, H.S., and Cho, K. (2007). Interface engineering in organic transistors. *Mater. Today* 10 (3): 46.

637 Braga, D. and Horowitz, G. (2009). High-performance organic field-effect transistors. *Adv. Mater.* 21 (14–15): 1473.

638 Khassanov, A., Steinruck, H.G., Schmaltz, T. et al. (2015). Structural investigations of self-assembled monolayers for organic electronics: results from X-ray reflectivity. *Acc. Chem. Res.* 48 (7): 1901.

639 Peters, R.D., Nealey, P.F., Crain, J.N., and Himpsel, F.J. (2002). A near edge X-ray absorption fine structure spectroscopy investigation of the structure of self-assembled films of octadecyltrichlorosilane. *Langmuir* 18 (4): 1250.

640 Wang, Y.L. and Lieberman, M. (2003). Growth of ultrasmooth octadecyltrichlorosilane self-assembled monolayers on SiO_2. *Langmuir* 19 (4): 1159.

641 Lee, D.H., Oh, T., and Cho, K. (2005). Combined effect of chain length and phase state on adhesion/friction behavior of self-assembled monolayers. *J. Phys. Chem. B* 109 (22): 11301.

642 Cho, J.H., Lee, D.H., Shin, H.S. et al. (2004). Exploiting poly(dimethylsiloxane)-modified tips to evaluate frictional behavior by friction force microscopy. *Langmuir* 20 (26): 11499.

643 Soliman, A.I., Ichii, T., Utsunomiya, T., and Sugimura, H. (2015). Chemical conversion of self-assembled hexadecyl monolayers with active oxygen species generated by vacuum ultraviolet irradiation in an atmospheric environment. *Soft Matter* 11 (28): 5678.

644 Yang, H., Shin, T.J., Ling, M.M. et al. (2005). Conducting AFM and 2D GIXD studies on pentacene thin films. *J. Am. Chem. Soc.* 127 (33): 11542.

645 Kim, D.H., Park, Y.D., Jang, Y.S. et al. (2005). Enhancement of field-effect mobility due to surface-mediated molecular ordering in regioregular polythiophene thin film transistors. *Adv. Funct. Mater.* 15 (1): 77.

646 Kline, R.J., DeLongchamp, D.M., Fischer, D.A. et al. (2007). Significant dependence of morphology and charge carrier mobility on substrate surface chemistry in high performance polythiophene semiconductor films. *Appl. Phys. Lett.* 90 (6): 062117.

647 Umeda, T., Tokito, S., and Kumaki, D. (2007). High-mobility and air-stable organic thin-film transistors with highly ordered semiconducting polymer films. *J. Appl. Phys.* 101 (5): 054517.

648 Sarcletti, M., Schmaltz, T., and Halik, M. (2014). The mutual influence of surface energy and substrate temperature on the saturation mobility in organic semiconductors. *Org. Electron.* 15 (11): 3082.

649 Liang, Z., Tang, Q., Xu, J., and Miao, Q. (2011). Soluble and stable N-heteropentacenes with high field-effect mobility. *Adv. Mater.* 23 (13): 1535.

650 Das, S., Lee, B.H., Linstadt, R.T. et al. (2016). Molecularly smooth self-assembled monolayer for high-mobility organic field-effect transistors. *Nano Lett.* 16 (10): 6709.

651 Ahn, B.K., Das, S., Linstadt, R. et al. (2015). High-performance mussel-inspired adhesives of reduced complexity. *Nat. Commun.* 6: 8663.

652 Luo, C., Kyaw, A.K., Perez, L.A. et al. (2014). General strategy for self-assembly of highly oriented nanocrystalline semiconducting polymers with high mobility. *Nano Lett.* 14 (5): 2764.

653 Kim, G., Kang, S.J., Dutta, G.K. et al. (2014). A thienoisoindigo-naphthalene polymer with ultrahigh mobility of 14.4 $cm^2/V.s$ that substantially exceeds benchmark values for amorphous silicon semiconductors. *J. Am. Chem. Soc.* 136 (26): 9477.

654 Yoshida, M., Uemura, S., Kodzasa, T. et al. (2003). Surface potential control of an insulator layer for the high performance organic FET. *Synth. Met.* 137 (1–3): 967.

655 Ohashi, N., Tomii, H., Matsubara, R. et al. (2007). Conductivity fluctuation within a crystalline domain and its origin in pentacene thin-film transistors. *Appl. Phys. Lett.* 91 (16): 162105.

656 Matsubara, R., Ohashi, N., Sakai, M. et al. (2008). Analysis of barrier height at crystalline domain boundary and in-domain mobility in pentacene polycrystalline films on SiO_2. *Appl. Phys. Lett.* 92 (24): 242108.

657 Wang, S.D., Miyadera, T., Minari, T. et al. (2008). Correlation between grain size and device parameters in pentacene thin film transistors. *Appl. Phys. Lett.* 93 (4): 043311.

658 Edura, T., Takahashi, H., Nakata, M. et al. (2006). Electrical characterization of single grain and single grain boundary of pentacene thin film by nano-scale electrode array. *Curr. Appl Phys.* 6 (1): 109.

659 Hu, Y.Y., Wang, L.M., Qi, Q. et al. (2011). Charge transport model based on single-layered grains and grain boundaries for polycrystalline pentacene thin-film transistors. *J. Phys. Chem. C* 115 (47): 23568.

660 Jo, S.J., Kim, C.S., Kim, J.B. et al. (2008). Surface property controllable multi-layered gate dielectric for low voltage organic thin film transistors. *Appl. Phys. Lett.* 93 (8): 083504.

661 Lu, Y.X., Lee, W.H., Lee, H.S. et al. (2009). Low-voltage organic transistors with titanium oxide/polystyrene bilayer dielectrics. *Appl. Phys. Lett.* 94 (11): 113303.

662 Wang, Y., Acton, O., Ting, G. et al. (2009). Low-voltage high-performance organic thin film transistors with a thermally annealed polystyrene/hafnium oxide dielectric. *Appl. Phys. Lett.* 95 (24): 243302.

663 Steudel, S., De Vusser, S., De Jonge, S. et al. (2004). Influence of the dielectric roughness on the performance of pentacene transistors. *Appl. Phys. Lett.* 85 (19): 4400.

664 Ling, M.M. and Bao, Z.N. (2004). Thin film deposition, patterning, and printing in organic thin film transistors. *Chem. Mater.* 16 (23): 4824.

665 Tang, M.L., Reichardt, A.D., Wei, P., and Bao, Z. (2009). Correlating carrier type with frontier molecular orbital energy levels in organic thin film transistors of functionalized acene derivatives. *J. Am. Chem. Soc.* 131 (14): 5264.

666 Zhang, X., Jie, J., Deng, W. et al. (2016). Alignment and patterning of ordered small-molecule organic semiconductor micro-/nanocrystals for device applications. *Adv. Mater.* 28 (13): 2475.

667 Heil, H., Finnberg, T., von Malm, N. et al. (2003). The influence of mechanical rubbing on the field-effect mobility in polyhexylthiophene. *J. Appl. Phys.* 93 (3): 1636.

668 Yang, C.Y., Soci, C., Moses, D., and Heeger, A.J. (2005). Aligned rrP3HT film: structural order and transport properties. *Synth. Met.* 155 (3): 639.

669 Brinkmann, M. and Rannou, P. (2007). Effect of molecular weight on the structure and morphology of oriented thin films of regioregular poly(3-hexylthiophene) grown by directional epitaxial solidification. *Adv. Funct. Mater.* 17 (1): 101.

670 Sirringhaus, H., Wilson, R.J., Friend, R.H. et al. (2000). Mobility enhancement in conjugated polymer field-effect transistors through chain alignment in a liquid-crystalline phase. *Appl. Phys. Lett.* 77 (3): 406.

671 van de Craats, A.M., Stutzmann, N., Bunk, O. et al. (2003). Meso-epitaxial solution-growth of self-organizing discotic liquid-crystalline semiconductors. *Adv. Mater.* 15 (6): 495.

672 Knipp, D., Street, R.A., Volkel, A., and Ho, J. (2003). Pentacene thin film transistors on inorganic dielectrics: morphology, structural properties, and electronic transport. *J. Appl. Phys.* 93 (1): 347.

673 Fritz, S.E., Kelley, T.W., and Frisbie, C.D. (2005). Effect of dielectric roughness on performance of pentacene TFTs and restoration of performance with a polymeric smoothing layer. *J. Phys. Chem. B* 109 (21): 10574.

674 Shin, K., Yang, C.W., Yang, S.Y. et al. (2006). Effects of polymer gate dielectrics roughness on pentacene field-effect transistors. *Appl. Phys. Lett.* 88 (7): 072109.

675 Possanner, S.K., Zojer, K., Pacher, P. et al. (2009). Threshold voltage shifts in organic thin-film transistors due to self-assembled monolayers at the dielectric surface. *Adv. Funct. Mater.* 19 (6): 958.

676 Lin, G.Q., Wang, Q.H., Peng, L. et al. (2015). Impact of the lateral length scales of dielectric roughness on pentacene organic field-effect transistors. *J. Phys. D: Appl. Phys.* 48 (10): 105103.

677 Chabinyc, M.L., Lujan, R., Endicott, F. et al. (2007). Effects of the surface roughness of plastic-compatible inorganic dielectrics on polymeric thin film transistors. *Appl. Phys. Lett.* 90 (23): 233508.

678 Meng, H.F., Liu, C.C., Jiang, C.J. et al. (2006). Effect of gate metal on polymer transistor with glass substrate. *Appl. Phys. Lett.* 89 (24): 243503.

679 Jung, Y., Kline, R.J., Fischer, D.A. et al. (2008). The effect of interfacial roughness on the thin film morphology and charge transport of high-performance polythiophenes. *Adv. Funct. Mater.* 18 (5): 742.

680 Li, M., An, C., Marszalek, T. et al. (2016). Impact of interfacial microstructure on charge carrier transport in solution-processed conjugated polymer field-effect transistors. *Adv. Mater.* 28 (11): 2245.

681 Li, M., Hinkel, F., Mullen, K., and Pisula, W. (2016). Self-assembly and charge carrier transport of solution-processed conjugated polymer monolayers on dielectric surfaces with controlled sub-nanometer roughness. *Nanoscale* 8 (17): 9211.

682 Li, M., Marszalek, T., Mullen, K., and Pisula, W. (2016). Role of sub-nanometer dielectric roughness on microstructure and charge carrier transport in alpha,omega-dihexylsexithiophene field-effect transistors. *ACS Appl. Mater. Interfaces* 8 (25): 16200.

683 Menard, E., Meitl, M.A., Sun, Y. et al. (2007). Micro- and nanopatterning techniques for organic electronic and optoelectronic systems. *Chem. Rev.* 107 (4): 1117.

684 Xu, Y., Zhang, F., and Feng, X. (2011). Patterning of conjugated polymers for organic optoelectronic devices. *Small* 7 (10): 1338.

685 Carlson, A., Bowen, A.M., Huang, Y. et al. (2012). Transfer printing techniques for materials assembly and micro/nanodevice fabrication. *Adv. Mater.* 24 (39): 5284.

686 Zheng, Z., Yim, K.H., Saifullah, M.S. et al. (2007). Uniaxial alignment of liquid-crystalline conjugated polymers by nanoconfinement. *Nano Lett.* 7 (4): 987.

687 Park, B.J., Fang, F.F., and Choi, H.J. (2010). Magnetorheology: materials and application. *Soft Matter* 6 (21): 5246.

688 Aryal, M., Trivedi, K., and Hu, W.W. (2009). Nano-confinement induced chain alignment in ordered P3HT nanostructures defined by nanoimprint lithography. *ACS Nano* 3 (10): 3085.

689 Ding, G.Z., Wang, K.X., Li, X.H. et al. (2017). Nanoimprinting-induced molecular orientation in poly(3-hexylthiophene) nanogratings and its extraordinary retention after thermal annealing. *Polym. Chem.* 8 (17): 2666.

690 Wei, S.H., Zhang, Y.Y., Liu, J. et al. (2015). Large modulation of charge transport anisotropy by controlling the alignment of pi-pi stacks in Diketopyrrolopyrrole-based polymers. *Adv. Mater. Interfaces* 2 (13): 1500153.

691 Kim, K., Rho, Y., Kim, Y. et al. (2016). A lattice-strained organic single-crystal nanowire array fabricated via solution-phase nanograting-assisted pattern transfer for use in high-mobility organic field-effect transistors. *Adv. Mater.* 28 (16): 3209.

692 Tseng, H.R., Ying, L., Hsu, B.B. et al. (2012). High mobility field effect transistors based on macroscopically oriented regioregular copolymers. *Nano Lett.* 12 (12): 6353.

693 Ward, J.W., Loth, M.A., Kline, R.J. et al. (2012). Tailored interfaces for self-patterning organic thin-film transistors. *J. Mater. Chem.* 22 (36): 19047.

694 Tseng, H.R., Phan, H., Luo, C. et al. (2014). High-mobility field-effect transistors fabricated with macroscopic aligned semiconducting polymers. *Adv. Mater.* 26 (19): 2993.

695 Kline, R.J. and McGehee, M.D. (2006). Morphology and charge transport in conjugated polymer. *Polym. Rev.* 46 (1): 27.

696 Gasperini, A. and Sivula, K. (2013). Effects of molecular weight on microstructure and carrier transport in a semicrystalline poly(thieno)thiophene. *Macromolecules* 46 (23): 9349.

697 Himmelberger, S., Vandewal, K., Fei, Z.P. et al. (2014). Role of molecular weight distribution on charge transport in semiconducting polymers. *Macromolecules* 47 (20): 7151.

698 Lee, B.H., Hsu, B.B., Patel, S.N. et al. (2016). Flexible organic transistors with controlled nanomorphology. *Nano Lett.* 16 (1): 314.

699 Roberts, M.E., Queralto, N., Mannsfeld, S.C.B. et al. (2009). Cross-linked polymer gate dielectric films for low-voltage organic transistors. *Chem. Mater.* 21 (11): 2292.

700 Yang, F., Jin, L., Sun, L. et al. (2018). Free-standing 2D hexagonal aluminum nitride dielectric crystals for high-performance organic field-effect transistors. *Adv. Mater.* 30: 1801891.

701 Ji, D., Xu, X., Jiang, L. et al. (2017). Surface polarity and self-structured nanogrooves collaboratively oriented molecular packing for high crystallinity toward efficient charge transport. *J. Am. Chem. Soc.* 139 (7): 2734.

702 Zhang, F., Dai, X., Zhu, W. et al. (2017). Large modulation of charge carrier mobility in doped nanoporous organic transistors. *Adv. Mater.* 29 (27): 1700411.

703 Jiang, Y.M., Huang, W., Zhuang, X.M. et al. (2017). Thickness modulation on semiconductor towards high performance gas sensors based on organic thin film transistors. *Mater. Sci. Eng., B* 226: 107.

704 Ji, D., Li, T., Zou, Y. et al. (2018). Copolymer dielectrics with balanced chain-packing density and surface polarity for high-performance flexible organic electronics. *Nat. Commun.* 9 (1): 2339.

705 Zhang, F.J., Qu, G., Mohammadi, E. et al. (2017). Solution-processed nanoporous organic semiconductor thin films: toward health and environmental monitoring of volatile markers. *Adv. Funct. Mater.* 27 (23): 1701117.

706 Mendez, H., Heimel, G., Opitz, A. et al. (2013). Doping of organic semiconductors: impact of dopant strength and electronic coupling. *Angew. Chem. Int. Ed.* 52 (30): 7751.

707 Scheinert, S., Paasch, G., Schrodner, M. et al. (2002). Subthreshold characteristics of field effect transistors based on poly(3-dodecylthiophene) and an organic insulator. *J. Appl. Phys.* 92 (1): 330.

708 Gu, G., Kane, M.G., Doty, J.E., and Firester, A.H. (2005). Electron traps and hysteresis in pentacene-based organic thin-film transistors. *Appl. Phys. Lett.* 87 (24): 243512.

709 Gu, G. and Kane, M.G. (2008). Moisture induced electron traps and hysteresis in pentacene-based organic thin-film transistors. *Appl. Phys. Lett.* 92 (5): 053305.

710 Koo, J.B., Kang, S.Y., You, I.K., and Suh, K.S. (2009). Effect of UV/ozone treatment on hysteresis of pentacene thin-film transistor with polymer gate dielectric. *Solid State Electron.* 53 (6): 621.

711 Olthuis, W. and Bergveld, P. (1992). On the charge storage and decay mechanism in silicon dioxide electrets. *IEEE Electr. Insul. Mag.* 27 (4): 691.

712 Yoon, M.H., Kim, C., Facchetti, A., and Marks, T.J. (2006). Gate dielectric chemical structure-organic field-effect transistor performance correlations for electron, hole, and ambipolar organic semiconductors. *J. Am. Chem. Soc.* 128 (39): 12851.

713 Goldmann, C., Gundlach, D.J., and Batlogg, B. (2006). Evidence of water-related discrete trap state formation in pentacene single-crystal field-effect transistors. *Appl. Phys. Lett.* 88 (6): 063501.

714 Suárez, S.P., Fleischli, F.D., Schaer, M., and Zuppiroli, L. (2010). From oxide surface to organic transistor properties: the nature and the role of oxide gate surface defects. *J. Phys. Chem. C* 114 (15): 7153.

715 Barra, M., Girolamo, F.V.D., Chiarella, F. et al. (2010). Transport property and charge trap comparison for N-channel perylene diimide transistors with different air-stability. *J. Phys. Chem. C* 114 (48): 20387.

716 Ahles, M., Schmechel, R., and von Seggern, H. (2004). n-Type organic field-effect transistor based on interface-doped pentacene. *Appl. Phys. Lett.* 85 (19): 4499.

717 Benson, N., Gassmann, A., Mankel, E. et al. (2008). The role of Ca traces in the passivation of silicon dioxide dielectrics for electron transport in pentacene organic field effect transistors. *J. Appl. Phys.* 104 (5): 054505.

718 Benson, N., Melzer, C., Schmechel, R., and von Seggern, H. (2008). Electronic states at the dielectric/semiconductor interface in organic field effect transistors. *Phys. Status Solidi A* 205 (3): 475.

719 Angst, D.L. and Simmons, G.W. (1991). Moisture absorption characteristics of organosiloxane self-assembled monolayers. *Langmuir* 7 (10): 2236.

720 Wasserman, S.R., Tao, Y.T., and Whitesides, G.M. (1989). Structure and reactivity of alkylsiloxane monolayers formed by reaction of alkyltrichlorosilanes on silicon substrates. *Langmuir* 5 (4): 1074.

721 Fadeev, A.Y. and McCarthy, T.J. (2000). Self-assembly is not the only reaction possible between alkyltrichlorosilanes and surfaces: monomolecular and oligomeric covalently attached layers of dichloro- and trichloroalkylsilanes on silicon. *Langmuir* 16 (18): 7268.

722 Chua, L.L., Zaumseil, J., Chang, J.F. et al. (2005). General observation of n-type field-effect behaviour in organic semiconductors. *Nature* 434 (7030): 194.

723 Singh, T.B., Marjanovic, N., Matt, G.J. et al. (2004). Nonvolatile organic field-effect transistor memory element with a polymeric gate electret. *Appl. Phys. Lett.* 85 (22): 5409.

724 Lee, S., Koo, B., Shin, J. et al. (2006). Effects of hydroxyl groups in polymeric dielectrics on organic transistor performance. *Appl. Phys. Lett.* 88 (16): 162109.

725 Veres, J., Ogier, S.D., Leeming, S.W. et al. (2003). Low-k insulators as the choice of dielectrics in organic field-effect transistors. *Adv. Funct. Mater.* 13 (3): 199.

726 Jang, Y., Kim, D.H., Park, Y.D. et al. (2005). Influence of the dielectric constant of a polyvinyl phenol insulator on the field-effect mobility of a pentacene-based thin-film transistor. *Appl. Phys. Lett.* 87 (15): 152105.

727 Chang, J.W., Hsu, W.L., Wu, C.Y. et al. (2010). The polymer gate dielectrics and source-drain electrodes on n-type pentacene-based organic field-effect transistors. *Org. Electron.* 11 (10): 1613.

728 Chua, L.L., Ho, P.K.H., Sirringhaus, H., and Friend, R.H. (2004). High-stability ultrathin spin-on benzocyclobutene gate dielectric for polymer field-effect transistors. *Appl. Phys. Lett.* 84 (17): 3400.

729 Umeda, T., Kumaki, D., and Tokito, S. (2008). High air stability of threshold voltage on gate bias stress in pentacene TFTs with a hydroxyl-free and amorphous fluoropolymer as gate insulators. *Org. Electron.* 9 (4): 545.

730 Willa, K., Hausermann, R., Mathis, T. et al. (2013). From organic single crystals to solution processed thin-films: charge transport and trapping with varying degree of order. *J. Appl. Phys.* 113 (13): 133707.

731 Diemer, P.J., Lamport, Z.A., Mei, Y.C. et al. (2015). Quantitative analysis of the density of trap states at the semiconductor-dielectric interface in organic field-effect transistors. *Appl. Phys. Lett.* 107 (10): 103303.

732 Hunter, S., Chen, J.H., and Anthopoulos, T.D. (2014). Microstructural control of charge transport in organic blend thin-film transistors. *Adv. Funct. Mater.* 24 (38): 5969.

733 Maddalena, F., Spijkman, M., Brondijk, J.J. et al. (2008). Device characteristics of polymer dual-gate field-effect transistors. *Org. Electron.* 9 (5): 839.

734 Spijkman, M.J., Myny, K., Smits, E.C. et al. (2011). Dual-gate thin-film transistors, integrated circuits and sensors. *Adv. Mater.* 23 (29): 3231.

735 Dao, T.T., Matsushima, T., Friedlein, R., and Murata, H. (2013). Controllable threshold voltage of a pentacene field-effect transistor based on a double-dielectric structure. *Org. Electron.* 14 (8): 2007.

736 Spijkman, M., Mathijssen, S.G.J., Smits, E.C.P. et al. (2010). Monolayer dual gate transistors with a single charge transport layer. *Appl. Phys. Lett.* 96 (14): 143304.

737 Ahmed, R., Kadashchuk, A., Simbrunner, C. et al. (2014). Geometrical structure and interface dependence of bias stress induced threshold voltage shift in C60-based OFETs. *ACS Appl. Mater. Interfaces* 6 (17): 15148.

738 Menard, E., Podzorov, V., Hur, S.H. et al. (2004). High-performance n- and p-type single-crystal organic transistors with free-space gate dielectrics. *Adv. Mater.* 16 (23–24): 2097.

739 Xia, Y., Kalihari, V., Frisbie, C.D. et al. (2007). Tetracene air-gap single-crystal field-effect transistors. *Appl. Phys. Lett.* 90 (16): 162106.

740 Hulea, I.N., Fratini, S., Xie, H. et al. (2006). Tunable Frohlich polarons in organic single-crystal transistors. *Nat. Mater.* 5 (12): 982.

741 Tang, Q.X., Tong, Y.H., Li, H.X. et al. (2008). High-performance air-stable bipolar field-effect transistors of organic single-crystalline ribbons with an air-gap dielectric. *Adv. Mater.* 20 (8): 1511.

742 Tang, Q., Tong, Y., Li, H., and Hu, W. (2008). Air/vacuum dielectric organic single crystalline transistors of copper-hexadecafluorophthalocyanine ribbons. *Appl. Phys. Lett.* 92 (8): 083309.

743 Uno, M., Uemura, T., Kanaoka, Y. et al. (2013). High-speed organic single-crystal transistors gated with short-channel air gaps: efficient hole and electron injection in organic semiconductor crystals. *Org. Electron.* 14 (6): 1656.

744 Chen, Y. and Podzorov, V. (2012). Bias stress effect in "air-gap" organic field-effect transistors. *Adv. Mater.* 24 (20): 2679.

745 Xia, Y., Cho, J.H., Lee, J. et al. (2009). Comparison of the mobility-carrier density relation in polymer and single-crystal organic transistors employing vacuum and liquid gate dielectrics. *Adv. Mater.* 21 (21): 2174.

746 Ono, S., Seki, S., Hirahara, R. et al. (2008). High-mobility, low-power, and fast-switching organic field-effect transistors with ionic liquids. *Appl. Phys. Lett.* 92 (10): 103313.

747 He, Y., Boswell, P.G., Buhlmann, P., and Lodge, T.P. (2007). Ion gels by self-assembly of a triblock copolymer in an ionic liquid. *J. Phys. Chem. B* 111 (18): 4645.

748 Panzer, M.J. and Frisbie, C.D. (2008). Exploiting ionic coupling in electronic devices: electrolyte-gated organic field-effect transistors. *Adv. Mater.* 20 (16): 3177.

749 Cho, J.H., Lee, J., Xia, Y. et al. (2008). Printable ion-gel gate dielectrics for low-voltage polymer thin-film transistors on plastic. *Nat. Mater.* 7 (11): 900.

750 Lin, P. and Yan, F. (2012). Organic thin-film transistors for chemical and biological sensing. *Adv. Mater.* 24 (1): 34.

751 Liao, C., Zhang, M., Yao, M.Y. et al. (2015). Flexible organic electronics in biology: materials and devices. *Adv. Mater.* 27 (46): 7493.

752 Chou, Y.H., Chang, H.C., Liu, C.L., and Chen, W.C. (2015). Polymeric charge storage electrets for non-volatile organic field effect transistor memory devices. *Polym. Chem.* 6 (3): 341.

753 Lin, Y.Y., Gundlach, D.J., Nelson, S.F., and Jackson, T.N. (1997). Stacked pentacene layer organic thin-film transistors with improved characteristics. *IEEE Electron. Device Lett.* 18 (12): 606.

754 Steinruck, H.G., Magerl, A., Deutsch, M., and Ocko, B.M. (2014). Pseudorotational epitaxy of self-assembled octadecyltrichlorosilane monolayers on sapphire (0001). *Phys. Rev. Lett.* 113 (15): 156101.

755 Mutin, P.H., Guerrero, G., and Vioux, A. (2005). Hybrid materials from organophosphorus coupling molecules. *J. Mater. Chem.* 15 (35–36): 3761.

756 Guerrero, G., Mutin, P.H., and Vioux, A. (2001). Organically modified aluminas by grafting and sol–gel processes involving phosphonate derivatives. *J. Mater. Chem.* 11 (12): 3161.

757 Vega, A., Thissen, P., and Chabal, Y.J. (2012). Environment-controlled tethering by aggregation and growth of phosphonic acid monolayers on silicon oxide. *Langmuir* 28 (21): 8046.

758 Lin, Y., Wang, L., Krumpfer, J.W. et al. (2013). Hydrophobization of inorganic oxide surfaces using dimethylsilanediol. *Langmuir* 29 (5): 1329.

759 Fadeev, A.Y. and McCarthy, T.J. (1999). Trialkylsilane monolayers covalently attached to silicon surfaces: wettability studies indicating that molecular topography contributes to contact angle hysteresis. *Langmuir* 15 (11): 3759.

760 Anac, I. and McCarthy, T.J. (2009). Chemical modification of chromium oxide surfaces using organosilanes. *J. Colloid Interface Sci.* 331 (1): 138.

761 Tian, R., Seitz, O., Li, M. et al. (2010). Infrared characterization of interfacial Si-O bond formation on silanized flat SiO_2/Si surfaces. *Langmuir* 26 (7): 4563.

762 Stassen, A.F., de Boer, R.W.I., Iosad, N.N., and Morpurgo, A.F. (2004). Influence of the gate dielectric on the mobility of rubrene single-crystal field-effect transistors. *Appl. Phys. Lett.* 85 (17): 3899.

763 Martinelli, N.G., Savini, M., Muccioli, L. et al. (2009). Modeling polymer dielectric/pentacene interfaces: on the role of electrostatic energy disorder on charge carrier mobility. *Adv. Funct. Mater.* 19 (20): 3254.

764 Salleo, A., Chabinyc, M.L., Yang, M.S., and Street, R.A. (2002). Polymer thin-film transistors with chemically modified dielectric interfaces. *Appl. Phys. Lett.* 81 (23): 4383.

765 Yagi, I., Tsukagoshi, K., and Aoyagi, Y. (2005). Modification of the electric conduction at the pentacene/SiO_2 interface by surface termination of SiO_2. *Appl. Phys. Lett.* 86 (10): 103502.

766 Kumaki, D., Yahiro, M., Inoue, Y., and Tokito, S. (2007). Air stable, high performance pentacene thin-film transistor fabricated on SiO_2 gate insulator treated with beta-phenethyltrichlorosilane. *Appl. Phys. Lett.* 90 (13): 133511.

767 Boulas, C., Davidovits, J.V., Rondelez, F., and Vuillaume, D. (1996). Suppression of charge carrier tunneling through organic self-assembled monolayers. *Phys. Rev. Lett.* 76 (25): 4797.

768 Collet, J., Tharaud, O., Chapoton, A., and Vuillaume, D. (2000). Low-voltage, 30 nm channel length, organic transistors with a self-assembled monolayer as gate insulating films. *Appl. Phys. Lett.* 76 (14): 1941.

769 Halik, M., Klauk, H., Zschieschang, U. et al. (2004). Low-voltage organic transistors with an amorphous molecular gate dielectric. *Nature* 431 (7011): 963.

770 Richards, T., Bird, M., and Sirringhaus, H. (2008). A quantitative analytical model for static dipolar disorder broadening of the density of states at organic heterointerfaces. *J. Chem. Phys.* 128 (23): 234905.

771 Bassler, H. (1993). Charge transport in disordered organic photoconductors – a Monte-Carlo simulation study. *Phys. Status Solidi B* 175 (1): 15.

772 Dimitrakopoulos, C.D., Veres, J., Ogier, S. et al. (2003). Gate insulators influencing electronic transport in organic FETs. *Proc. SPIE* 5217: 147.

773 Kobayashi, S., Nishikawa, T., Takenobu, T. et al. (2004). Control of carrier density by self-assembled monolayers in organic field-effect transistors. *Nat. Mater.* 3 (5): 317.

774 Jang, Y., Cho, J.H., Kim, D.H. et al. (2007). Effects of the permanent dipoles of self-assembled monolayer-treated insulator surfaces on the field-effect mobility of a pentacene thin-film transistor. *Appl. Phys. Lett.* 90 (13): 132104.

775 Zhang, Y., Ziegler, D., and Salmeron, M. (2013). Charge trapping states at the SiO_2-oligothiophene monolayer interface in field effect transistors studied by Kelvin probe force microscopy. *ACS Nano* 7 (9): 8258.

776 Guo, M., Hayakawa, T., Kakimoto, M.A. et al. (2011). Organic macromolecular high dielectric constant materials: synthesis, characterization, and applications. *J. Phys. Chem. B* 115 (46): 13419.

777 Li, J.H., Liu, D.Q., Miao, Q., and Yan, F. (2012). The application of a high-k polymer in flexible low-voltage organic thin-film transistors. *J. Mater. Chem.* 22 (31): 15998.

778 Choi, Y., Kim, I.D., Tuller, H.L., and Akinwande, A.I. (2005). Low-voltage organic transistors and depletion-load inverters with high-K pyrochlore BZN gate dielectric on polymer substrate. *IEEE Trans. Electron Devices* 52 (12): 2819.

779 Klauk, H., Zschieschang, U., Pflaum, J., and Halik, M. (2007). Ultralow-power organic complementary circuits. *Nature* 445 (7129): 745.

780 Ono, S., Hausermann, R., Chiba, D. et al. (2014). High performance organic field-effect transistors with ultra-thin HfO_2 gate insulator deposited directly onto the organic semiconductor. *Appl. Phys. Lett.* 104 (1): 013307.

781 Majewski, L.A., Schroeder, R., Grell, M. et al. (2004). High capacitance organic field-effect transistors with modified gate insulator surface. *J. Appl. Phys.* 96 (10): 5781.

782 Hofer, R., Textor, M., and Spencer, N.D. (2001). Alkyl phosphate monolayers, self-assembled from aqueous solution onto metal oxide surfaces. *Langmuir* 17 (13): 4014.

783 Chen, X., Luais, E., Darwish, N. et al. (2012). Studies on the effect of solvents on self-assembled monolayers formed from organophosphonic acids on indium tin oxide. *Langmuir* 28 (25): 9487.

784 McDowell, M., Hill, I.G., McDermott, J.E. et al. (2006). Improved organic thin-film transistor performance using novel self-assembled monolayers. *Appl. Phys. Lett.* 88 (7): 073505.

785 Fukuda, K., Hamamoto, T., Yokota, T. et al. (2009). Effects of the alkyl chain length in phosphonic acid self-assembled monolayer gate dielectrics on the performance and stability of low-voltage organic thin-film transistors. *Appl. Phys. Lett.* 95 (20): 203301.

786 Spori, D.M., Venkataraman, N.V., Tosatti, S.G. et al. (2007). Influence of alkyl chain length on phosphate self-assembled monolayers. *Langmuir* 23 (15): 8053.

787 Tao, Y.T. (1993). Structural comparison of self-assembled monolayers of n-alkanoic acids on the surfaces of silver, copper, and aluminum. *J. Am. Chem. Soc.* 115 (10): 4350.

788 Motte, L. and Pileni, M.P. (2000). Self-assemblies of silver sulfide nanocrystals: influence of length of thio-alkyl chains used as coating agent. *Appl. Surf. Sci.* 164 (1–4): 60.

789 Sekitani, T., Zschieschang, U., Klauk, H., and Someya, T. (2010). Flexible organic transistors and circuits with extreme bending stability. *Nat. Mater.* 9 (12): 1015.

790 Reeder, J., Kaltenbrunner, M., Ware, T. et al. (2014). Mechanically adaptive organic transistors for implantable electronics. *Adv. Mater.* 26 (29): 4967.

791 Park, Y.M., Daniel, J., Heeney, M., and Salleo, A. (2011). Room-temperature fabrication of ultrathin oxide gate dielectrics for low-voltage operation of organic field-effect transistors. *Adv. Mater.* 23 (8): 971.

792 Wong, W. S.; Salleo, A. *Flexible Electronics*, 2009.

793 Wagner, S. and Bauer, S. (2012). Materials for stretchable electronics. *MRS Bull.* 37 (3): 207.

794 Nomura, K., Ohta, H., Takagi, A. et al. (2004). Room-temperature fabrication of transparent flexible thin-film transistors using amorphous oxide semiconductors. *Nature* 432 (7016): 488.

795 Mardare, A.I., Kaltenbrunner, M., Sariciftci, N.S. et al. (2012). Ultra-thin anodic alumina capacitor films for plastic electronics. *Phys. Status Solidi A* 209 (5): 813.

796 Lohrengel, M.M. (1993). Thin anodic oxide layers on aluminium and other valve metals: high field regime. *Mater. Sci. Eng.: R: Rep.* 11 (6): 243.

797 Sekitani, T., Yokota, T., Kuribara, K. et al. (2016). Ultraflexible organic amplifier with biocompatible gel electrodes. *Nat. Commun.* 7: 11425.

798 Gelinck, G.H., Huitema, H.E., van Veenendaal, E. et al. (2004). Flexible active-matrix displays and shift registers based on solution-processed organic transistors. *Nat. Mater.* 3 (2): 106.

799 Sekitani, T., Takamiya, M., Noguchi, Y. et al. (2007). A large-area wireless power-transmission sheet using printed organic transistors and plastic MEMS switches. *Nat. Mater.* 6 (6): 413.

800 Bettinger, C.J. and Bao, Z. (2010). Organic thin-film transistors fabricated on resorbable biomaterial substrates. *Adv. Mater.* 22 (5): 651.

801 Granstrom, J. and Katz, H.E. (2011). Pentacene transistors with polymer gate dielectrics on metallized optical fibers. *J. Mater. Res.* 19 (12): 3540.

802 Lee, J.B. and Subramanian, V. (2005). Weave patterned organic transistors on fiber for e-textiles. *IEEE Trans. Electron Devices* 52 (2): 269.

803 Tobjork, D. and Osterbacka, R. (2011). Paper electronics. *Adv. Mater.* 23 (17): 1935.

804 Hyun, W.J., Secor, E.B., Rojas, G.A. et al. (2015). All-printed, foldable organic thin-film transistors on glassine paper. *Adv. Mater.* 27 (44): 7058.

805 Khan, M.A., Bhansali, U.S., and Alshareef, H.N. (2012). High-performance non-volatile organic ferroelectric memory on banknotes. *Adv. Mater.* 24 (16): 2165.

806 Zhang, L., Wang, H., Zhao, Y. et al. (2013). Substrate-free ultra-flexible organic field-effect transistors and five-stage ring oscillators. *Adv. Mater.* 25 (38): 5455.

807 Myny, K., Steudel, S., Smout, S. et al. (2010). Organic RFID transponder chip with data rate compatible with electronic product coding. *Org. Electron.* 11 (7): 1176.

808 Jager, C.M., Schmaltz, T., Novak, M. et al. (2013). Improving the charge transport in self-assembled monolayer field-effect transistors: from theory to devices. *J. Am. Chem. Soc.* 135 (12): 4893.

809 Huang, J., Sun, J., and Katz, H.E. (2008). Monolayer-dimensional 5,5'-Bis(4-hexylphenyl)-2,2′-bithiophene transistors and chemically responsive heterostructures. *Adv. Mater.* 20 (13): 2567.

810 Sokolov, A.N., Roberts, M.E., and Bao, Z.A. (2009). Fabrication of low-cost electronic biosensors. *Mater. Today* 12 (9): 12.

811 Mannebach, E.M., Spalenka, J.W., Johnson, P.S. et al. (2013). High hole mobility and thickness-dependent crystal structure in alpha,omega-dihexylsexithiophene single-monolayer field-effect transistors. *Adv. Funct. Mater.* 23 (5): 554.

812 Jiang, L., Dong, H., Meng, Q. et al. (2011). Millimeter-sized molecular monolayer two-dimensional crystals. *Adv. Mater.* 23 (18): 2059.

813 Defaux, M., Gholamrezaie, F., Wang, J. et al. (2012). Solution-processable septithiophene monolayer transistor. *Adv. Mater.* 24 (7): 973.

814 Zhang, X., Hou, L., and Samori, P. (2016). Coupling carbon nanomaterials with photochromic molecules for the generation of optically responsive materials. *Nat. Commun.* 7: 11118.

815 Shallcross, R.C., Zacharias, P., Kohnen, A. et al. (2013). Photochromic transduction layers in organic memory elements. *Adv. Mater.* 25 (3): 469.

816 Fu, L.N., Leng, B., Li, Y.S., and Gao, X.K. (2016). Photoresponsive organic field-effect transistors involving photochromic molecules. *Chin. Chem. Lett.* 27 (8): 1319.

817 Zhang, H., Guo, X., Hui, J. et al. (2011). Interface engineering of semiconductor/dielectric heterojunctions toward functional organic thin-film transistors. *Nano Lett.* 11 (11): 4939.

818 Shen, Q., Wang, L., Liu, S. et al. (2010). Photoactive gate dielectrics. *Adv. Mater.* 22 (30): 3282.

819 Suda, M., Kato, R., and Yamamoto, H.M. (2015). Superconductivity: light-induced superconductivity using a photoactive electric double layer. *Science* 347 (6223): 743.

820 Chen, H., Cheng, N., Ma, W. et al. (2016). Design of a photoactive hybrid bilayer dielectric for flexible nonvolatile organic memory transistors. *ACS Nano* 10 (1): 436.

821 Ueno, K., Nakamura, S., Shimotani, H. et al. (2008). Electric-field-induced superconductivity in an insulator. *Nat. Mater.* 7 (11): 855.

822 Ye, J.T., Inoue, S., Kobayashi, K. et al. (2010). Liquid-gated interface superconductivity on an atomically flat film. *Nat. Mater.* 9 (2): 125.

823 Bollinger, A.T., Dubuis, G., Yoon, J. et al. (2011). Superconductor-insulator transition in $La_{2-x}Sr_xCuO_4$ at the pair quantum resistance. *Nature* 472 (7344): 458.

824 Ueno, K., Nakamura, S., Shimotani, H. et al. (2011). Discovery of superconductivity in $KTaO_3$ by electrostatic carrier doping. *Nat. Nanotechnol.* 6 (7): 408.

825 Ye, J.T., Zhang, Y.J., Akashi, R. et al. (2012). Superconducting dome in a gate-tuned band insulator. *Science* 338 (6111): 1193.

826 Irie, M. (2000). Diarylethenes for memories and switches. *Chem. Rev.* 100 (5): 1685.

827 Jia, C., Wang, J., Yao, C. et al. (2013). Conductance switching and mechanisms in single-molecule junctions. *Angew. Chem. Int. Ed.* 52 (33): 8666.

828 Di, C.A., Zhang, F., and Zhu, D. (2013). Multi-functional integration of organic field-effect transistors (OFETs): advances and perspectives. *Adv. Mater.* 25 (3): 313.

829 Trung, T.Q. and Lee, N.E. (2016). Flexible and stretchable physical sensor integrated platforms for wearable human-activity monitoringand personal healthcare. *Adv. Mater.* 28 (22): 4338.

830 Maheshwari, V. and Saraf, R. (2008). Tactile devices to sense touch on a par with a human finger. *Angew. Chem. Int. Ed.* 47 (41): 7808.

831 Hammock, M.L., Chortos, A., Tee, B.C. et al. (2013). 25th anniversary article: the evolution of electronic skin (e-skin): a brief history, design considerations, and recent progress. *Adv. Mater.* 25 (42): 5997.

832 Sokolov, A.N., Tee, B.C., Bettinger, C.J. et al. (2012). Chemical and engineering approaches to enable organic field-effect transistors for electronic skin applications. *Acc. Chem. Res.* 45 (3): 361.

833 Chortos, A. and Bao, Z.N. (2014). Skin-inspired electronic devices. *Mater. Today* 17 (7): 321.

834 Zang, Y.P., Zhang, F.J., Di, C.A., and Zhu, D.B. (2015). Advances of flexible pressure sensors toward artificial intelligence and health care applications. *Mater. Horiz.* 2 (2): 140.

835 Someya, T., Sekitani, T., Iba, S. et al. (2004). A large-area, flexible pressure sensor matrix with organic field-effect transistors for artificial skin applications. *Proc. Natl. Acad. Sci. U.S.A.* 101 (27): 9966.

836 Someya, T., Kato, Y., Sekitani, T. et al. (2005). Conformable, flexible, large-area networks of pressure and thermal sensors with organic transistor active matrixes. *Proc. Natl. Acad. Sci. U.S.A.* 102 (35): 12321.
837 Mannsfeld, S.C., Tee, B.C., Stoltenberg, R.M. et al. (2010). Highly sensitive flexible pressure sensors with microstructured rubber dielectric layers. *Nat. Mater.* 9 (10): 859.
838 Schwartz, G., Tee, B.C., Mei, J. et al. (1859). Flexible polymer transistors with high pressure sensitivity for application in electronic skin and health monitoring. *Nat. Commun.* 2013: 4.
839 Zang, Y., Zhang, F., Huang, D. et al. (2015). Flexible suspended gate organic thin-film transistors for ultra-sensitive pressure detection. *Nat. Commun.* 6: 6269.
840 Yokota, T., Inoue, Y., Terakawa, Y. et al. (2015). Ultraflexible, large-area, physiological temperature sensors for multipoint measurements. *Proc. Natl. Acad. Sci. U.S.A.* 112 (47): 14533.
841 Triet, N.M., Trung, T.Q., Hien, N.T.D. et al. (2015). A flexible magnetoelectric field-effect transistor with magnetically responsive nanohybrid gate dielectric layer. *Nano Res.* 8 (10): 3421.
842 Zirkl, M., Haase, A., Fian, A. et al. (2007). Low-voltage organic thin-film transistors with high-*k* nanocomposite gate dielectrics for flexible electronics and optothermal sensors. *Adv. Mater.* 19 (17): 2241.
843 Podzorov, V., Menard, E., Borissov, A. et al. (2004). Intrinsic charge transport on the surface of organic semiconductors. *Phys. Rev. Lett.* 93 (8): 086602.
844 Sakanoue, T. and Sirringhaus, H. (2010). Band-like temperature dependence of mobility in a solution-processed organic semiconductor. *Nat. Mater.* 9 (9): 736.
845 Xie, W., Willa, K., Wu, Y. et al. (2013). Temperature-independent transport in high-mobility dinaphtho-thieno-thiophene (DNTT) single crystal transistors. *Adv. Mater.* 25 (25): 3478.
846 Ostroverkhova, O. (2016). Organic optoelectronic materials: mechanisms and applications. *Chem. Rev.* 116 (22): 13279.
847 Wu, X.H., Ma, Y., Zhang, G.Q. et al. (2015). Thermally stable, biocompatible, and flexible organic field-effect transistors and their application in temperature sensing arrays for artificial skin. *Adv. Funct. Mater.* 25 (14): 2138.
848 Zang, Y., Zhang, F., Huang, D. et al. (2015). Sensitive flexible magnetic sensors using organic transistors with magnetic-functionalized suspended gate electrodes. *Adv. Mater.* 27 (48): 7979.
849 Graz, I., Krause, M., Bauer-Gogonea, S. et al. (2009). Flexible active-matrix cells with selectively poled bifunctional polymer-ceramic nanocomposite for pressure and temperature sensing skin. *J. Appl. Phys.* 106 (3): 034503.
850 Tien, N.T., Trung, T.Q., Seoul, Y.G. et al. (2011). Physically responsive field-effect transistors with giant electromechanical coupling induced by nanocomposite gate dielectrics. *ACS Nano* 5 (9): 7069.
851 Trung, T.Q., Tien, N.T., Seol, Y.G., and Lee, N.E. (2012). Transparent and flexible organic field-effect transistor for multi-modal sensing. *Org. Electron.* 13 (4): 533.

852 Tien, N.T., Jeon, S., Kim, D.I. et al. (2014). A flexible bimodal sensor array for simultaneous sensing of pressure and temperature. *Adv. Mater.* 26 (5): 796.

853 Han, S.T., Zhou, Y., and Roy, V.A. (2013). Towards the development of flexible non-volatile memories. *Adv. Mater.* 25 (38): 5425.

854 Scott, J.C. and Bozano, L.D. (2007). Nonvolatile memory elements based on organic materials. *Adv. Mater.* 19 (11): 1452.

855 Cho, B., Song, S., Ji, Y. et al. (2011). Organic resistive memory devices: performance enhancement, integration, and advanced architectures. *Adv. Funct. Mater.* 21 (15): 2806.

856 Burkhardt, M., Jedaa, A., Novak, M. et al. (2010). Concept of a molecular charge storage dielectric layer for organic thin-film memory transistors. *Adv. Mater.* 22 (23): 2525.

857 Mok, S.M., Yan, F., and Chan, H.L.W. (2008). Organic phototransistor based on poly(3-hexylthiophene)/TiO_2 nanoparticle composite. *Appl. Phys. Lett.* 93 (2): 023310.

858 Sekitani, T., Yokota, T., Zschieschang, U. et al. (2009). Organic nonvolatile memory transistors for flexible sensor arrays. *Science* 326 (5959): 1516.

859 Shaw, J., Zhong, Y.W., Hughes, K.J. et al. (2011). Integration of self-assembled redox molecules in flash memory devices. *IEEE Trans. Electron Devices* 58 (3): 826.

860 Zhu, H., Hacker, C.A., Pookpanratana, S.J. et al. (2013). Non-volatile memory with self-assembled ferrocene charge trapping layer. *Appl. Phys. Lett.* 103 (5): 053102.

861 Po-Chiang, C., Guozhen, S., and Chongwu, Z. (2008). Chemical sensors and electronic noses based on 1-D metal oxide nanostructures. *IEEE Trans. Nanotechnol.* 7 (6): 668.

862 Seol, M.L., Choi, S.J., Kim, C.H. et al. (2012). Porphyrin-silicon hybrid field-effect transistor with individually addressable top-gate structure. *ACS Nano* 6 (1): 183.

863 Tans, S.J., Verschueren, A.R.M., and Dekker, C. (1998). Room-temperature transistor based on a single carbon nanotube. *Nature* 393 (6680): 49.

864 Balliou, A., Papadimitropoulos, G., Skoulatakis, G. et al. (2016). Low-dimensional polyoxometalate molecules/tantalum oxide hybrids for non-volatile capacitive memories. *ACS Appl. Mater. Interfaces* 8 (11): 7212.

865 Busche, C., Vila-Nadal, L., Yan, J. et al. (2014). Design and fabrication of memory devices based on nanoscale polyoxometalate clusters. *Nature* 515 (7528): 545.

866 Ling, Q.D., Liaw, D.J., Zhu, C.X. et al. (2008). Polymer electronic memories: materials, devices and mechanisms. *Prog. Polym. Sci.* 33 (10): 917.

867 Heremans, P., Gelinck, G.H., Müller, R. et al. (2011). Polymer and organic nonvolatile memory devices. *Chem. Mater.* 23 (3): 341.

868 Lee, J.S. (2011). Progress in non-volatile memory devices based on nanostructured materials and nanofabrication. *J. Mater. Chem.* 21 (37): 14097.

869 Kang, M., Baeg, K.J., Khim, D. et al. (2013). Printed, flexible, organic nano-floating-gate memory: effects of metal nanoparticles and blocking dielectrics on memory characteristics. *Adv. Funct. Mater.* 23 (28): 3503.

870 Zhang, J.Y., Liu, L.M., Su, Y.J. et al. (2015). Synergistic effect in organic field-effect transistor nonvolatile memory utilizing bimetal nanoparticles as nano-floating-gate. *Org. Electron.* 25: 324.

871 Kim, B.J., Ko, Y., Cho, J.H., and Cho, J. (2013). Organic field-effect transistor memory devices using discrete ferritin nanoparticle-based gate dielectrics. *Small* 9 (22): 3784.

872 Aimi, J., Lo, C.T., Wu, H.C. et al. (2016). Phthalocyanine-cored star-shaped polystyrene for nano floating gate in nonvolatile organic transistor memory device. *Adv. Electron. Mater.* 2 (2): 1500300.

873 Chen, C.M., Liu, C.M., Tsai, M.C. et al. (2013). A nanostructured micellar diblock copolymer layer affects the memory characteristics and packing of pentacene molecules in non-volatile organic field-effect transistor memory devices. *J. Mater. Chem. C* 1 (12): 2328.

874 Baeg, K.J., Noh, Y.Y., Sirringhaus, H., and Kim, D.Y. (2010). Controllable shifts in threshold voltage of top-gate polymer field-effect transistors for applications in organic nano floating gate memory. *Adv. Funct. Mater.* 20 (2): 224.

875 Jung, J.H., Kim, S., Kim, H. et al. (2015). High-performance flexible organic nano-floating gate memory devices functionalized with cobalt ferrite nanoparticles. *Small* 11 (37): 4976.

876 Kang, M., Chung, K., Baeg, K.J. et al. (2015). Multi-layered nanocomposite dielectrics for high density organic memory devices. *Appl. Phys. Lett.* 106 (4): 043302.

877 Chang, H.C., Lu, C., Liu, C.L., and Chen, W.C. (2015). Single-crystal C_{60} needle/CuPc nanoparticle double floating-gate for low-voltage organic transistors based non-volatile memory devices. *Adv. Mater.* 27 (1): 27.

878 She, X.J., Liu, C.H., Sun, Q.J. et al. (2012). Morphology control of tunneling dielectric towards high-performance organic field-effect transistor nonvolatile memory. *Org. Electron.* 13 (10): 1908.

879 Han, S.T., Zhou, Y., Xu, Z.X. et al. (2011). Nanoparticle size dependent threshold voltage shifts in organic memory transistors. *J. Mater. Chem.* 21 (38): 14575.

880 Park, Y.S. and Lee, J.S. (2015). Design of an efficient charge-trapping layer with a built-in tunnel barrier for reliable organic-transistor memory. *Adv. Mater.* 27 (4): 706.

881 Shih, C.C., Lee, W.Y., Chiu, Y.C. et al. (2016). High performance transparent transistor memory devices using nano-floating gate of polymer/ZnO nanocomposites. *Sci. Rep.* 6: 20129.

882 Tseng, C.W. and Tao, Y.T. (2009). Electric bistability in pentacene film-based transistor embedding gold nanoparticles. *J. Am. Chem. Soc.* 131 (34): 12441.

883 Chang, H.C., Liu, C.L., and Chen, W.C. (2013). Flexible nonvolatile transistor memory devices based on one-dimensional electrospun P3HT:Au hybrid nanofibers. *Adv. Funct. Mater.* 23 (39): 4960.

884 Chang, H.C., Liu, C.L., and Chen, W.C. (2013). Nonvolatile organic thin film transistor memory devices based on hybrid nanocomposites of semiconducting polymers: gold nanoparticles. *ACS Appl. Mater. Interfaces* 5 (24): 13180.

885 Leong, W.L., Mathews, N., Tan, B. et al. (2011). Towards printable organic thin film transistor based flash memory devices. *J. Mater. Chem.* 21 (14): 5203.

886 Baeg, K.J., Noh, Y.Y., Ghim, J. et al. (2006). Organic non-volatile memory based on pentacene field-effect transistors using a polymeric gate electret. *Adv. Mater.* 18 (23): 3179.

887 Chung, F.J., Liu, H.Y., Jiang, B.Y. et al. (2016). Random styrenic copolymers with pendant pyrene moieties: synthesis and applications in organic field-effect transistor memory. *J. Polym. Sci., Part A: Polym. Chem.* 54 (7): 910.

888 Debucquoy, M., Rockele, M., Genoe, J. et al. (2009). Charge trapping in organic transistor memories: on the role of electrons and holes. *Org. Electron.* 10 (7): 1252.

889 Baeg, K.J., Noh, Y.Y., Ghim, J. et al. (2008). Polarity effects of polymer gate electrets on non-volatile organic field-effect transistor memory. *Adv. Funct. Mater.* 18 (22): 3678.

890 Chou, Y.H., Takasugi, S., Goseki, R. et al. (2014). Nonvolatile organic field-effect transistor memory devices using polymer electrets with different thiophene chain lengths. *Polym. Chem.* 5 (3): 1063.

891 Dong, L., Chiu, Y.C., Chueh, C.C. et al. (2014). Semi-conjugated acceptor-based polyimides as electrets for nonvolatile transistor memory devices. *Polym. Chem.* 5 (23): 6834.

892 Yu, A.D., Tung, W.Y., Chiu, Y.C. et al. (2014). Multilevel nonvolatile flexible organic field-effect transistor memories employing polyimide electrets with different charge-transfer effects. *Macromol. Rapid Commun.* 35 (11): 1039.

893 Chou, Y.-H., You, N.-H., Kurosawa, T. et al. (2012). Thiophene and selenophene donor–acceptor polyimides as polymer electrets for nonvolatile transistor memory devices. *Macromolecules* 45 (17): 6946.

894 Chiu, Y.C., Sun, H.S., Lee, W.Y. et al. (2015). Oligosaccharide carbohydrate dielectrics toward high-performance non-volatile transistor memory devices. *Adv. Mater.* 27 (40): 6257.

895 Li, Y.Q., Song, Y., Zhang, X.Y. et al. (2015). Programmable polymer memory device based on hydrophilic polythiophene and poly(ionic liquid) electrolyte. *Macromol. Chem. Phys.* 216 (1): 113.

896 Sun, H.S., Chen, Y., Lee, W.Y. et al. (2016). Synthesis, morphology, and electrical memory application of oligosaccharide-based block copolymers with pi-conjugated pyrene moieties and their supramolecules. *Polym. Chem.* 7 (6): 1249.

897 Wu, W.P., Zhang, H.L., Wang, Y. et al. (2008). High-performance organic transistor memory elements with steep flanks of hysteresis. *Adv. Funct. Mater.* 18 (17): 2593.

898 Chiu, Y.C., Chen, T.Y., Chen, Y. et al. (2014). High-performance nonvolatile organic transistor memory devices using the electrets of semiconducting blends. *ACS Appl. Mater. Interfaces* 6 (15): 12780.

899 Chiu, Y.C., Otsuka, I., Halila, S. et al. (2014). High-performance nonvolatile transistor memories of pentacence using the green electrets of sugar-based block copolymers and their supramolecules. *Adv. Funct. Mater.* 24 (27): 4240.

900 Chiu, Y.-C., Chen, T.-Y., Chueh, C.-C. et al. (2014). High performance non-volatile transistor memories of pentacene using the electrets of star-branched p-type polymers and their donor–acceptor blends. *J. Mater. Chem. C* 2 (8): 1436.

901 Chiu, Y.C., Shih, C.C., and Chen, W.C. (2015). Nonvolatile memories using the electrets of conjugated rod-coil block copolymer and its nanocomposite with single wall carbon nanotubes. *J. Mater. Chem. C* 3 (3): 551.

902 Wei, Q.S., Lin, Y., Anderson, E.R. et al. (2012). Additive-driven assembly of block copolymer-nanoparticle hybrid materials for solution processable floating gate memory. *ACS Nano* 6 (2): 1188.

903 Chen, C.M., Liu, C.M., Wei, K.H. et al. (2012). Non-volatile organic field-effect transistor memory comprising sequestered metal nanoparticles in a diblock copolymer film. *J. Mater. Chem.* 22 (2): 454.

904 Cho, I., Kim, B.J., Ryu, S.W. et al. (2014). Transistor memory devices with large memory windows, using multi-stacking of densely packed, hydrophobic charge trapping metal nanoparticle array. *Nanotechnology* 25 (50): 505604.

905 Zhou, Y., Han, S.T., Xu, Z.X., and Roy, V.A.L. (2012). Low voltage flexible nonvolatile memory with gold nanoparticles embedded in poly(methyl methacrylate). *Nanotechnology* 23 (34): 344014.

906 Wang, H., Ji, Z.Y., Shang, L.W. et al. (2011). Nonvolatile nano-crystal floating gate OFET memory with light assisted program. *Org. Electron.* 12 (7): 1236.

907 Tseng, C.W., Huang, D.C., and Tao, Y.T. (2013). Azobenzene-functionalized gold nanoparticles as hybrid double-floating-gate in pentacene thin-film transistors/memories with enhanced response, retention, and memory windows. *ACS Appl. Mater. Interfaces* 5 (19): 9528.

908 Liu, J., Liu, C.H., She, X.J. et al. (2014). Organic field-effect transistor non-volatile memories utilizing sputtered C nanoparticles as nano-floating-gate. *Appl. Phys. Lett.* 105 (16): 163302.

909 Sun, H.S., Chiu, Y.C., Lee, W.Y. et al. (2015). Synthesis of oligosaccharide-based block copolymers with pendent pi-conjugated oligofluorene moieties and their electrical device applications. *Macromolecules* 48 (12): 3907.

910 Wang, Y.F., Tsai, M.R., Lin, Y.S. et al. (2015). High-response organic thin-film memory transistors based on dipole-functional polymer electret layers. *Org. Electron.* 26: 359.

911 Dong, L., Sun, H.S., Wang, J.T. et al. (2015). Fluorene based donor–acceptor polymer electrets for nonvolatile organic transistor memory device applications. *J. Polym. Sci. Polym. Chem.* 53 (4): 602.

912 Hsu, J.C., Lee, W.Y., Wu, H.C. et al. (2012). Nonvolatile memory based on pentacene organic field-effect transistors with polystyrene para-substituted oligofluorene pendent moieties as polymer electrets. *J. Mater. Chem.* 22 (12): 5820.

913 Chi, H.Y., Hsu, H.W., Tung, S.H., and Liu, C.L. (2015). Nonvolatile organic field-effect transistors memory devices using supramolecular block copolymer/functional small molecule nanocomposite electret. *ACS Appl. Mater. Interfaces* 7 (10): 5663.

914 Magliulo, M., Manoli, K., Macchia, E. et al. (2015). Tailoring functional interlayers in organic field-effect transistor biosensors. *Adv. Mater.* 27 (46): 7528.

915 Erouel, M., Diallo, K., Tardy, J. et al. (2008). Stability and 2,4-dinitrotoluene response of organic field effect transistors based on pi-conjugated thiophene oligomers. *Mater. Sci. Eng., C* 28 (5–6): 965.

916 Tiwari, S., Singh, A.K., Joshi, L. et al. (2012). Poly-3-hexylthiophene based organic field-effect transistor: detection of low concentration of ammonia. *Sens. Actuators, B* 171: 962.

917 Marinelli, F., Dell'Aquila, A., Torsi, L. et al. (2009). An organic field effect transistor as a selective NO_x sensor operated at room temperature. *Sens. Actuators, B* 140 (2): 445.

918 Yamazoe, N. (1991). New approaches for improving semiconductor gas sensors. *Sens. Actuators B* 5 (1–4): 7.

919 Wang, Z., Huang, L., Zhu, X. et al. (2017). An ultrasensitive organic semiconductor NO_2 sensor based on crystalline TIPS-pentacene films. *Adv. Mater.* 29 (38): 1703192.

920 Shaymurat, T., Tang, Q., Tong, Y. et al. (2013). Gas dielectric transistor of CuPc single crystalline nanowire for SO_2 detection down to sub-ppm levels at room temperature. *Adv. Mater.* 25 (16): 2269.

921 Mirza, M., Wang, J.W., Wang, L. et al. (2015). Response enhancement mechanism of NO_2 gas sensing in ultrathin pentacene field-effect transistors. *Org. Electron.* 24: 96.

922 Zhang, F., Di, C.A., Berdunov, N. et al. (2013). Ultrathin film organic transistors: precise control of semiconductor thickness via spin-coating. *Adv. Mater.* 25 (10): 1401.

923 Li, L., Gao, P., Wang, W. et al. (2013). Growth of ultrathin organic semiconductor microstripes with thickness control in the monolayer precision. *Angew. Chem. Int. Ed.* 52 (48): 12530.

924 Chew, D.J., Zhu, L., Delivopoulos, E. et al. (2013). A microchannel neuroprosthesis for bladder control after spinal cord injury in rat. *Sci. Transl. Med.* 5 (210): 210ra155.

925 Ji, S., Wang, H., Wang, T., and Yan, D. (2013). A high-performance room-temperature NO_2 sensor based on an ultrathin heterojunction film. *Adv. Mater.* 25 (12): 1755.

926 Zhang, B., Tai, H.L., Xie, G.Z. et al. (2013). The investigation of a new NO_2 OTFT sensor based on heterojunction $F_{16}CuPc$/CuPc thin films. *Adv. Mater. Res.* 721: 159.

927 Fan, H.D., Shi, W., Yu, X.G., and Yu, J.S. (2016). High performance nitrogen dioxide sensor based on organic field-effect transistor utilizing ultrathin CuPc/PTCDI-C8 heterojunction. *Synth. Met.* 211: 161.

928 Lai, S., Demelas, M., Casula, G. et al. (2013). Ultralow voltage, OTFT-based sensor for label-free DNA detection. *Adv. Mater.* 25 (1): 103.

929 Lai, S., Barbaro, M., and Bonfiglio, A. (2016). Tailoring the sensing performances of an OFET-based biosensor. *Sens. Actuators, B* 233: 314.

930 Caboni, A., Orgiu, E., Barbaro, M., and Bonfiglio, A. (2009). Flexible organic thin-film transistors for pH monitoring. *IEEE Sens. J.* 9 (12): 1963.

931 Caboni, A., Orgiu, E., Scavetta, E. et al. (2009). Organic-based sensor for chemical detection in aqueous solution. *Appl. Phys. Lett.* 95 (12): 123304.

932 Park, M.H., Hang, D., Chand, R. et al. (2016). Mechanism of label-free DNA detection using the floating electrode on pentacene thin film transistor. *J. Phys. Chem. C* 120 (9): 4854.

933 Minami, T., Sasaki, Y., Minamiki, T. et al. (2015). A mercury(II) ion sensor device based on an organic field effect transistor with an extended-gate modified by dipicolylamine. *Chem. Commun.* 51 (100): 17666.

934 Minami, T., Minamiki, T., and Tokito, S. (2016). Electric detection of phosphate anions in water by an extended-gate-type organic field-effect transistor functionalized with a zinc(II)-dipicolylamine derivative. *Chem. Lett.* 45 (3): 371.

935 Song, J., Dailey, J., Li, H. et al. (2017). Extended solution gate OFET-based biosensor for label-free glial fibrillary acidic protein detection with polyethylene glycol-containing bioreceptor layer. *Adv. Funct. Mater.* 27 (20): 1606506.

936 Torsi, L., Farinola, G.M., Marinelli, F. et al. (2008). A sensitivity-enhanced field-effect chiral sensor. *Nat. Mater.* 7 (5): 412.

937 Mulla, M.Y., Tuccori, E., Magliulo, M. et al. (2015). Capacitance-modulated transistor detects odorant binding protein chiral interactions. *Nat. Commun.* 6: 6010.

938 Kergoat, L., Herlogsson, L., Braga, D. et al. (2010). A water-gate organic field-effect transistor. *Adv. Mater.* 22 (23): 2565.

939 Kim, T.H., Lee, S.H., Lee, J. et al. (2009). Single-carbon-atomic-resolution detection of odorant molecules using a human olfactory receptor-based bioelectronic nose. *Adv. Mater.* 21 (1): 91.

940 Yoon, H., Lee, S.H., Kwon, O.S. et al. (2009). Polypyrrole nanotubes conjugated with human olfactory receptors: high-performance transducers for FET-type bioelectronic noses. *Angew. Chem. Int. Ed.* 48 (15): 2755.

941 Jang, M., Kim, H., Lee, S. et al. (2015). Highly sensitive and selective biosensors based on organic transistors functionalized with cucurbit[6]uril derivatives. *Adv. Funct. Mater.* 25 (30): 4882.

942 Lee, M.Y., Kim, H.J., Jung, G.Y. et al. (2015). Highly sensitive and selective liquid-phase sensors based on a solvent-resistant organic-transistor platform. *Adv. Mater.* 27 (9): 1540.

943 Ji, T., Rai, P., Jung, S., and Varadan, V.K. (2008). In vitro evaluation of flexible pH and potassium ion-sensitive organic field effect transistor sensors. *Appl. Phys. Lett.* 92 (23): 233304.

944 Bernards, D.A., Malliaras, G.G., Toombes, G.E.S., and Gruner, S.M. (2006). Gating of an organic transistor through a bilayer lipid membrane with ion channels. *Appl. Phys. Lett.* 89 (5): 053505.

945 Schmoltner, K., Kofler, J., Klug, A., and List-Kratochvil, E.J. (2013). Electrolyte-gated organic field-effect transistor for selective reversible ion detection. *Adv. Mater.* 25 (47): 6895.

946 Mousavi, Z., Ekholm, A., Bobacka, J., and Ivaska, A. (2009). Ion-selective organic electrochemical junction transistors based on poly(3,4-ethylenedioxythiophene) doped with poly(styrene sulfonate). *Electroanalysis* 21 (3–5): 472.

947 Saxena, V., Shirodkar, V., and Prakash, R. (2000). Copper(II) ion-selective microelectrochemical transistor. *J. Solid State Electrochem.* 4 (4): 234.

948 Dezieck, A., Acton, O., Leong, K. et al. (2010). Threshold voltage control in organic thin film transistors with dielectric layer modified by a genetically engineered polypeptide. *Appl. Phys. Lett.* 97 (1): 013307.

949 Angione, M.D., Cotrone, S., Magliulo, M. et al. (2012). Interfacial electronic effects in functional biolayers integrated into organic field-effect transistors. *Proc. Natl. Acad. Sci. U.S.A.* 109 (17): 6429.

950 Huang, W., Besar, K., LeCover, R. et al. (2012). Highly sensitive NH3 detection based on organic field-effect transistors with tris(pentafluorophenyl)borane as receptor. *J. Am. Chem. Soc.* 134 (36): 14650.

951 Klug, A., Denk, M., Bauer, T. et al. (2013). Organic field-effect transistor based sensors with sensitive gate dielectrics used for low-concentration ammonia detection. *Org. Electron.* 14 (2): 500.

952 Zang, Y., Zhang, F., Huang, D. et al. (2014). Specific and reproducible gas sensors utilizing gas-phase chemical reaction on organic transistors. *Adv. Mater.* 26 (18): 2862.

953 Shi, W., Yu, X.E., Zheng, Y.F., and Yu, J.S. (2016). DNA based chemical sensor for the detection of nitrogen dioxide enabled by organic field-effect transistor. *Sens. Actuators, B* 222: 1003.

954 Yang, R.D., Park, J., Colesniuc, C.N. et al. (2009). Analyte chemisorption and sensing on n- and p-channel copper phthalocyanine thin-film transistors. *J. Chem. Phys.* 130 (16): 164703.

955 Huang, J., Miragliotta, J., Becknell, A., and Katz, H.E. (2007). Hydroxy-terminated organic semiconductor-based field-effect transistors for phosphonate vapor detection. *J. Am. Chem. Soc.* 129 (30): 9366.

956 Hammock, M.L., Sokolov, A.N., Stoltenberg, R.M. et al. (2012). Organic transistors with ordered nanoparticle arrays as a tailorable platform for selective, in situ detection. *ACS Nano* 6 (4): 3100.

957 Knopfmacher, O., Hammock, M.L., Appleton, A.L. et al. (2014). Highly stable organic polymer field-effect transistor sensor for selective detection in the marine environment. *Nat. Commun.* 5: 2954.

958 Maddalena, F., Kuiper, M.J., Poolman, B. et al. (2010). Organic field-effect transistor-based biosensors functionalized with protein receptors. *J. Appl. Phys.* 108 (12): 124501.

959 Lienerth, P., Fall, S., Leveque, P. et al. (2016). Improving the selectivity to polar vapors of OFET-based sensors by using the transfer characteristics hysteresis response. *Sens. Actuators, B* 225: 90.

960 Werkmeister, F.X., Koide, T., and Nickel, B.A. (2016). Ammonia sensing for enzymatic urea detection using organic field effect transistors and a semipermeable membrane. *J. Mater. Chem. B* 4 (1): 162.

961 Gannot, Y., Hertzog-Ronen, C., Tessler, N., and Eichen, Y. (2010). Sensing of alkylating agents using organic field-effect transistors. *Adv. Funct. Mater.* 20 (1): 105.

962 Park, Y.D., Kang, B., Lim, H.S. et al. (2013). Polyelectrolyte interlayer for ultra-sensitive organic transistor humidity sensors. *ACS Appl. Mater. Interfaces* 5 (17): 8591.

963 Yang, G., Di, C.A., Zhang, G.X. et al. (2013). Highly sensitive chemical-vapor sensor based on thin-film organic field-effect transistors with benzothiadiazole-fused-tetrathiafulvalene. *Adv. Funct. Mater.* 23 (13): 1671.

964 Minami, T., Sasaki, Y., Minamiki, T. et al. (2016). Selective nitrate detection by an enzymatic sensor based on an extended-gate type organic field-effect transistor. *Biosens. Bioelectron.* 81: 87.

965 Minami, T., Sato, T., Minamiki, T. et al. (2015). A novel OFET-based biosensor for the selective and sensitive detection of lactate levels. *Biosens. Bioelectron.* 74: 45.

966 Royer, J.E., Kappe, E.D., Zhang, C.Y. et al. (2012). Organic thin-film transistors for selective hydrogen peroxide and organic peroxide vapor detection. *J. Phys. Chem. C* 116 (46): 24566.

967 Huang, W.G., Sinha, J., Yeh, M.L. et al. (2013). Diverse organic field-effect transistor sensor responses from two functionalized naphthalenetetracarboxylic diimides and copper phthalocyanine semiconductors distinguishable over a wide analyte range. *Adv. Funct. Mater.* 23 (33): 4094.

968 Rivnay, J., Owens, R.M., and Malliaras, G.G. (2014). The rise of organic bioelectronics. *Chem. Mater.* 26 (1): 679.

969 Berggren, M. and Richter-Dahlfors, A. (2007). Organic bioelectronics. *Adv. Mater.* 19 (20): 3201.

970 Alivisatos, A.P., Andrews, A.M., Boyden, E.S. et al. (2013). Nanotools for neuroscience and brain activity mapping. *ACS Nano* 7 (3): 1850.

971 Poo, M.M., Du, J.L., Ip, N.Y. et al. (2016). China brain project: basic neuroscience, brain diseases, and brain-inspired computing. *Neuron* 92 (3): 591.

972 White, H.S., Kittlesen, G.P., and Wrighton, M.S. (1984). Chemical derivatization of an array of three gold microelectrodes with polypyrrole: fabrication of a molecule-based transistor. *J. Am. Chem. Soc.* 106 (18): 5375.

973 Fu, Y., Wang, N., Yang, A. et al. (2017). Highly sensitive detection of protein biomarkers with organic electrochemical transistors. *Adv. Mater.* 29 (41): 1703787.

974 Hammock, M.L., Knopfmacher, O., Ng, T.N. et al. (2014). Electronic readout enzyme-linked immunosorbent assay with organic field-effect transistors as a preeclampsia prognostic. *Adv. Mater.* 26 (35): 6138.

975 Hai, W., Goda, T., Takeuchi, H. et al. (2017). Specific recognition of human influenza virus with PEDOT bearing sialic acid-terminated trisaccharides. *ACS Appl. Mater. Interfaces* 9 (16): 14162.

976 Zhang, M., Liao, C.Z., Yao, Y.L. et al. (2014). High- dopamine sensors based on WholeGraphene solution- gated transistors. *Adv. Funct. Mater.* 24 (7): 978.

977 Liao, C., Mak, C., Zhang, M. et al. (2015). Flexible organic electrochemical transistors for highly selective enzyme biosensors and used for saliva testing. *Adv. Mater.* 27 (4): 676.

978 Liao, C.Z. and Yan, F. (2013). Organic semiconductors in organic thin-film transistor-based chemical and biological sensors. *Polym. Rev.* 53 (3): 352.

979 Park, S., Kang, Y.J., and Majd, S. (2015). A review of patterned organic bioelectronic materials and their biomedical applications. *Adv. Mater.* 27 (46): 7583.

980 Irimia-Vladu, M. (2014). "Green" electronics: biodegradable and biocompatible materials and devices for sustainable future. *Chem. Soc. Rev.* 43 (2): 588.

981 Strakosas, X., Bongo, M., and Owens, R.M. (2015). The organic electrochemical transistor for biological applications. *J. Appl. Polym. Sci.* 132 (15): 41735.

982 Wong, J.Y., Langer, R., and Ingber, D.E. (1994). Electrically conducting polymers can noninvasively control the shape and growth of mammalian-cells. *Proc. Natl. Acad. Sci. U.S.A.* 91 (8): 3201.

983 Wan, A.M., Chandler, E.M., Madhavan, M. et al. (2013). Fibronectin conformation regulates the proangiogenic capability of tumor-associated adipogenic stromal cells. *Biochim. Biophys. Acta* 1830 (9): 4314.

984 Herland, A., Persson, K.M., Lundin, V. et al. (2011). Electrochemical control of growth factor presentation to steer neural stem cell differentiation. *Angew. Chem. Int. Ed.* 50 (52): 12529.

985 Bolin, M.H., Svennersten, K., Nilsson, D. et al. (2009). Active control of epithelial cell-density gradients grown along the channel of an organic electrochemical transistor. *Adv. Mater.* 21 (43): 4379.

986 Lin, P., Yan, F., Yu, J. et al. (2010). The application of organic electrochemical transistors in cell-based biosensors. *Adv. Mater.* 22 (33): 3655.

987 Yao, C., Xie, C., Lin, P. et al. (2013). Organic electrochemical transistor array for recording transepithelial ion transport of human airway epithelial cells. *Adv. Mater.* 25 (45): 6575.

988 Ramuz, M., Hama, A., Huerta, M. et al. (2014). Combined optical and electronic sensing of epithelial cells using planar organic transistors. *Adv. Mater.* 26 (41): 7083.

989 Curto, V.F., Marchiori, B., Hama, A. et al. (2017). Organic transistor platform with integrated microfluidics for in-line multi-parametric in vitro cell monitoring. *Microsyst. Nanoeng.* 3: 17028.

990 Green, R. and Abidian, M.R. (2015). Conducting polymers for neural prosthetic and neural interface applications. *Adv. Mater.* 27 (46): 7620.

991 Khodagholy, D., Gelinas, J.N., Thesen, T. et al. (2015). NeuroGrid: recording action potentials from the surface of the brain. *Nat. Neurosci.* 18 (2): 310.

992 Park, S.I., Brenner, D.S., Shin, G. et al. (2015). Soft, stretchable, fully implantable miniaturized optoelectronic systems for wireless optogenetics. *Nat. Biotechnol.* 33 (12): 1280.

993 Buzsaki, G. (2004). Large-scale recording of neuronal ensembles. *Nat. Neurosci.* 7 (5): 446.

994 Bennett, P.B., Yazawa, K., Makita, N., and George, A.L. Jr., (1995). Molecular mechanism for an inherited cardiac arrhythmia. *Nature* 376: 683.

995 Rubehn, B., Bosman, C., Oostenveld, R. et al. (2009). A MEMS-based flexible multichannel ECoG-electrode array. *J. Neural Eng.* 6 (3): 036003.

996 Viventi, J., Kim, D.H., Vigeland, L. et al. (2011). Flexible, foldable, actively multiplexed, high-density electrode array for mapping brain activity in vivo. *Nat. Neurosci.* 14 (12): 1599.

997 Minev, I.R., Musienko, P., Hirsch, A. et al. (2015). Electronic dura mater for long-term multimodal neural interfaces. *Science* 347 (6218): 159.

998 Harrison, R.R., Watkins, P.T., Kier, R.J. et al. (2007). A low-power integrated circuit for a wireless 100-electrode neural recording system. *IEEE J. Solid-State Circuits* 42 (1): 123.

999 Campana, A., Cramer, T., Simon, D.T. et al. (2014). Electrocardiographic recording with conformable organic electrochemical transistor fabricated on resorbable bioscaffold. *Adv. Mater.* 26 (23): 3874.

1000 Khodagholy, D., Doublet, T., Quilichini, P. et al. (2013). In vivo recordings of brain activity using organic transistors. *Nat. Commun.* 4: 1575.

1001 Rogers, J.A., Someya, T., and Huang, Y. (2010). Materials and mechanics for stretchable electronics. *Science* 327 (5973): 1603.

1002 Jeong, J.W., McCall, J.G., Shin, G. et al. (2015). Wireless optofluidic systems for programmable in vivo pharmacology and optogenetics. *Cell* 162 (3): 662.

1003 Viventi, J., Kim, D.H., Moss, J.D. et al. (2010). A conformal, bio-interfaced class of silicon electronics for mapping cardiac electrophysiology. *Sci. Transl. Med.* 2 (24): 24ra22.

1004 Vetter, R.J., Williams, J.C., Hetke, J.F. et al. (2004). Chronic neural recording using silicon-substrate microelectrode arrays implanted in cerebral cortex. *IEEE Trans. Biomed. Eng.* 51 (6): 896.

1005 Stieglitz, T., Beutel, H., Schuettler, M., and Meyer, J.U. (2000). Micromachined, polyimide-based devices for flexible neural interfaces. *Biomed. Microdevices* 2 (4): 283.

1006 Tyler, D.J. and Durand, D.M. (2002). Functionally selective peripheral nerve stimulation with a flat interface nerve electrode. *IEEE Trans. Neural Syst. Rehabil. Eng.* 10 (4): 294.

1007 Benfenati, V., Toffanin, S., Bonetti, S. et al. (2013). A transparent organic transistor structure for bidirectional stimulation and recording of primary neurons. *Nat. Mater.* 12 (7): 672.

1008 Stavrinidou, E., Leleux, P., Rajaona, H. et al. (2013). A simple model for ion injection and transport in conducting polymers. *J. Appl. Phys.* 113 (24): 244501.

1009 Leleux, P., Rivnay, J., Lonjaret, T. et al. (2015). Organic electrochemical transistors for clinical applications. *Adv. Healthcare Mater.* 4 (1): 142.

1010 Moschner, C. and Baloh, R.W. (1994). Age-related changes in visual tracking. *J. Gerontol.* 49 (5): M235.

1011 McDonagh, J., Stephen, L.J., Dolan, F.M. et al. (2003). Peripheral retinal dysfunction in patients taking vigabatrin. *Neurology* 61 (12): 1690.

1012 Lam, R.W., Beattie, C.W., Buchanan, A. et al. (1991). Low electrooculographic ratios in patients with seasonal affective disorder. *Am. J. Psychiatry* 148 (11): 1526.

1013 Kotilahti, K., Nissila, I., Nasi, T. et al. (2010). Hemodynamic responses to speech and music in newborn infants. *Hum. Brain Mapp.* 31 (4): 595.

1014 Gu, X., Yao, C., Liu, Y., and Hsing, I.M. (2016). 16-channel organic electrochemical transistor array for in vitro conduction mapping of cardiac action potential. *Adv. Healthcare Mater.* 5 (18): 2345.

1015 Jasper, H.H., Arfel-Capdeville, G., and Rasmussen, T. (1961). Evaluation of EEG and cortical electrographic studies for prognosis of seizures following surgical excision of epileptogenic lesions. *Epilepsia* 2 (2): 130.

1016 Gilletti, A. and Muthuswamy, J. (2006). Brain micromotion around implants in the rodent somatosensory cortex. *J. Neural Eng.* 3 (3): 189.

1017 Williamson, A., Ferro, M., Leleux, P. et al. (2015). Localized neuron stimulation with organic electrochemical transistors on delaminating depth probes. *Adv. Mater.* 27 (30): 4405.

1018 Kim, T.I., McCall, J.G., Jung, Y.H. et al. (2013). Injectable, cellular-scale optoelectronics with applications for wireless optogenetics. *Science* 340 (6129): 211.

1019 Kim, B.J., Kuo, J.T., Hara, S.A. et al. (2013). 3D Parylene sheath neural probe for chronic recordings. *J. Neural Eng.* 10 (4): 045002.

1020 Kim, D.H., Viventi, J., Amsden, J.J. et al. (2010). Dissolvable films of silk fibroin for ultrathin conformal bio-integrated electronics. *Nat. Mater.* 9 (6): 511.

1021 Boyden, E.S., Zhang, F., Bamberg, E. et al. (2005). Millisecond-timescale, genetically targeted optical control of neural activity. *Nat. Neurosci.* 8 (9): 1263.

1022 Deisseroth, K. (2011). Optogenetics. *Nat. Methods* 8 (1): 26.

1023 Deisseroth, K., Feng, G., Majewska, A.K. et al. (2006). Next-generation optical technologies for illuminating genetically targeted brain circuits. *J. Neurosci.* 26 (41): 10380.

1024 Antognazza, M.R., Martino, N., Ghezzi, D. et al. (2015). Shedding light on living cells. *Adv. Mater.* 27 (46): 7662.

1025 Deisseroth, K. (2015). Optogenetics: 10 years of microbial opsins in neuroscience. *Nat. Neurosci.* 18 (9): 1213.

1026 Kang, S.K., Murphy, R.K., Hwang, S.W. et al. (2016). Bioresorbable silicon electronic sensors for the brain. *Nature* 530 (7588): 71.

1027 Antognazza, M.R., Ghezzi, D., Musitelli, D. et al. (2009). A hybrid solid-liquid polymer photodiode for the bioenvironment. *Appl. Phys. Lett.* 94 (24): 243501.

1028 Ghezzi, D., Antognazza, M.R., Dal Maschio, M. et al. (2011). A hybrid bioorganic interface for neuronal photoactivation. *Nat. Commun.* 2: 166.

1029 Ghezzi, D., Antognazza, M.R., Maccarone, R. et al. (2013). A polymer optoelectronic interface restores light sensitivity in blind rat retinas. *Nat. Photonics* 7 (5): 400.

1030 Lee, W., Kim, D., Rivnay, J. et al. (2016). Integration of organic electrochemical and field-effect transistors for ultraflexible, high temporal resolution electrophysiology arrays. *Adv. Mater.* 28 (44): 9722.

1031 Doyen, S., Klein, O., Pichon, C.L., and Cleeremans, A. (2012). Behavioral priming: it's all in the mind, but whose mind? *PLoS One* 7 (1): e29081.

1032 Ishizuka, T., Kakuda, M., Araki, R., and Yawo, H. (2006). Kinetic evaluation of photosensitivity in genetically engineered neurons expressing green algae light-gated channels. *Neurosci. Res.* 54 (2): 85.

1033 Lee, W., Kim, D., Matsuhisa, N. et al. (2017). Transparent, conformable, active multielectrode array using organic electrochemical transistors. *Proc. Natl. Acad. Sci. U.S.A.* 114 (40): 10554.

1034 Tang, W., Huang, Y.K., Han, L. et al. (2019). Recent progress in printable organic field effect transistors. *J. Mater. Chem. C* 7 (4): 790.

1035 Bariya, M., Nyein, H.Y.Y., and Javey, A. (2018). Wearable sweat sensors. *Nat. Electron.* 1 (3): 160.

1036 Choi, S., Lee, H., Ghaffari, R. et al. (2016). Recent advances in flexible and stretchable bio-electronic devices integrated with nanomaterials. *Adv. Mater.* 28 (22): 4203.

Index

Interface Engineering in Organic Field-Effect Transistors, First Edition. Xuefeng Guo and Hongliang Chen.
© 2023 WILEY-VCH GmbH. Published 2023 by WILEY-VCH GmbH.

s

t